The Practical Xilinx® Designer Lab Book, Version 1.5

David Van den Bout

Prentice Hall
Upper Saddle River, New Jersey 07458

Library of Congress Cataloging-in-Publication Data
Van Den Bout, David E.
    The practical XILINX designer lab book, version 1.5 / David Van
den Bout.
        p. cm.
    Includes bibliographical references.
    ISBN 0-13-021617-8
    1. Digital integrated circuits--Design and construction 2. Logic
design. 3. Microcomputers--Design and construction I. Title.
TK7874.65.V36   1999
621.3815--dc21   99–12146
                                                        CIP

Publisher: Tom Robbins
Editorial / Production Supervision: Rose Kernan
Editor-in-Chief: Marcia Horton
Managing Editor: Eileen Clark
Vice President of Production and Manufacturing:
    David W. Riccardi
Manufacturing Buyer: Diane Hynes
Manufacturing Manager: Trudy Pisciotti
Marketing Manager: Danny Hoyt
Editorial Assistant: Dan DePasquale
Book Cover, Box, and CD Holder Design: Paul
    Gourhan
Composition: RDD Consultants, Inc.

Printed in the United States of America

10 9 8 7 6 5 4 3 2 1

ISBN 0-13-021617-8

Prentice-Hall International (UK) Limited, *London*
Prentice-Hall of Australia Pty. Limited, *Sydney*
Prentice-Hall Canada, Inc., *Toronto*
Prentice-Hall Hispanoamericana, S.A., *Mexico*
Prentice-Hall of India Private Limited, *New Delhi*
Prentice-Hall of Japan, Inc., *Tokyo*
Simon & Schuster Asia Pte. Ltd., *Singapore*
Editora Prentice-Hall do Brasil, Ltda., *Rio de Janeiro*

Trademarks

ΣXILINX®

The Xilinx logo shown above is a registered trademark of Xilinx, Inc.

XILINX, XACT, XC2064, XC3090, XC4005, XC5210, XC-DS501, FPGA Architect, FPGA Foundry, NeoCAD, NeoCAD EPIC, NeoCAD PRISM, NeoROUTE, Plus Logic, Plustran, P+, Timing Wizard, and TRACE are registered trademarks of Xilinx, Inc.

Σ

The shadow X shown above is a trademark of Xilinx, Inc.

All XC-prefix product designations, XACT *step*, XACT *step* Advanced, XACT *step* Foundry, XACT-Floorplanner, XACT-Performance, XAPP, XAM, X-BLOX, X-BLOX plus, XChecker, XDM, XDS, XEPLD, XPP, XSI, BITA, Configurable Logic Cell, CLC, Dual Block, FastCLK, FastCONNECT, FastFLASH, FastMap, Foundation, HardWire, LCA, LogiBLOX, Logic Cell, LogiCORE, LogicProfessor, MicroVia, PLUSASM, PowerGuide, PowerMaze, Select-RAM, SMARTswitch, TrueMap, UIM, VectorMaze, VersaBlock, VersaRing, XABEL, Xilinx Foundation Series, and ZERO+ are trademarks of Xilinx, Inc. The Programmable Logic Company and The Programmable Gate Array Company are service marks of Xilinx, Inc.

Aldec is a trademark of Aldec, Inc. IBM is a registered trademark and PC/AT, PC/XT, PS/2 and Micro Channel are trademarks of International Business Machines Corporation. DASH, Data I/O and FutureNet are registered trademarks and ABEL, ABEL-HDL and ABEL-PLA are trademarks of Data I/O Corporation. SimuCad and Silos are registered trademarks and P-Silos and P/C-Silos are trademarks of SimuCad Corporation. Microsoft is a registered trademark and MS-DOS is a trademark of Microsoft Corporation. Centronics is a registered trademark of Centronics Data Computer Corporation. PALASM is a registered trademark of Advanced Micro Devices, Inc. UNIX is a trademark of AT&T Technologies, Inc. CUPL, PROLINK, and MAKEPRG are trademarks of Logical Devices, Inc. Apollo and AEGIS are registered trademarks of Hewlett-Packard Corporation. Mentor and IDEA are registered trademarks and NETED, Design Architect, System Architect, QuickSim, QuickSim II, and EXPAND are trademarks of Mentor Graphics, Inc. Sun is a registered trademark of Sun Microsystems, Inc. SCHEMA II+ and SCHEMA III are trademarks of Omation Corporation. OrCAD is a registered trademark of OrCAD Systems Corporation. Viewlogic, Viewsim, and Viewdraw are registered trademarks of Viewlogic Systems, Inc. CASE Technology is a trademark of CASE Technology, a division of the Teradyne Electronic Design Automation Group. DECstation is a trademark of Digital Equipment Corporation. Synopsys is a registered trademark of Synopsys, Inc. Verilog is a registered trademark of Cadence Design Systems, Inc. FLEXlm is a trademark of Globetrotter, Inc. DynaText is a registered trademark of Inso Corporation. All other trademarks are owned by their respective companies.

Xilinx, Inc. does not assume any liability arising out of the application or use of any product described or shown herein; nor does it convey any license under its patents, copyrights, or maskwork rights or any rights of others. Xilinx, Inc. reserves the right to make changes, at any time, in order to improve reliability, function or design and to supply the best product possible. Xilinx, Inc. will not assume responsibility for the use of any circuitry described herein other than circuitry entirely embodied in its products. Xilinx, Inc. devices and products are protected under one or more of the following U.S. Patents: 4,642,487; 4,695,740; 4,706,216; 4,713,557; 4,746,822; 4,750,155; 4,758,985; 4,820,937; 4,821,233; 4,835,418; 4,853,626; 4,855,619; 4,855,669; 4,902,910; 4,940,909; 4,967,107; 5,012,135; 5,023,606; 5,028,821; 5,047,710; 5,068,603; 5,140,193; 5,148,390; 5,155,432; 5,166,858; 5,224,056; 5,243,238; 5,245,277; 5,267,187; 5,291,079; 5,295,090; 5,302,866; 5,319,252; 5,319,254; 5,321,704; 5,329,174; 5,329,181; 5,331,220; 5,331,226; 5,332,929; 5,337,255; 5,343,406; 5,349,248; 5,349,249; 5,349,250; 5,349,691; 5,357,153; 5,360,747; 5,361,229; 5,362,999; 5,365,125; 5,367,207; 5,386,154; 5,394,104; 5,399,924; 5,399,925; 5,410,189; 5,410,194; 5,414,377; 5,422,833; 5,426,378; 5,426,379; 5,430,687; 5,432,719; 5,448,181; 5,448,493; 5,450,021; 5,450,022; 5,453,706; 5,466,117; 5,469,003; 5,475,253; 5,477,414; 5,481,206; 5,483,478; 5,486,707; 5,486,776; 5,488,316; 5,489,858; 5,489,866; 5,491,353; 5,495,196; 5,498,979; 5,498,989; 5,499,192; 5,500,608; 5,500,609; 5,502,000; 5,502,440; RE 34,363, RE 34,444, and RE 34,808. Other U.S. and foreign patents pending.

Xilinx, Inc. does not represent that devices shown or products described herein are free from patent infringement or from any other third party right. Xilinx, Inc. assumes no obligation to correct any errors contained herein or to advise any user of this text of any correction if such be made. Xilinx, Inc. will not assume any liability for the accuracy or correctness of any engineering or software support or assistance provided to a user.

Xilinx products are not intended for use in life support appliances, devices, or systems. Use of a Xilinx product in such applications without the written consent of the appropriate Xilinx officer is prohibited.

Contents

5. Electrical Characteristics, 141

6. Flip-Flops, 171

7. State Machine Design, 203

8. Memories, 259

9. The GNOME Microcomputer, 313

10. The DWARF Microcomputer, 359

Final Word, 400

Appendix A. Building the XS40 and XS95 Lite Boards, 401

Introduction

So, what is this book about? It all started in July of 1993 when my associate department head in the EE department at NCSU asked me to teach the lab session for the introductory sophomore digital design course. This is the academic equivalent of being asked to clean the grease trap, but assistant professors don't say that to associate department heads.

Since I was going to have to teach the lab session, I tried to think of ways to make it interesting. To me, at least! The idea of spending three hours trying to help twenty students debug ten protoboards with spaghetti-wiring threatening to engulf us all was not attractive. Then we would have to rip all the work apart at the end of each lab session to make room for the next group. That meant there wouldn't be a chance to build upon what was accomplished from week to week.

I considered the use of FPGAs in the sophomore lab. FPGAs would let the students write "software-like" hardware descriptions instead of cutting and clipping wires. Granted, student software can look as much like spaghetti as their wiring, but at least they could be handed a working description of a circuit that they could modify, compile, and load into an FPGA. And writing a hardware description takes less time than wiring together chips (especially when you add in debugging time) so the students could do more challenging designs while they were in lab. It would also be easier for them to redesign portions of their circuit if they found they were taking the wrong approach. And they might do more iterations to improve their designs since they wouldn't have to build each one from scratch. They could even do their designs outside of the lab session and bring in their hardware descriptions on floppy disks so the circuits could be tested in an actual FPGA.

From the instructor's viewpoint, FPGAs lower lab operating costs since a single chip replaces a cabinet of TTL parts. And the time to clean-up lab stations between sessions is eliminated because the FPGA can be erased in seconds. A student's design could even span multiple sessions since the FPGA design can be saved and recalled from disk as often as needed. These factors would help to improve the size and challenge of designs that students can work on. It seemed like a good idea.

We had plenty of software tools for FPGA design on our system of networked workstations. However, after the students finished their lab course, what then?

How would a student build a digital circuit if he/she didn't have access to a workstation and a bunch of tools costing $10,000? If this forced a student back to designing with TTL chips, then teaching FPGA-based design in the labs wouldn't be very useful. The FPGA design environment had to be affordable so that a student could own one and use it once they finished the lab course.

Luckily, I had attended an FPGA conference and saw a $40 FPGA with 3,500 gates and a set of free software tools. The FPGA was not cheap, but it was less than what a drawer full of TTL chips would cost. The 84-pin FPGA could be mounted on a standard breadboard along with some interface circuitry for about $120. That's pretty cheap for a complete FPGA design environment.

The FPGA-based protoboards were used for twelve different labs throughout the semester. Every lab did not go smoothly, of course, but it was relatively painless when you consider the magnitude of a switchover from using TTL devices to FPGAs. And we only burned up one FPGA during the semester so the operating costs were low.

After several years, the company that provided the FPGAs for the labs exited the programmable logic market and the chip was eventually discontinued. Luckily, I managed to connect with Jason Feinsmith at XILINX and reached an agreement to re-write the labs based on their XC9500 CPLDs and XC4000 FPGAs and their Foundation Series software package. The prices of their chips have fallen into the $10-$20 range. And they agreed to include the Foundation tools on a CDROM with every lab book so the students could each have their own copy.

This book contains a complete set of lab sessions on using both CPLDs and FPGAs for beginning logic design. The material is organized into the following chapters and appendices:

The Digital Design Process discusses the steps involved in designing a digital circuit and shows how to build a circuit using TTL chips.

Programmable Logic Design Techniques lists some advantages of using programmable logic, discusses the evolution of programmable logic, and demonstrates the use of the XILINX F1 design tools.

Combinational Logic discusses several commonly-used logic circuits and demonstrates how to design and implement an LED decoder circuit.

Modular Designs and Hierarchy discusses the advantages of step-wise refinement and encapsulation in the design process and shows how the F1 package supports these concepts.

Electrical Characteristics describes various I/O drivers and timing/delay models of CPLDs and FPGAs.

Flip-Flops introduces sequential logic by building several types of level-sensitive and edge-triggered flip-flop circuits.

State Machine Design shows design examples for a counter and several versions of a controller for a drink machine.

Memories discusses how random-access memories are organized and shows how to build a small memories using flip-flops and logic gates, how to use built-in FPGA memories, and how to use external memory chips to build a FIFO.

The GNOME Microcomputer describes the architecture, instruction set, and construction of a simple, four-bit microcomputer.

The DWARF Microcomputer describes how to extend the GNOME microcomputer to eight bits and adds more addressing modes, subroutine support, and interrupts.

Appendix A shows how to build a CPLD or FPGA-based prototyping board. It also shows the connections between the programmable logic device and other components. You will need this information to do the examples in each chapter.

Appendix B gives an overview on using the ABEL hardware description language for FPGA designs.

Each chapter begins with a discussion of the logic design principles that will be applied in that chapter. This is followed by an experimental section where you can build and test logic circuits that demonstrate those principles. My guiding principle is that expertise can only be developed by building and testing logic designs. In order for this book to teach you how to design with FPGAs and CPLDs, you have to actively work the examples. This is not the kind of book that you can read in bed and get anything out of it. It's the kind of book that belongs by the computer while you go through the examples, or on the lab bench while you're testing an FPGA or CPLD to find out if it's working. By the time you are done with all the chapters, the book should be scuffed and marked-up beyond belief. Practice makes perfect, but it's rarely pretty.

Each chapter ends with a list of additional projects that give you some ideas for extending and building upon the designs described in each chapter. Hopefully, you will find doing designs with FPGAs and CPLDs so easy that you will tackle more challenging projects than you would otherwise.

After perusing the book, you will notice many important topics that are not covered in these chapters. If you pointed out these omissions to me, I would even agree that they are important topics. But any book is a trade-off between what we want to do and what we have time to do well. This book is intended to be the start of your education in the use of FPGAs and CPLDs, but not the end. If this book imparts enough knowledge so you can extend your expertise through further study of the FPGA/CPLD datasheets, application examples, and case-studies, then it has achieved its purpose.

There are two intended audiences for this book:

Beginning electrical and computer engineering students. This audience is learning the theoretical and practical aspects of digital logic design. This book is oriented toward the practical portion. It can be understood by someone with a knowledge of Boolean arithmetic, Boolean operations, and binary number systems. For a formal digital logic design course, the initial chapters of this book serve as a practical supplement to any of the great number of standard texts on digital design. The final two chapters present two simple, extendible microcomputer architectures that students can experiment with in a beginning computer design course.

Practicing engineers and hobbyists. This audience is already well-versed in designing and building digital logic circuits, but may not have had the money or time to get acquainted with programmable logic. The directions for creating a prototyping environment for around $100 will get them past the first obstacle. And while they will not need the digital logic fundamentals discussed in the first few chapters, the worked-out examples will be helpful for getting started quickly. The final chapters on the GNOME and DWARF microcomputers should also give them a good grasp of the size of designs that can be housed in FPGAs and CPLDs.

No book gets done in isolation, and this one is no different. I would like to thank Jason Feinsmith of XILINX for providing me with the opportunity to translate my labs to the XILINX programmable logic family. And Tom Robbins, Nancy Garcia, and Rose Kernan of Prentice Hall were inspirational as they guided this manuscript to publication in record time. Also, thanks to all the students, faculty, and reviewers at various universities who suffered through previous incarnations of this material and made helpful suggestions and corrections. In particular, I'd like to thank Jim Harris at Cal. Polytechnic Institute, Bob Reese at Mississippi State, Jan Van der Spiegel at the University of Pennsylvania, Edward Stabler at Syracuse University, John Froehlich, University of Harford, Jack Sandell, Central Queensland University, and Don Bouldin at the University of Tennessee. Finally, I thank my wife, Liz, who supported me through the entire effort.

The Digital Design Process

Objectives

- Discuss the steps involved in designing a digital circuit.
- Demonstrate the design steps on a small problem.
- Demonstrate the use of transistor-transistor logic small-scale integrated circuits.

Discussion

The Digital Design Process

Entire books are written about digital design techniques and the process of creating a digital circuit. A single chapter can only brush over this topic lightly; the remaining chapters will flesh out the skeleton that will be given here.

The digital design flow shown in Figure 1.1 has eight steps. A design always starts with specifications (or "specs") such as "the circuit must be able to add two 4-bit numbers in less than 12 nanoseconds." Specifications describe what the circuit must do, but not how it is done. The specifications are usually set by the customer who wants the final circuit, but the designer can set specs as well since the customer does not always know what he or she wants.

Once the designer knows what the circuit must do, he or she can begin to determine how it is done. Defining what the circuit receives as inputs and the outputs it generates is a good first step. In the preceding example, for instance, the adder circuit needs to receive two 4-bit numbers as inputs and it delivers one 5-bit number as an output. So the adder would have eight binary inputs and five binary outputs.

Once the inputs and outputs are known, the designer has to create truth tables, which list what values the outputs will have for each possible combination of input values. To continue with the adder example, if the two 4-bit inputs were 0111 and 1011 (7 and 11 in decimal, respectively), then the output would

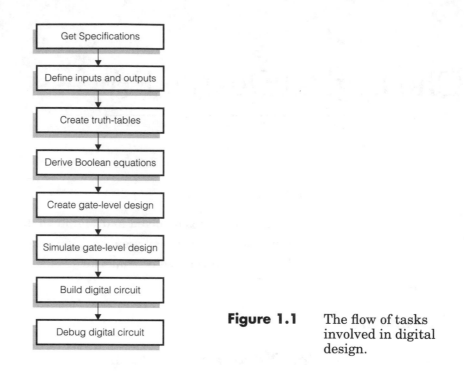

Figure 1.1 The flow of tasks involved in digital design.

be 10010 (18 in decimal). This would be one of many entries in a truth table like this:

Inputs								Outputs				
A3	**A2**	**A1**	**A0**	**B3**	**B2**	**B1**	**B0**	**O4**	**O3**	**O2**	**O1**	**O0**
...	...	...	...	...	...	...	...	...	...	...	...	...
0	1	1	1	1	0	1	1	1	0	0	1	0
...	...	...	...	...	...	...	...	...	...	...	...	...

Once the truth table is written down, the designer has to derive Boolean equations that describe how each binary output can be computed from the binary inputs using the logical operations of AND, OR, and NOT. There are a variety of manual and computer-assisted methods to accomplish this step. For the adder example, the result of this step would be five Boolean equations (one for each binary output) employing the eight binary inputs.

Next, the Boolean equations derived in the previous step are transformed into a gate-level circuit schematic drawing. Each AND, OR, and NOT operation in the Boolean equations is replaced with a corresponding AND gate, OR gate, or inverter symbol in the schematic. Lines connect the inputs and outputs of these symbols to represent the passage of binary results between logical operations.

Before building the circuit, it is a good thing to check and make sure that all the previous steps have been completed correctly. Therefore, the gate-level design of the previous step is simulated to check its operation. In its most primitive form, simulation is performed by the designer, who traces various input combinations through the gates of the schematic to determine if the values generated on the outputs match what he or she listed in the truth table. If so, great! If not, then the designer must go back through the previous steps and find where the error was made. Manual simulation is very tedious, so computers usually take care of this step.

At this point it looks like the design might actually work, so the designer can go ahead and build a physical circuit. One way to do this is by plugging digital integrated circuits (ICs) that contain individual inverters, AND gates, and OR gates into a prototyping board and then connecting the gates together with wires, as indicated by the gate-level schematic drawing. This is a time-consuming operation that is prone to errors.

Once the circuit is built, the fun begins! The designer applies real inputs to the circuit and sees if the outputs match what is listed in the truth table. If so, great! If not, then the designer made an error in a previous step. Assuming that the simulation performed was complete and correct, then the designer made the error while building the circuit. Therefore, he or she can concentrate on looking for a physical problem in the circuit. If a simulation had not been performed, however, then the problem could have been caused by a mistake in any of the preceding steps. This makes the search for the mistake more difficult and increases the likelihood that, once found, the remedy for correcting the error will be expensive. If, for example, the designer made an incorrect entry in the truth table, then this mistake would ripple through the Boolean equations, the gate-level schematic, and the physical construction of the circuit. Doing a simulation would have taken less time than it will take to find the incorrect entry and disassemble and reassemble the circuit to correct the problem.

Experimental

Building a Transistor-Transistor Logic Circuit

Design flows and design frameworks always sound great, but you never really notice their strengths and weaknesses until you actually try to use one. Therefore, the design flow that was just described will be used for a simple product: a router for traveling salespeople.

Specifications

Specifications usually come from customers, and in this case the customer is the owner of a small company employing a group of traveling salespeople who service four cities: Hooterville, Mayberry, Mt. Pilot and Siler City. The owner's problem is that his salespeople often make poor choices on routes between cities and spend more time on the road than they do selling his products. The owner has checked the travel time between cities for the various highways that connect them and has entered this information on a map (Figure 1.2). What he

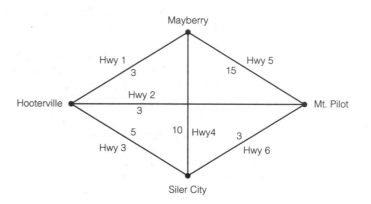

Figure 1.2   The four cities and the connecting highways traveled by the salespeople.

wants is a small, hand-held device that can be given to each salesperson and that will indicate the fastest route between any two cities. He even has an "artist's conception" of his Trip-Genie™ (Figure 1.3). The Trip-Genie has four

Figure 1.3   The Trip-Genie showing the six LED outputs for indicating the highways on the best route between the two cities selected by two of the four toggle switches.

switches (one for each city) and six light-emitting-diodes (LEDs) (one for each highway). The salesperson would flip two switches to indicate the starting and ending cities, after which the Trip-Genie would light the LEDs for those highways that are on the fastest route connecting the two cities.

It should be obvious that what the owner really needs is a smarter sales force or, at least, some printed cards that would show the salespeople the fastest route between each pair of cities. But the owner has already given you the specifications for what he wants. And it is usually the case that you will be more successful if you give others what they want instead of what they need. So the following subsections will demonstrate the design of the Trip-Genie.

Inputs and Outputs

The logic circuitry for the Trip-Genie will have four inputs for indicating the cities at the beginning and end of the route (see Figure 1.4). The circuitry will also have six outputs to drive LEDs, which indicate the highways that are on the fastest route between the cities.

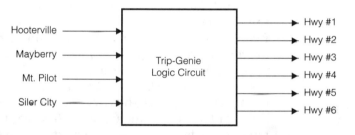

Figure 1.4   A very high-level view of the Trip-Genie logic circuit with its 4 binary inputs and 6 binary outputs.

The Boolean equations that will describe the logic circuitry operate on the binary values 0 and 1. The physical logic circuitry, however, uses voltages to represent binary logic values. As part of defining the inputs and outputs of the Trip-Genie, you also get to define the correspondence between binary logic values and voltages. We will keep it simple in this book and assign logic 0 to any voltage less than 0.8 V and logic 1 to any voltage greater than 2.0 V. (These are standard thresholds for many digital logic families that operate from 5 V power supplies.)

Finally, you have to decide on the meaning of a logic 0 or 1 for each given input and output. For the Trip-Genie, a logic 1 on any input means that the city corresponding to that input is either the beginning or the ending city on the route. But if an input is zero, then the corresponding city is not on the route. For the outputs, if any output is at logic 1, then the highway corresponding to that output is part of the fastest route between the beginning and ending city. If an output is at logic 0, then the corresponding highway is not part of the fastest route.

Truth Tables

Now that the inputs and outputs are defined, you have to determine the relationship between them so that you can write down the truth table. This boils down to answering the question, Given any two cities, what highways make up the fastest route between them? You could imagine building a very smart circuit that analyzes all possible paths between the pair of cities and picks the fastest one. That is a bit much, especially for a chapter near the beginning of the book. Since there are only four cities, you can manually find the fastest route for every possible situation. There are $4 \times 3 = 12$ possible combinations of beginning and ending cities, and this can be cut down to six combinations because the fastest route is the same regardless of the direction the path is traversed. The six possible combinations of beginning and ending cities are shown in Figure 1.5 along with the fastest route between the cities.

The truth table for the Trip-Genie logic (see Table 1.1) can be filled in based on the information shown in Figure 1.5. Examining the first route shown in Figure 1.5, we can see that the beginning and ending cities are Hooterville and Mayberry. Therefore, in row 1 of the truth table the logic values of the Hooterville and Mayberry inputs are set to 1 while the Mt. Pilot and Siler City inputs are set to logic 0. The fastest route between Hooterville and Mayberry consists of only Highway 1, so the Hwy1 output in row 1 is set to logic 1 while the Hwy2 through Hwy6 outputs are set to 0 since the corresponding highways are not part of the fastest route. The next five rows of the truth table are filled by applying the same reasoning to the remainder of Figure 1.5.

The remaining 10 rows of Table 1.1 define what happens when the Trip-Genie user enters less than or more than two cities as endpoints of the route. In these cases, all the outputs are held at logic 0, which indicates an error condition.

Boolean Equations

Once the truth table is available, a Boolean equation has to be written for each output. For some of the outputs, such as Hwy4 and Hwy5, which are never used, this is very easy:

Hwy4 = 0

Hwy5 = 0

For other outputs, such as Hwy3, the equation can be found by first stating under what conditions the output will be a logic 1:

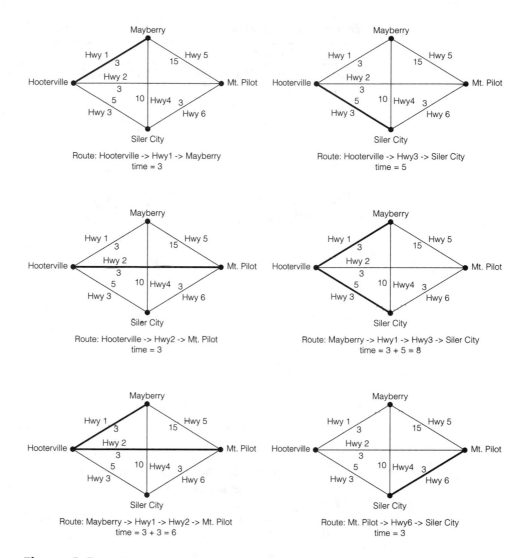

Figure 1.5 The fastest routes between each pair of cities.

Highway 3 will be taken if you travel on the fastest route between Hooterville and Siler City or Mayberry and Siler City.

We can begin replacing pieces of this sentence with language that uses Boolean operations:

(Highway 3 is taken) IF

> (one end of the route is Hooterville AND
> the other end is Siler City)

Table 1.1 The truth table for the Trip-Genie logic.

Inputs				Outputs					
Hoot.	May.	Siler	Pilot	Hwy1	Hwy2	Hwy3	Hwy4	Hwy5	Hwy6
1	1	0	0	1	0	0	0	0	0
1	0	1	0	0	0	1	0	0	0
1	0	0	1	0	1	0	0	0	0
0	1	1	0	1	0	1	0	0	0
0	1	0	1	1	1	0	0	0	0
0	0	1	1	0	0	0	0	0	1
0	0	0	0	0	0	0	0	0	0
1	0	0	0	0	0	0	0	0	0
0	1	0	0	0	0	0	0	0	0
0	0	1	0	0	0	0	0	0	0
0	0	0	1	0	0	0	0	0	0
1	1	1	0	0	0	0	0	0	0
1	1	0	1	0	0	0	0	0	0
1	0	1	1	0	0	0	0	0	0
0	1	1	1	0	0	0	0	0	0
1	1	1	1	0	0	0	0	0	0

> OR
>
> (one end of the route is Mayberry AND
> the other end is Siler City).

The fact that the other cities are not endpoints of the route also has to be included in the sentence:

> (Highway 3 is taken) IF
>
>> (one end of the route is Hooterville AND
>> the other end is Siler City AND
>> Mayberry is NOT one end of the route AND
>> Mt. Pilot is NOT one end of the route)
>
>> OR
>
>> (one end of the route is Mayberry AND
>> the other end is Siler City AND
>> Hooterville is NOT one end of the route AND
>> Mt. Pilot is NOT one end of the route.)

The sentence can be restated using more algebraic notation. If a city is an endpoint of the route, then its associated input is set to logic 1. If the city is not an endpoint of the route, then (obviously) its input is set to logic 0. If a highway is part of the fastest route, then its associated output is set to logic 1. This gives

(Hwy3 = 1) IF

   (Hooterville = 1 AND Siler City = 1 AND
   Mayberry = 0 AND Mt. Pilot = 0)

   OR

   (Mayberry = 1 AND Siler City = 1 AND
   Hooterville = 0 AND Mt. Pilot = 0)

We will introduce some notation to shorten the previous statement. If an input or an output is equal to a logic 1, we will replace the equality with the name of the input or output. On the other hand, if an input or output is equal to logic zero, we will replace the equality with the name of the input or output with a line drawn over it. (This is called the NOT or inversion operation.) For example,

Hwy3 = 1 → Hwy3

Hooterville = 0 → $\overline{\text{Hooterville}}$

We will also replace the words IF, AND, and OR with the =, ·, and + Boolean operators, respectively. Making these changes finally gives a Boolean equation:

Hwy3 = Hooterville · Siler City · $\overline{\text{Mayberry}}$ · $\overline{\text{Mt. Pilot}}$ +
    Mayberry · Siler City · $\overline{\text{Hooterville}}$ · $\overline{\text{Mt. Pilot}}$

This equation can be simplified slightly by factoring out a common subexpression as follows:

Hwy3 = (Hooterville · $\overline{\text{Mayberry}}$ + $\overline{\text{Hooterville}}$ · Mayberry) ·
    (Siler City · $\overline{\text{Mt. Pilot}}$)

That is all we can do right now to simplify this equation. You can read more about logic minimization techniques in any standard text on digital design. The procedure we just went through can be applied to Hwy1, Hwy2, and Hwy6 to get the Boolean equations for these outputs as well.

Gate-level Design

Armed with the Boolean equations, we can create the gate-level schematic for the Trip-Genie logic circuitry. We do this by replacing the AND, OR, and inversion operators in the Boolean equation with the logic gate schematic symbols shown in Figure 1.6. The NOT gate takes a single binary input and produces the logical inversion on its output. The AND and OR gates output the logical

AND or OR of their two inputs, respectively. The NAND and NOR gates are the same as the AND and OR gates except that their outputs are inverted (this is represented by the "bubble" on their outputs).

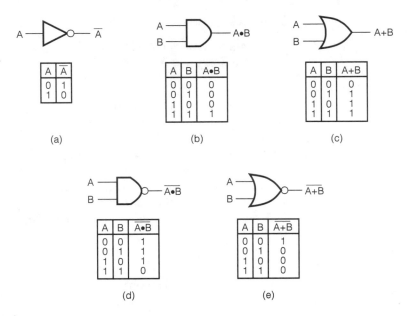

Figure 1.6   Logic gate schematic symbols for the (a) NOT, (b) AND, (c) OR, (d) NAND, and (e) NOR Boolean operations.

Of course, it is one thing to say that the Boolean operators will be replaced with schematic symbols, but it is another to do it. As an example, let's generate the schematic drawing of the Boolean equation for Hwy3. We start by noting that the Hwy3 output is the logical AND of two terms:

$$(\text{Hooterville} \cdot \overline{\text{Mayberry}} + \overline{\text{Hooterville}} \cdot \text{Mayberry}) \tag{1.1}$$

$$(\text{Siler City} \cdot \overline{\text{Mt. Pilot}}) \tag{1.2}$$

Therefore, the Hwy3 output is generated by an AND gate that has the previous two terms as its inputs (see Figure 1.7a).

We can continue the process by noting that Eq. (1.1) is the logical OR of Hooterville · $\overline{\text{Mayberry}}$ and $\overline{\text{Hooterville}}$ · Mayberry. So an OR gate can be added to the drawing that has these two terms as inputs and whose output is connected to one of the inputs of the first AND gate (see Figure 1.7b). In addition, Eq. (1.2) is the logical AND of Siler City and $\overline{\text{Mt. Pilot}}$. The output of this AND gate is connected to the other input of our first AND gate.

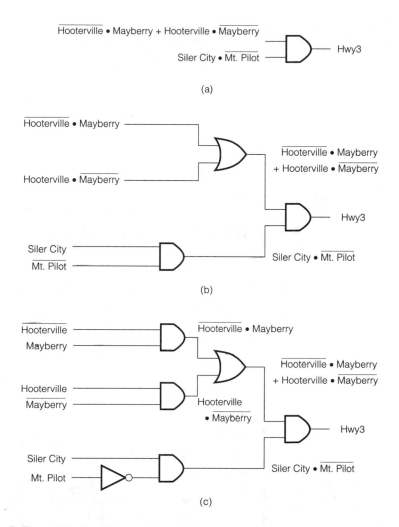

Figure 1.7 The build-up of a gate-level schematic for the Hwy3 output of the Trip Genie logic.

The inputs of the OR gate are seen to be outputs of AND gates, so these can be added as well (see Figure 1.7c). Finally, all the logically inverted inputs can be replaced with inverter gates, as shown in Figure 1.8.

In this chapter, the logic circuits will be built using small-scale integration chips (SSI chips) that contain only a few logic gates. This will give you a better appreciation of the advantages of programmable logic when it is presented in the following chapters. SSI chips have been around since the 1960s, and there are many different functions they can perform. We will only use the simple SSI chips shown in Figure 1.9. (These are examples of chips built using older transistor-transistor logic (TTL) technology, but complementary metal-oxide semi-

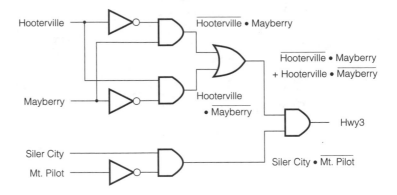

Figure 1.8   The schematic for the AND-OR-NOT logic which computes whether highway #3 is in a route or not.

conductor (CMOS) technology is used in most modern chips.) Each chip is contained in a dual-inline package (DIP), which has 14 pins that allow you to connect wires to the inputs and outputs of the logic gates and to supply power and ground connections.

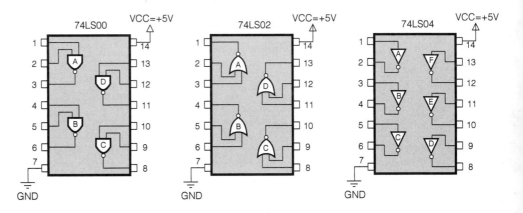

Figure 1.9   Some common TTL ICs: (a) the quad 2-input NAND chip (74LS00); (b) the quad 2-input NOR chip (74LS02); (c) the hex inverter chip (74LS04).

Unfortunately, the schematic of Figure 1.8 uses AND and OR gates, but the TTL chips only provide NAND and NOR gates. A fitting operation must be performed to change the types of gates without changing the function of the design. In this example, the fitting can be done using DeMorgan's theorem which, loosely stated, says that

Inverting all the inputs and the output of an AND gate makes it operate just like an OR gate. Inverting all the inputs and the output of an OR gate makes it operate just like an AND gate.

A picture of what DeMorgan's theorem means is given in Figure 1.10.

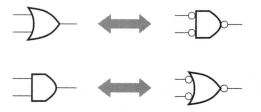

Figure 1.10 DeMorgan's theorem.

Figure 1.11 demonstrates how we can use DeMorgan's theorem to replace the AND and OR gates in our original schematic with NAND and NOR gates. The process starts by replacing the final AND gate in Figure 1.8 with a NOR gate that has its inputs inverted (Figure 1.11a). Then the inverters on the inputs of this NOR gate can be moved back to the outputs of the preceding OR and AND gates to transform them into NOR and NAND gates, respectively (Figure 1.11b). The remaining two AND gates in the schematic are replaced with NOR gates having inverted inputs and outputs (Figure 1.11c). Now we note that each of these NOR gates receives an input that has been previously inverted. These dual inverters can be removed. Finally, the remaining inverted input of each NOR gate can be redrawn as an individual inverter gate symbol. This gives us the final schematic of Figure 1.11d, which contains only NAND gates, NOR gates, and inverters.

Before building the logic circuit, it makes sense to check its operation to ensure that we did not introduce an error during all the changes. We can perform a manual simulation by picking input patterns and tracing the effect they have as they propagate through the circuit and reach the output. For example, suppose we choose the input pattern, or vector,

Hooterville = 1
Mayberry = 0
Siler City = 1
Mt. Pilot = 0

that corresponds to the problem of selecting the the fastest route between Hooterville and Siler City. Highway 3 should be part of this route, so Hwy3 should be at logic 1 if this input vector is applied to the logic circuit.

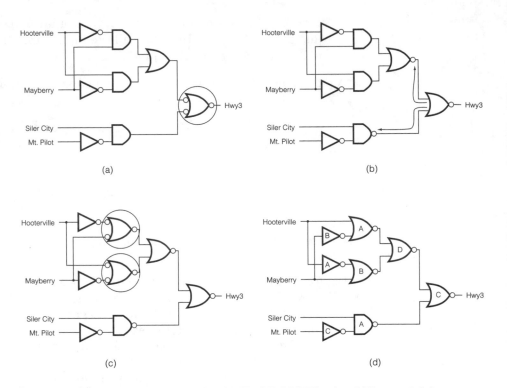

Figure 1.11 Transforming the AND-OR-NOT logic of Figure 1.8 into NAND-NOR-NOT logic.

Figure 1.12 shows the reaction of the logic circuit of Figure 1.11d to the input vector just given. In Figure 1.12a the inputs have traveled through the initial inverters and reached the first set of NAND and NOR gates. Using the truth tables shown in Figure 1.6, the outputs of these NAND and NOR gates can be computed (Figure 1.12b). Continuing this process leads to the final output (Figure 1.12c), which shows that the circuit operates correctly for this single input pattern.

The procedure just described has to be repeated with each possible input pattern for every output of the logic circuit in order to validate the entire circuit. The Trip-Genie has four binary inputs, so there are $2^4 = 16$ possible input patterns. With six distinct outputs, this means we have to perform $16 \times 6 = 96$ simulation traces. That is why computers are used for simulation.

Building the Circuit

Assuming we checked the Hwy3 logic circuit with all 16 possible input vectors, now it is time to build the circuit. Figure 1.13 shows how the gates in the TTL chips can be connected to build the circuit shown in Figure 1.11d.

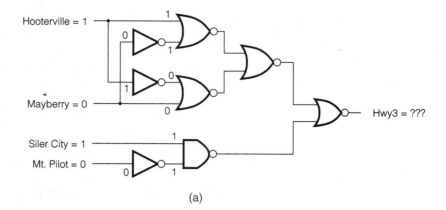

(a)

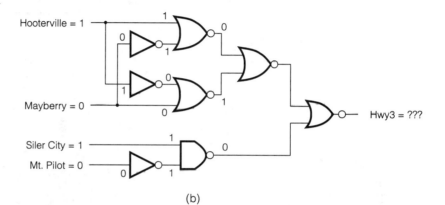

(b)

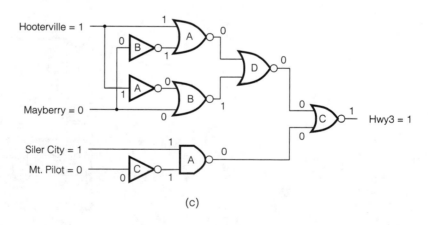

(c)

Figure 1.12 The reaction of the Trip-Genie logic to an input pattern.

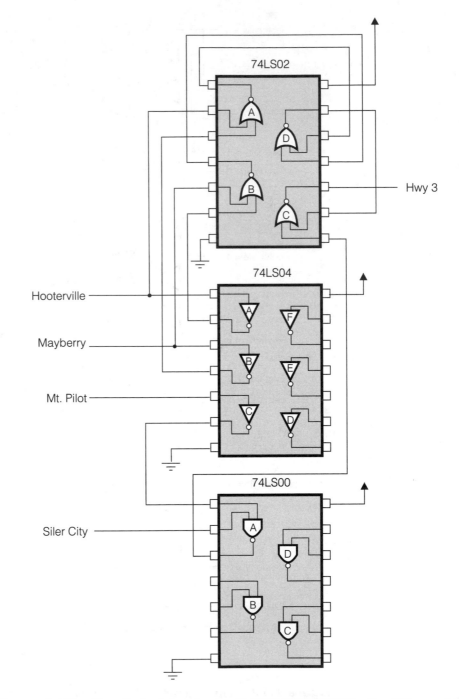

Figure 1.13 Detailed wiring for the NAND-NOR-NOT logic of Figure 1.11.

Figure 1.14 shows how this circuit would actually look when constructed on a solderless breadboard. The solderless breadboard provides a "sea of holes" that chips and wires can be plugged into to make connections. The holes are connected internally within the breadboard as follows:

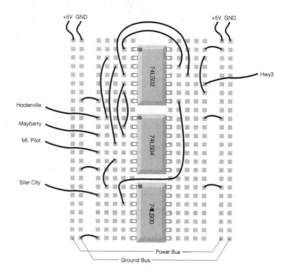

Figure 1.14   Physical layout of the logic from Figure 1.13.

1. All the holes in the leftmost column of Figure 1.14 are electrically connected. This is also true for the next column and the two rightmost columns.

2. There are two columns of five-hole rows in the center of Figure 1.14. Each row of five holes is electrically connected.

The steps for wiring a circuit should be completed in the following order:

1. Turn the power off before you begin to build anything.

2. Make sure the power is off before you begin to build anything.

3. Connect the +5 V and ground leads of your power supply to the power and ground bus bars of your breadboard. (These are the leftmost and rightmost columns of the breadboard.)

4. Plug the chips you will be using into the breadboard. Point all the chips in the same direction with pin 1 at the upperleft corner. (Pin 1 is often identified by a dot next to it on the chip package.)

5. Connect the +5 V and GND pins of each chip to the power and ground bus bars on the breadboard.

6. Select a connection on your schematic and identify the beginning and ending pins on the chips. Then place a piece of 22 to 26 gauge hookup wire between the corresponding pins of the chips on your breadboard. One end of the wire should be plugged into one of the four remaining holes in the row containing the beginning pin. The other end should be plugged into the row holding the ending pin. (Remember to strip the insulation off the ends of the wire before inserting them.) It usually is best to make the short connections first and then add the longer wires later. Highlight each connection on your schematic as you go so you do not forget and make the same connection again at a later time. And keep the wiring neat and close to the breadboard. There is nothing worse than accidentally snagging a long wire and pulling it from the breadboard and then not being able to tell where it should be reinserted. Remember: Short connections first, keep it neat, and highlight what you connect.

Debugging the Circuit

Once the circuit is built, the debugging begins. This involves placing a pattern of logic ones and zeros on the inputs and observing the output with a logic probe. In Figure 1.15, the input vector used to do the simulation in Figure 1.12 has been placed on the inputs of the circuit by attaching the Hooterville and Siler City inputs to +5 V while connecting the Mt. Pilot and Mayberry inputs to ground. The logic probe is touched to pin 10 of the 74LS02, which is the Hwy3 output from the C NOR gate. This pin should have a high voltage corresponding to a logic 1, but the probe senses a low voltage, or logic 0, instead.

Obviously, there is some error in the circuit. Therefore, we have to trace backward from the output to find the error. We first check the inputs to the C NOR gate. Probing the input that is driven by the A NAND gate (pin 8 of the 74LS02) shows a logic 0 that is correct according to the previous simulation (Figure 1.16). Probing the other input of the C NOR gate that is driven by the D NOR gate shows a logic 1 instead of the correct logic 0 level (Figure 1.17). Therefore, we trace back and directly probe the output of the D NOR gate and observe that it truly is high (Figure 1.18). Continuing our trace to the input of the D NOR gate that is driven by the B NOR gate, we find a logic 0 when it should be a logic 1 (Figure 1.19). Directly probing the output of the B NOR gate shows that it is low instead of high (Figure 1.20). As we begin to probe the inputs of the B NOR gate, we can see an obvious mistake—the wire connecting the output of inverter A to the input of the B NOR gate is missing. Leaving an input of a TTL gate unconnected allows it to float to a logic 1 level. This would cause the output of the B NOR gate to go low, which ripples through the rest of the circuit to give an incorrect result.

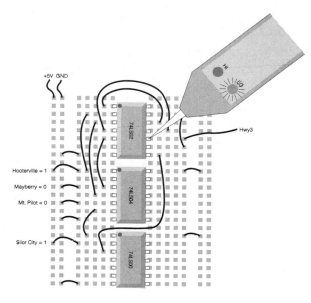

Figure 1.15   Testing the Hwy3 output of the Trip-Genie logic circuit. The output is low (logic 0) when it should be high (logic 1).

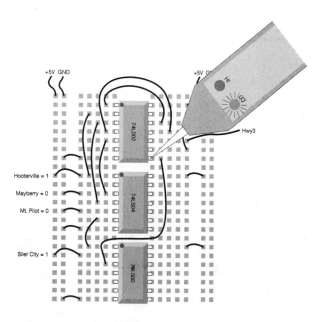

Figure 1.16   Debugging the circuit by tracing backward to find the error. The input to the C NOR gate from the A NAND gate is correct.

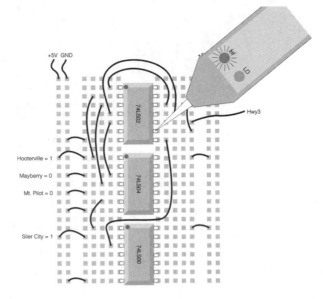

Figure 1.17   Continuing the trace backward to find the error, the input to the C NOR gate from the D NOR gate is incorrect.

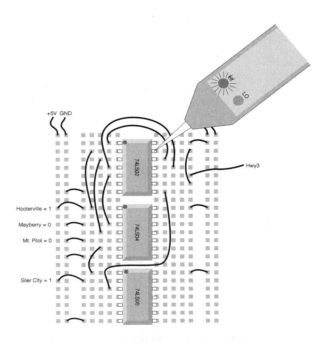

Figure 1.18   Continuing the backward trace, the output of the D NOR gate is high when it should be low.

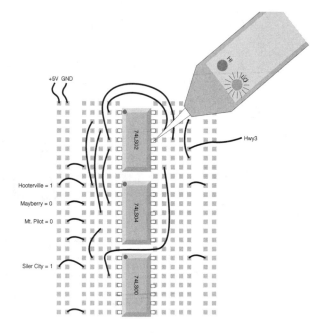

Figure 1.19 Checking the input to the D NOR gate driven by the output of the B NOR gate. The input is low when it should be high.

The problem can be corrected as follows:

1. Turn off the power to the circuit.

2. Make sure the power to the circuit is off.

3. Connect a wire from pin 2 of the 74LS04 (the output of inverter A) to pin 6 of the 74LS02 (the input to the B NOR gate).

4. Apply power to the circuit.

After this, a check of the Hwy3 output shows that it is high (logic 1) as it should be. The circuit has passed for this particular input pattern, but we have to check it on the remaining 15 patterns before we can say the circuit is completely correct.

It is worth mentioning here some of the most common causes of problems:

1. Not connecting power and/or ground pins for every chip

2. Not turning on the power before checking the operation of the circuit

3. Leaving out wires

4. Plugging wires into the wrong hole (usually one above or one below the hole you wanted)

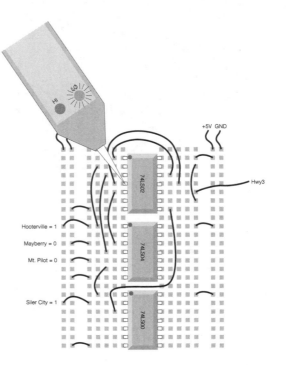

Figure 1.20 Checking the B NOR gate for errors. The output is low when it should be high. You can see that the input wire to the B NOR gate on pin 6 is missing.

5. Driving a single gate input with the outputs of two or more gates

6. Modifying the circuit while the power is still on.

Projects

1. Design the logic circuits for the other outputs of the Trip-Genie.

2. Build the logic circuit for the Hwy1 output and test its operation.

3. Out of your sight, have a lab partner remove one of the wires from your Hwy1 circuit. Then go through the debugging process to detect the problem.

Programmable Logic Design Techniques

Objectives

- Discuss the advantages of designing with programmable logic.
- Discuss the evolution of programmable logic.
- Introduce the use of the XILINX Foundation Series 1.5 Software tools for programmable logic design.

Discussion

The Advantages of Computer-aided Design and Programmable Logic

You can build digital circuits of up to several hundred gates using SSI and medium-scale integration (MSI) ICs. As you saw in the last chapter, these ICs are typically plugged into a breadboard and connections are made between them using 22- or 24-gauge hookup wire. Then switches and LEDs are connected to the inputs and outputs, respectively, and the circuit is exercised to see if it operates correctly. As you have probably already realized, there are a lot of things wrong with this method of building circuits:

- Cutting and stripping wires to make connections between ICs is time consuming.
- Wires are often plugged into the wrong place, so a lengthy check must be made to find the error in order to make the circuit work.
- With a limited budget, you may not have every type of TTL IC immediately available to you.
- Once your circuit works, you have to tear it apart to make room for the next design. Using a previously built circuit as a part of your current design is difficult to do.

In addition to the problems involved in physically building the circuit, you are also forced to use primitive tools when designing the circuit. For combinatorial circuits, you begin with a truth table and derive an efficient logic circuit from it by inspection or by formal logic minimization techniques. For sequential circuits that contain internal state information, you have to perform the additional steps of creating a state transition table and doing a state assignment (you will see this in Chapter 7). There is a good chance of making an error when doing all these steps manually. A manual simulation of the operation of the circuit is usually incomplete, so you will find additional (possibly serious) errors when the circuit is finally built and it fails to work correctly.

Practicing engineers use a completely different procedure when designing and building circuits. They may still begin by describing the truth table or state transition diagram for a logic circuit. But then the details of the logic circuit needed to realize the truth table are worked out by a logic-synthesis program. The operation of this logic circuit is checked using a simulation program. If the circuit simulates correctly, the gates and wires are mapped into a field programmable gate array (FPGA) or complex programmable logic device (CPLD) IC using specialized place&route or fitter programs. (We will combine FPGAs and CPLDs under the generic term *field programmable logic device,* or FPLD.) The FPLD contains logic gates and the means for interconnecting them within a single integrated circuit. The software programs determine how the gates in the device can be connected to build the logic circuit. The program's output is a bitstream configuration file that is downloaded into the FPLD to make it act like the logic circuit. The programmed FPLD can then be placed into a larger circuit where it will perform its functions (after some debugging, possibly). Note that the routine, mechanical chores of Figure 2.1 have been automated and are done for engineers by computer-aided design (CAD) programs and FPLDs.

Inexpensive computers and software now let the rest of us make use of computer-aided design programs and FPLDs to remove the drudgery from building digital circuits. This allows us to concentrate more fully on the creative parts of logic design. You can write logic programs in a hardware description language (HDL) using a text editor, or draw your circuit using a schematic editor. The HDL or schematic is compiled by CAD programs to create detailed logic circuits that perform the actions specified in your programs. A program on the PC simulates the circuit to make sure it operates correctly. Then the design is downloaded through the parallel port of the PC into an FPLD, where it can be debugged.

In summary, using CAD programs and FPLDs allows you to:

- Build designs faster because manual wiring is minimized.
- Avoid mistakes caused by errors in wiring.
- Save your designs for as long as you like in electronic files and recall them whenever you want.
- Experiment with many types of chips without having to stock them in your supply cabinet.

- Avoid or easily correct design errors.
- Design larger projects because the tedious manual procedures are automated.

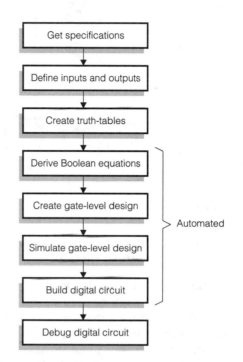

Figure 2.1　Phases of the digital design flow that can be automated.

Programmable Logic Architecture

It is natural to ask how FPGA and CPLD chips can be electronically programmed to perform any possible logic function. These devices actually evolved from the PLA™ (programmable logic array) devices of the early 1970s. The basic PLA structure is shown in Figure 2.2 and consists of a bunch of AND gates, OR gates, and inverters interconnected through some programmable switch arrays. In the PLA, every input and its logical inversion is passed into an AND array on the horizontal wires. The vertical wires in the AND array are inputs to a row of AND gates. The AND gates receive input signals by tying the horizontal and vertical wires together (as symbolized by the black dots in the AND array). Thus, the left-most AND gate receives the logical inverse of the C signal and ANDs it with the A signal.

The OR array has a function analogous to the AND array. The vertical wires carry the outputs from the AND gates into the OR array. There they can be connected to the horizontal wires which are inputs to a column of OR gates. By

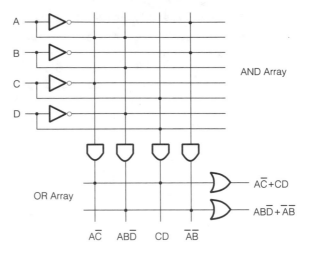

Figure 2.2   A PLA.

connecting the outputs of the AND gates to the inputs of the OR gates, a sum of products can be created at each output of the PLA.

The flexibility provided by both a programmable AND and OR array often went unused, so engineers came up with the simpler programmable array logic (PAL) structure shown in Figure 2.3. In the PAL, the programmable OR array has been replaced with a set of fixed connections from AND gates into the OR gates. So you can program the switches to form any product term you want. But the output from each OR gate is fixed to be the sum of only two product terms (in this example). In addition, the PAL outputs feedback into the AND array. The feedback terms are used to build multilevel logic functions.

PALs and PLAs are good for combinational logic, but they cannot be used for sequential logic without adding external flip-flops. (If you do not know what flip-flops are yet, you will find out in Chapter 6.) So, naturally, flip-flops were added to the PAL structure, as shown in Figure 2.4. At this point, the circuit is called an SPLD (simple programmable logic device). Multiplexers are added to each output to select either the flip-flop output or the combinational output as the actual output. The AND gates, OR gates, flip-flops, and multiplexers that drive each output are collectively known as a macrocell. Modern SPLDs have a variety of programmable circuit structures with many options that can be enabled to increase the usefulness of the chips.

The PLAs, PALs, and SPLDs still had to be placed on a printed circuit board and wired to each other and other components. If small errors were found on the board, programmable logic devices on the board could be replaced with reprogrammed devices that "wire around" the problem. However, large errors could only be corrected by manually changing the wiring pattern connecting the chips.

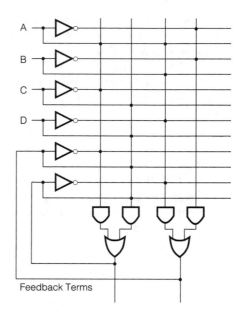

Feedback Terms

Figure 2.3   A PAL.

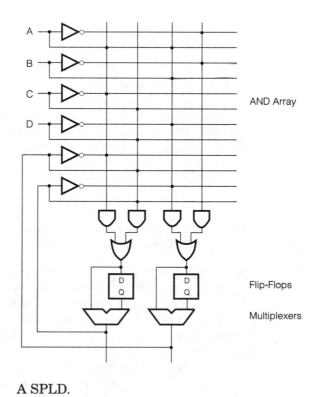

AND Array

Flip-Flops

Multiplexers

Figure 2.4   A SPLD.

It was not long before engineers got around that problem by combining several SPLDs into a single IC to create complex programmable logic devices (CPLDs). The XILINX XC9500 series of CPLDs is an example of such a CPLD (Figure 2.5). For example, the XC95108 contains six configurable function blocks (CFBs) that are each equivalent to an 18-macrocell SPLD with 36 inputs and 18 outputs. The outputs of the macrocells exit the chip through I/O pins, but they also feed back into a global interconnection matrix (XILINX calls this a "FastCONNECT Switch Matrix"). Each CFB accepts 36 inputs from the $6 \times 18 = 108$ macrocell feedback signals that enter the switch matrix. Very complex multilevel logic functions can be built by programming the individual logic functions of each macrocell in each CFB and then connecting them through the switch matrix. The result is a design where each pin on a CPLD is driven by a macrocell that implements a wide logic function of a combination of many inputs.

An alternative architecture is found in the field-programmable gate array (FPGA). The basic building block for the FPGA is the lookup table or LUT (Figure 2.6). A typical lookup table has only four inputs and a small memory containing 16 bits. Applying a binary combination to the inputs (such as 0110) will match the address of a particular memory bit and make it output its value. Any four-input logic function can be built by programming the lookup table memory with the appropriate bits. For example, a four-input AND gate is built by loading the entire memory with 0 bits except for a 1 bit that is placed in the cell that is activated when all the inputs are 1.

In FPGAs such as the XILINX XC4000 series, three LUTs are combined with two flip-flops and some additional steering circuitry to form a configurable logic block (CLB), as shown in Figure 2.7. Then the CLBs are arranged in an array with programmable switch matrices (PSMs) between the CLBs (Figure 2.8). The PSMs are analogous to the FastCONNECT switch matrix in the XC9500 CPLDs and are used to route outputs from neighboring CLBs to the inputs of a CLB. The FPGA I/O pins can attach to the PSMs and CLBs or may even be attached to their own large routing matrix. Most FPGAs have many more CLBs than I/O pins, so each CLB cannot have a direct connection to the outside world as do the macrocells in a CPLD. For example, the XILINX XC4005XL FPGA has 196 CLBs in a 14×14 array, but only 64 I/O pins when packaged in an 84-pin plastic-leaded chip carrier (PLCC).

The logical question now is, "How are all these switches set to make the connections in programmable devices?" All the wiring in PLAs, PALs, SPLDs, CPLDs, and FPGAs is internal to the IC, so there is no way an engineer can physically go in and wire the connections together. Instead, the connections are programmed electrically. In early versions of programmable logic, the switch arrays were manufactured with fuses at every crosspoint such that every input was connected to each logic gate. Specialized circuitry was added that allowed an engineer to place high voltages on selected vertical and horizontal wires.

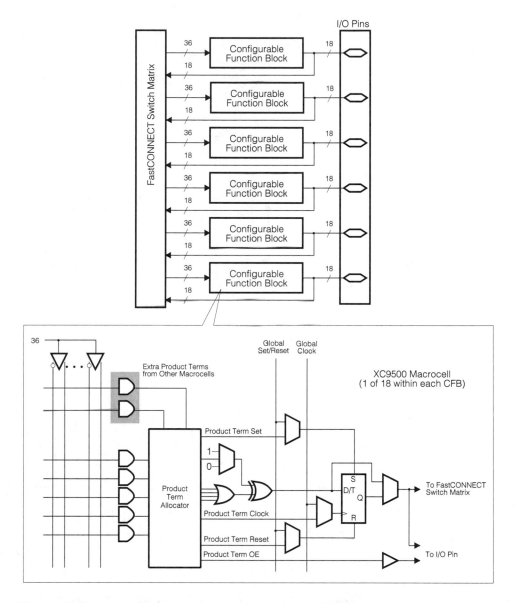

Figure 2.5   Internal organization of the XC95108 CPLD.

The high voltage burned out the fuse at the crosspoint between the two wires. The engineer performed this operation many times to knock out all the unwanted connections. At the end of the programming process, only the connections needed to build the desired logic functions would remain.

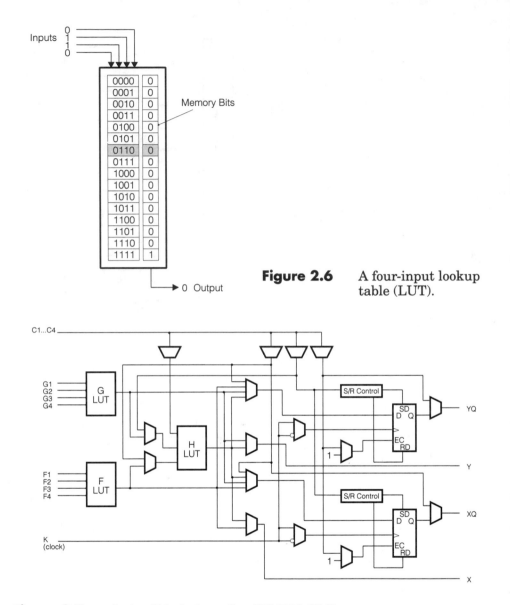

Figure 2.6   A four-input lookup table (LUT).

Figure 2.7   A simplified view of an XC4000 CLB.

The disadvantage of fuses is that once they are blown, they stay blown. So if a mistake is made, the programmable device has to be thrown out and a new one must be programmed. It is more convenient (and cheaper) if the connections can be erased and reprogrammed. This is a major advantage of the XILINX XC9500 and XC4000 programmable logic families: they contain reprogrammable switches where the fuses would normally be. Each switch is controlled

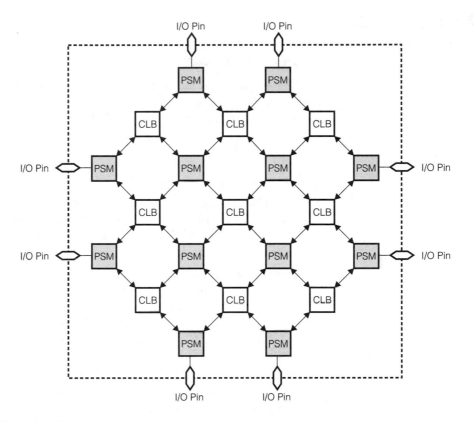

Figure 2.8   A generic FPGA architecture.

by a storage element that records whether the attached switch is opened or closed. Changing the values in these storage elements changes the state of the switches and alters the functions of the programmable device. These switches can be repeatedly programmed to implement new designs (or repair faulty designs), eliminating the need to buy a new device for each design modification. The XC9500 CPLDs use nonvolatile FLASH-based storage cells so the device retains its programming even if the power is turned off. The XC4000 FPGAs, however, employ static random access memory (RAM) storage cells so they need to be reprogrammed each time power is interrupted.

Now the question arises, "How do I tell the FPLD which wires to connect?" In the next section, I will show you how to use the XILINX Foundation Series 1.5 Software tools to create logic designs that you can download into the XC95108 CPLD and the XC4005XL FPGA.

Experimental

The Programmable Logic Design Flow

XILINX's Foundation Series Software is an environment for creating programs which describe logic designs. When using Foundation, your main design flow (shown in Figure 2.9) progresses as follows:

1. Digital designs are entered using the schematic editor, state machine editor, or by writing ABEL or VHDL programs with the HDL text editor. You can describe a design using just one or a mix of these methods.

2. A functional simulator checks the operation of the compiled design and lets you view the results to see if it is doing what you want. You can go back and edit the ABEL or VHDL file, schematic, and/or state machine diagram if any errors are found.

3. The Foundation Implementation tools compile the list of gates and connections, or netlist, into a bitstream which is used to program the FPLD. It is in this step that a particular device is targeted, such as the XC95108 or XC4005XL. For XC9500 devices, a fitter algorithm maps the netlist onto the CPLD architecture. For XC4000 devices, the implementation requires mapping the circuit to the FPGA architecture, placing the gates in specific CLBs, and then routing the wires using the PSMs.

4. A timing simulation of your design can be run after the Foundation Implementation tools have determined the gate and routing delays associated with a particular mapping to an FPLD architecture.

5. The XSLOAD program is used to download the bitstream into the XS95 Board or the XS40 Board (see Appendix A for a description of these boards).

6. Debugging is performed by forcing inputs into the XS95 or XS40 Boards via the parallel port cable. A 7-segment LED on the board shows the response of the FPLD.

These steps will be described in the following sections. As an introductory example of the modified design flow, we will recreate the design for the Hwy3 logic of the Trip-Genie shown in Chapter 1. Since the specification and I/O steps are unchanged, we can begin with the design-entry step.

The XILINX Foundation 1.5 Series Software has two distinct design flows:

Schematic mode is used if your top-level design is described using schematics, state diagrams, or ABEL.

HDL mode is used if your top-level design is described using VHDL.

We will show how to build the TripGenie using both of these design flows.

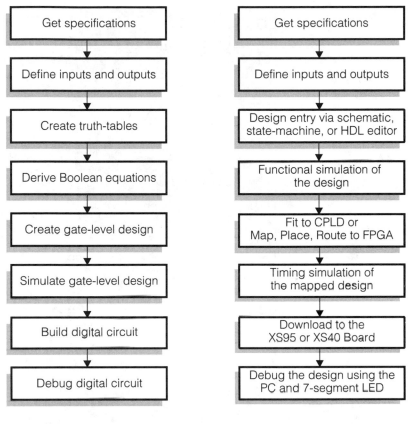

Generic Design Flow	F1 and XC9500/XC4000-Specific Design Flow
Get specifications	Get specifications
Define inputs and outputs	Define inputs and outputs
Create truth-tables	Design entry via schematic, state-machine, or HDL editor
Derive Boolean equations	Functional simulation of the design
Create gate-level design	Fit to CPLD or Map, Place, Route to FPGA
Simulate gate-level design	Timing simulation of the mapped design
Build digital circuit	Download to the XS95 or XS40 Board
Debug digital circuit	Debug the design using the PC and 7-segment LED

Figure 2.9   Changes in the digital design flow when using Xilinx Foundation Series software with the XC9500 CPLD or XC4000 FPGA.

Trip-Genie Project Design (Schematic Mode)

Starting a New Project

In the previous chapter, we could start a new design by getting a blank sheet of paper and clearing the old circuitry from our breadboard. In the modified design flow, we do the equivalent procedure when we start Foundation and create a new project.

To begin, create a main directory for all your projects. (In my case, I place all schematic mode designs in C:\XCPROJ.) Then click on the **Foundation Project Manager** icon. This will bring up the window shown in Figure 2.10.

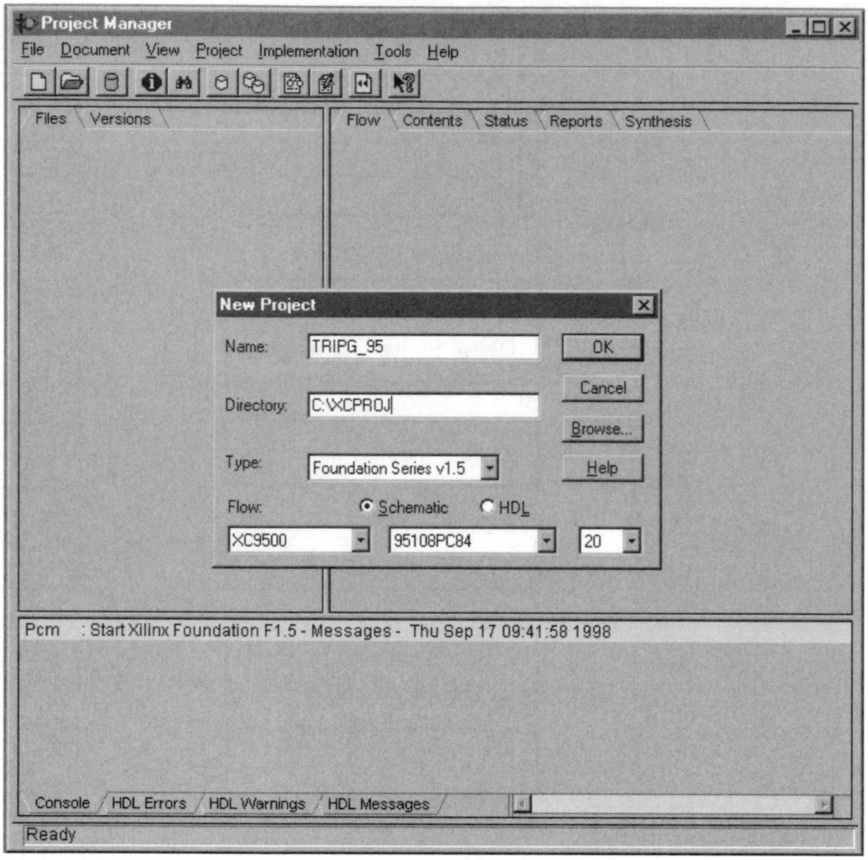

Figure 2.10   Foundation Series 1.5 Project Manager main window.

The **New Project** sub-window is displayed. (You can also bring up this sub-window by selecting the **File** → **New Project** menu item.) Then, enter the project name, project directory, type of design flow, chip family, chip part number, and device speed for either the XC95108 CPLD or the XC4005XL FPGA as shown. In Figure 2.10, I have created a project called TRIPG_95 in directory C:\XCPROJ which is targeted to the XC95108 CPLD in the 84-pin PLCC package and the slowest speed. (You could also create a project called TRIPG_40 in directory C:\XCPROJ which is targeted to the XC4005XL FPGA in the 84-pin PLCC package and the slowest speed.) I have also clicked on the **Schematic** radio button to select the schematic design flow mode of the Foundation Series software. Then click **OK** to return to the **Project Manager** window.

Now your window should look just like Figure 2.11. There are three main panes in the **Project Manager** window. The left-hand pane displays the source files and libraries for your design in the **Files** tab, while the **Versions** tab shows design implementation files generated by the XILINX Foundation Software for all the versions of your design. The right-hand pane has a **Flow** tab for controlling the activation of tools used in various phases of the design flow, a **Contents** tab which shows the status of the files in your project, a **Reports** tab which lists the report files generated by the tools as they do their work, and a **Synthesis** tab which activates the logic synthesizer for various HDL macros included in your design. The bottom pane lists diagnostic messages output by the Foundation software as it processes your design files. For now, you will spend most of your time working in the **Files** and **Flow** tabs of the left and right panes, respectively.

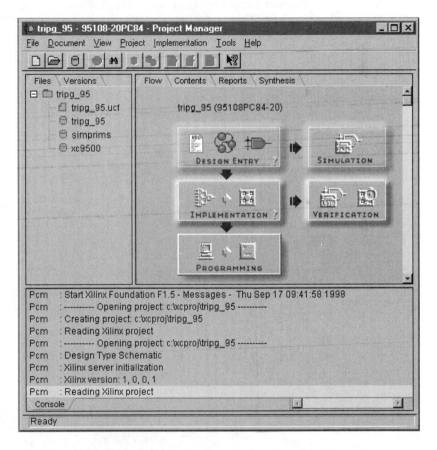

Figure 2.11   Initial Project Manager window for TripGenie design (schematic design flow).

Entering the Design Using the ABEL Hardware Description Language

In this section, I will show you how to enter the design for the Hwy3 output of the Trip-Genie using the ABEL HDL. (Go to the following section to see how the design is entered using the schematic editor.) ABEL is a simple design language that has matured and evolved as programmable logic progressed from PALs, PLAs, and SPLDs to FPLDs.

In the **Project Manager** window, select **Tools → Design Entry → HDL Editor** (or click on the ▓ button in the Flow tab of the **Project Manager** window). In the window that appears, click on the radio-button labeled **Use HDL Design Wizard** and then click on **OK**. Then click on **Next** in the **Design Wizard** window that appears to reach the **Design Wizard-Language** window. Here, select the **ABEL** radio-button and click on **Next** to move to the **Design Wizard-Name** window. Enter TRIPGENI as the name and click **Next**.

Now we should be in the **Design Wizard-Ports** window where we specify the inputs and outputs for our design (Figure 2.12). To add an input, click on **New** and then type **Hooterville** (for example) in the **Name** field. Click on the **Input** radio-button to set the port to be an input. Repeat these steps for the **Mayberry**, **SilerCity**, and **MtPilot** inputs.

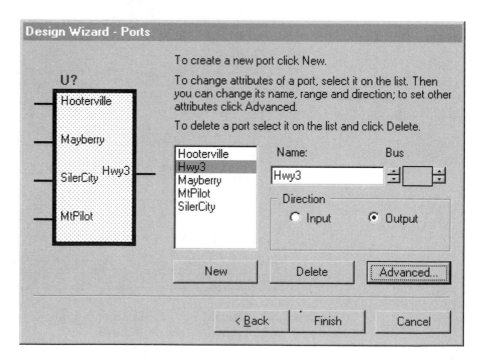

Figure 2.12 Adding inputs and outputs to the TRIPGENI design.

To add the output of our circuit click on **New**, enter **Hwy3** in the **Name** field, and click on the **Output** radio-button. Then click on the **Advanced** button to bring up the **Advanced Port Settings** window. Click on the **Combinatorial** radio-button and then the **OK** button. Now all the inputs and outputs are defined, so click on the **Finish** button.

At this point, an **HDL Editor** window will appear (Figure 2.13) with the skeleton of the Trip-Genie logic. You should see the module name and title at the top of the file. These are followed by declarations for the input pins and a single combinational output pin we added previously. Further down in the file is a section intended for logic equations. All we need to do is enter the Boolean equation we derived for the Hwy3 output in the previous chapter. This is typed on line 17 and 18 as follows:

```
Hwy3 = (Hooterville & !Mayberry # !Hooterville & Mayberry) &
       (SilerCity & !MtPilot);
```

We have used ABEL operators to express the following Boolean operations:

Boolean Operation	ABEL Operator
AND	&
OR	#
NOT	!

Now we have to check to make sure we have not made any simple mistakes. Select the **Synthesis** → **Check Syntax** menu item. A small pop-up window will appear informing you that the ABEL code is being examined for errors. Within a few seconds, this window should be replaced by another window stating **Check Successful**. Click the **OK** button in this window.

(You are likely to get syntax errors when you first begin to use ABEL for your own designs. Each error will be highlighted in the **HDL Editor** window and an error message will appear at the bottom of the window. You can get error-free ABEL code examples by selecting the **Tools** → **Language Assistant** menu item. A **Language Assistant-ABEL** window will appear which lists topics on the left-hand side. Click on the + symbols to drill-down to a particular topic. Click on the topic and an ABEL code example for that topic will appear on the right side of the window. You can cut-and-paste the example into your own code and then edit it for your particular design.)

Once we know there are no syntax errors, we can generate a netlist from the ABEL code. The netlist is a distillation of the ABEL code into a set of logic gates and the connections between them. First, click on the **Synthesis** → **Options...** menu item. In the **ABEL6** window that appears, click on the **Chip** radio-button and then click **OK**. This directs the synthesizer to create a netlist

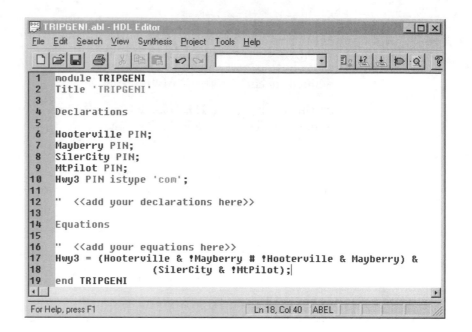

Figure 2.13 HDL Editor window with the filled-in Trip-Genie ABEL code skeleton.

for a stand-alone design. (The **Macro** radio-button is used when the ABEL code describes a logic circuit that is used as part of a larger design.) Then click on the **Synthesis** → **Synthesize** menu item to start the compilation process. A small pop-up window will appear informing you that the ABEL code is being synthesized. Within a minute, this window should be replaced by another window stating **Synthesis Successful**. Click the **OK** button in this window. The netlist for the Hwy3 output of the Trip-Genie now exists.

Now that the design entry is complete, select **File** → **Save** in the **HDL Editor** window. Then select **File** → **Exit**.

Upon returning to the **Project Manager** window, we must make the TRIP-GENI.ABL file a part of our project. Select the **Document** → **Add...** menu item and list items of type **ABEL6** (***.ABL**)[1] in the dialog window. Highlight TRIPGENI.ABL and click **Open**. You should then see TRIPGENI.ABL in the list of files making up the TRIPGENI project (Figure 2.14).

---

1. You can get all the design examples in this book from the enclosed CD-ROM if you want to avoid reentering them all.

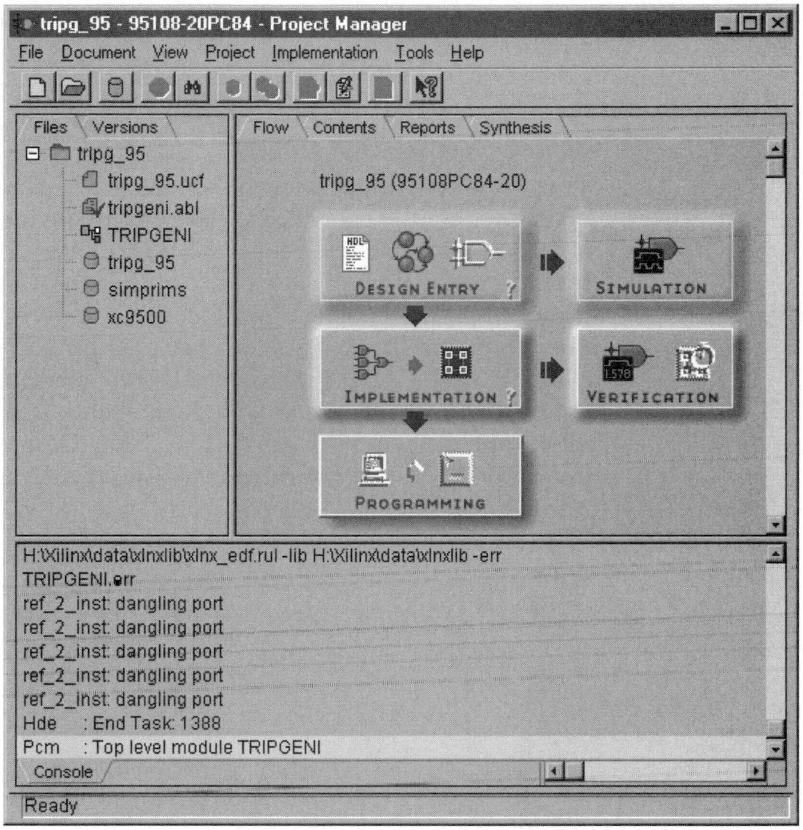

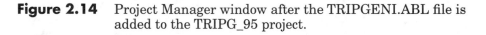

Figure 2.14   Project Manager window after the TRIPGENI.ABL file is added to the TRIPG_95 project.

Entering the Design Using the Schematic Editor

In this section, I will show you how to enter the design for the Hwy3 output of the Trip-Genie using the schematic editor. (You can skip this section if you have already described the design using an ABEL file. I have oriented this schematic-based design toward the XC4005XL FPGA, so you will find it in the TRIPG_40 project.) In the **Project Manager** window, select **Tools → Design Entry → Schematic Editor...** (or click on the ▦ button in the **Flow** tab of the **Project Manager** window) and a schematic editor window will appear. Then name the schematic TRIPGENI.SCH using the **File → Save As...** menu item.

The first thing we need to do is add the gates for the Hwy3 circuit. Select the **Mode → Symbols** menu item (or click on the ▫ button) and the **SC Symbols** window will appear with a list of all the types of components we can use. We will use the following Boolean Equation from Chapter 1 as the basis for our circuit:

$$\text{Hwy3} = (\text{Hooterville} \cdot \overline{\text{Mayberry}} + \overline{\text{Hooterville}} \cdot \text{Mayberry})$$
$$\cdot\ (\text{Siler City} \cdot \overline{\text{Mt. Pilot}})$$

This equation requires four 2-input AND gates, a 2-input OR gate, and 3 inverters. Scroll through the list of components in the **SC Symbols** window until you find **AND2**. Click on **AND2** to select it and then move the cursor into the schematic drawing area. You will see the symbol for a 2-input AND gate attached to the cursor. Simply click to drop the AND gate into the drawing area. You can attach another copy of the AND gate to your cursor simply by clicking on the AND gate you just dropped. Then you can drop the new AND gate onto the schematic as well. Repeat this process twice more to arrive at the schematic shown in Figure 2.15. The 2-input OR gate (**OR2**) and three inverters (**INV**) can be added in a similar manner.

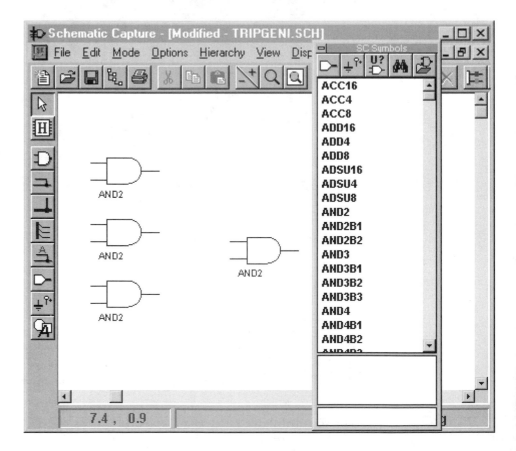

Figure 2.15 Placing AND gates in the schematic editor window.

We also need to get our inputs and outputs into the circuit. To do this, click on the ▣ button at the upper-left corner of the toolbar in the **SC Symbols** window. A dialog window will appear in which you can type the name and type of each input or output. First, enter the information for the Mayberry input as shown in Figure 2.16. Click **OK** and an input terminal symbol will be attached to your cursor. Simply click the mouse to drop the terminal into your schematic. Then add the other 3 inputs in the same manner. Finally, you can add the Hwy3 output in the same way with the only change being to select **OUTPUT** as the **Terminal Type**.

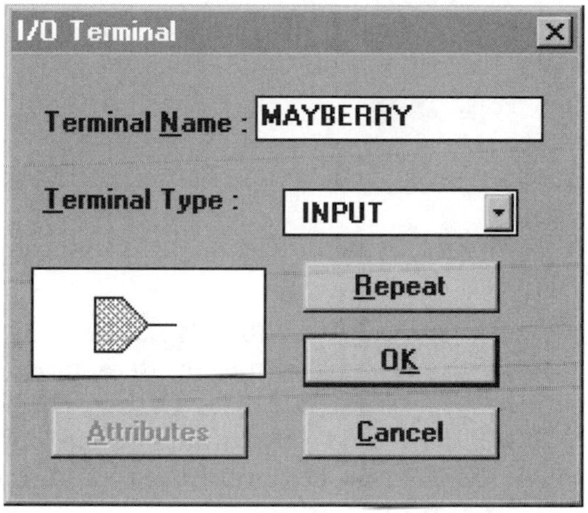

Figure 2.16 Adding a terminal to the schematic.

While we have entered the input and output terminals, we still need to add buffers between the terminals and the logic gates. These buffers indicate that the signals attached to them will actually enter and exit the FPLD chip via the I/O pins. To add the input buffers, select the **IBUF** symbol from the **SC Symbols** window and drop one near each input terminal. Then select the **OBUF** symbol for an output buffer and drop it near the output terminal. At this point, your placement of gates, terminals, and buffers should look something like Figure 2.17.

Now we have all the components we need so you can double-click on the upper-left corner of the **SC Symbols** window to get rid of it. The next step is to connect the gates. Select the **Mode → Draw Wires** menu item to begin the process. Then click on the output of the top-most AND gate followed by clicking on the upper input of the OR gate. A line will appear connecting the output to the input as shown in Figure 2.17.

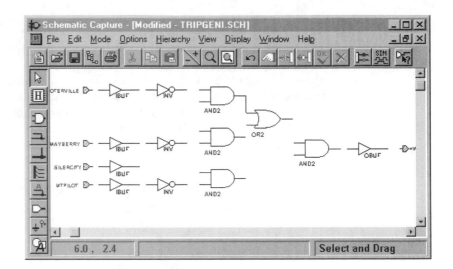

Figure 2.17   Wiring gates together in the schematic editor window.

You can continue in this manner until all the gates are connected as required by the Boolean equation. Remember to attach the input and output terminals only to the IBUF and OBUF symbols, respectively. Use the outputs of the IBUFs to drive the inputs of the logic gates and connect the input of the lone OBUF to the output of the final AND gate. The completed schematic is shown in Figure 2.18.

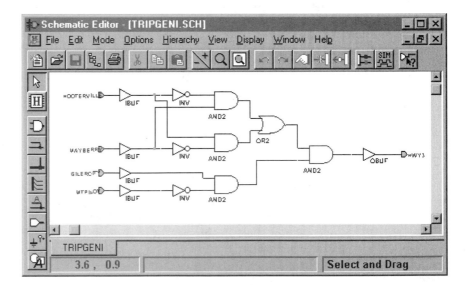

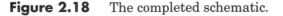

Figure 2.18   The completed schematic.

Now that the schematic is done, we need to check it for errors. First, select **Options → Create Netlist**. This will activate a program that examines our schematic drawing and generates a machine-readable netlist which describes what types of gates are used and how they are connected to each other. Once this is done, select **Options → Integrity Test** to initiate an error check on the netlist. The check should indicate that there are no errors in the schematic. Then save the schematic using the **File → Save** menu item.

Since there are no errors, we can go ahead and export the netlist for the design. The netlist is generated from the schematic and contains a machine-readable list of the logic gates and the connections between them. The netlist must be exported in a format that the other tools (like the simulator) understand. First, click on the **Options → Export Netlist...** menu item. An **Export Netlist** window will appear (Figure 2.19). Select Edif 200 [*.EDN] in the **Netlist Format** selection box. (EDIF 200 is the standard netlist format for the Foundation tools.) Click on the **Open** button. For several seconds, progress will be shown in a pop-up window as the internal netlist format of the schematic editor is exported as EDIF 200 and stored into the TRIPG_40.ALB file.

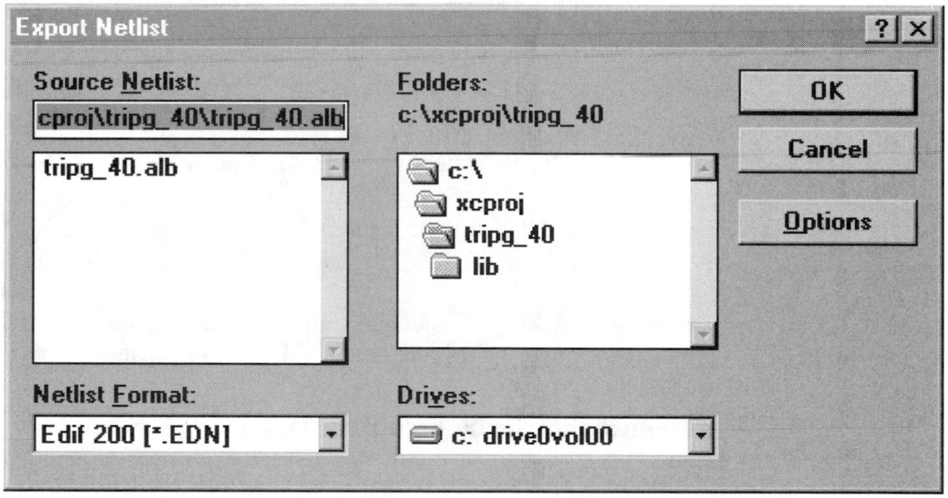

Figure 2.19   Window for exporting a netlist in EDIF 200 format.

Now select **File → Exit** to close the schematic editor. Upon returning to the **Project Manager** window, we must make the TRIPGENI.SCH file a part of our project. Select the **Document → Add...** menu item and list items of type **Schematic** (*.SCH) in the dialog window. Highlight TRIPGENI.SCH and click **Open**. You should then see TRIPGENI.SCH in the list of files making up the TRIPG_40 project (Figure 2.20).

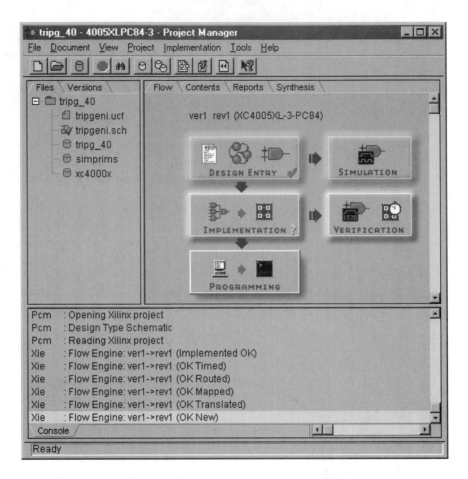

Figure 2.20 Project Manager window after the TRIPGENI.SCH file is added to the TRIPG_40 project

Functional Simulation of the Design

At this point, we have entered the logic for the Hwy3 output of the Trip-Genie. Now it is a good idea to use the functional simulator to see if what we have entered is working correctly.

Start the functional simulator by clicking the 🖼 button in the **Flow** tab of the **Project Manager** window. This will bring up the **Logic Simulator-Xilinx Foundation F1.5** window and a single, empty **Waveform Viewer** window (Figure 2.21).

The first thing we must do is add the inputs and output of the Hwy3 logic circuit to the **Waveform Viewer** so we can see what is happening as the circuit is simulated. Do this by selecting the **Signal → Add Signals...** menu item.

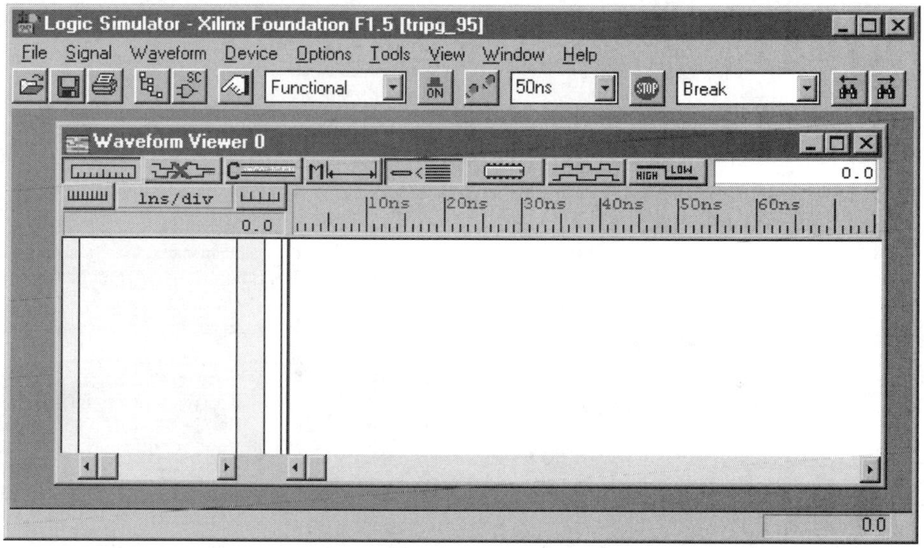

Figure 2.21 The **Logic Simulator** and **Waveform Viewer** windows.

The **Component Selection for Waveform Viewer** window will appear. Click on the **Hooterville** input to highlight it and then click on the **Add** button. A waveform labeled Hooterville will appear in the Waveform Viewer and a red checkmark will appear by the selected signal (Figure 2.22). We can repeat this procedure for the Mayberry, SilerCity and MtPilot inputs and the Hwy3 output. Then click on the **Close** button.

Now the inputs and outputs are displayed, but nothing interesting is happening because all the inputs are set to logic 0. We need to apply a stimulus to the circuit so, naturally, we select the **Signal → Add Stimulators...** menu item. This brings up the **Stimulator Selection** window (Figure 2.23). There are a frightening number of buttons in this thing, but we are only interested in a single item: a 16-bit binary counter labeled **Bc**. During a simulation, the right-most 4 bits of this counter will go through the following sequence: 0000, 0001, 0010, 0011, 0100, 0101, 0110, 0111, 1000, 1001, 1010, 1011, 1100, 1101, 1110, 1111. We can completely test the response of our Hwy3 circuit to every possible combination of inputs by attaching the 4 inputs of our logic circuit to these bits of the counter. Do this by clicking on the name of an input in the **Waveform Viewer** window and then clicking on one of the bit-circles in the **Bc** section of the **Stimulator Selection** window. The label of the counter bit attached to the circuit input will appear to the right of the input name in the **Waveform Viewer**. Once all the inputs are attached to the counter bits, we can click on **Close** to leave the **Stimulator Selection** window. Here is the set of attachements that I used:

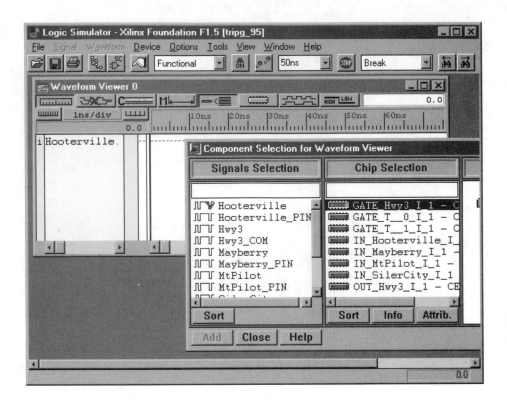

Figure 2.22 Adding signals to the **Waveform Viewer**.

```
Hooterville    B3
Mayberry       B2
SilerCity      B1
MtPilot        B0
```

Now we need to setup a parameter that controls the speed of the simulation. First, select the **Options → Preferences...** menu item. Then in the **Preferences** window (Figure 2.24), set the frequency of the B0 bit of the binary counter at 50 MHz. Then click on **OK**.

Now we can run the simulation. Set the simulation mode to **Functional** in the small, drop-down menu in the Logic simulator window tool-bar. This indicates we are doing a functional simulation that checks only the logical operation of our circuit and ignores detailed timing issues that result from the physical implementation of the circuit. Then click on the ![button] button to initiate the functional simulation. Within seconds, the results of the simulation appear as shown in the **Waveform Viewer** window in Figure 2.25. Note that the Hwy3 output goes high only when the inputs [Hooterville, Mayberry, SilerCity,

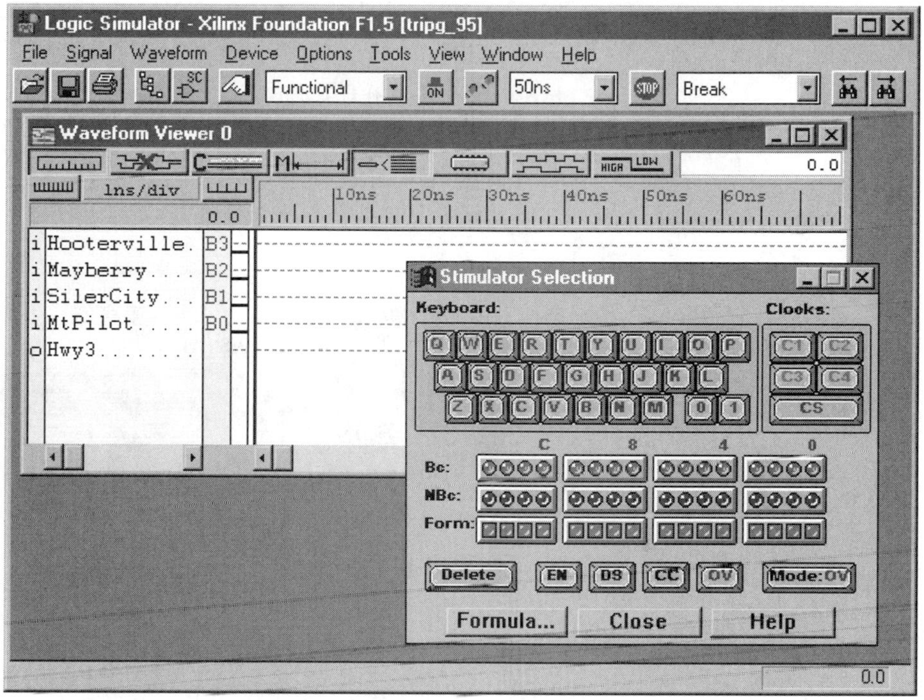

Figure 2.23   The Stimulator Selection window.

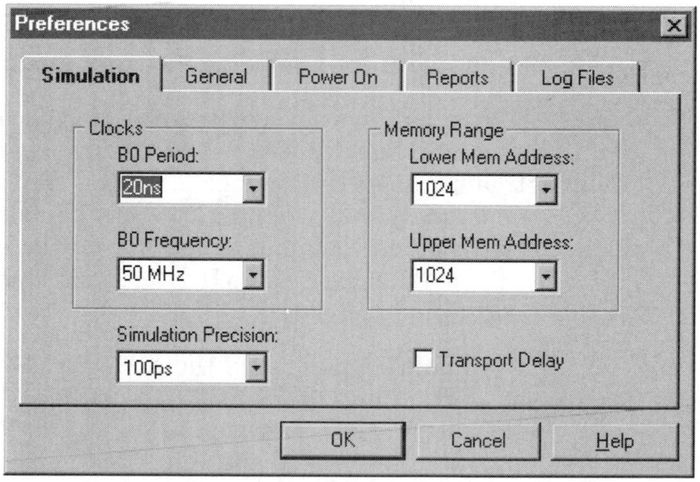

Figure 2.24   Window for setting simulator control parameters.

MtPilot] = [1010] or [0110]. This matches the truth-table shown in Chapter 1 for the Hwy3 output. So it looks like our circuit functions correctly.

After going through all the set-up of the simulator, you might want to keep it. Select the **File → Save Simulation State...** menu item and save the simulation in TRIPGENI.DES. Then select **File → Exit** to close the **Logic Simulator** window.

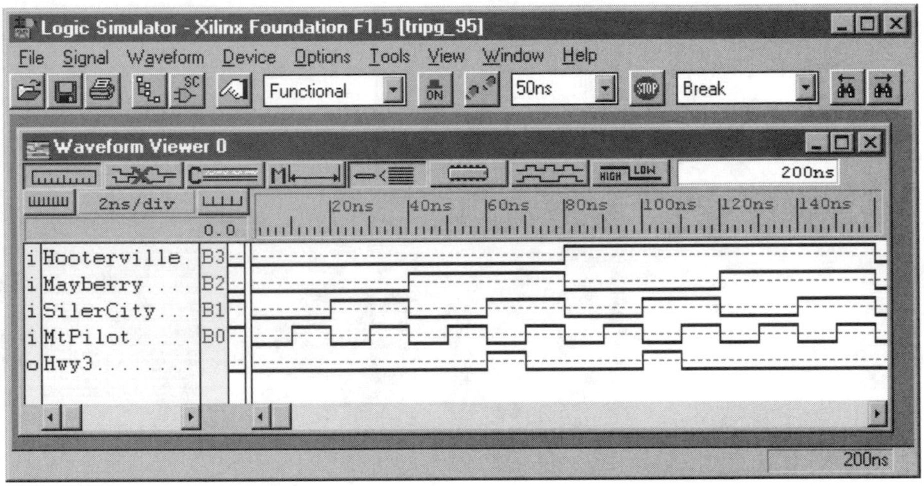

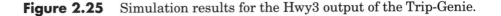

Figure 2.25   Simulation results for the Hwy3 output of the Trip-Genie.

Preparations for Compiling the Logic Design

At this point, we have entered the Hwy3 logic circuit and we have seen that it functions correctly. Now we are almost ready to begin compiling it into a bit-stream that we can load into an FPLD chip and test. But physically testing the design means we will have to apply signals to certain pins of the FPLD that correspond to the inputs of our Trip-Genie circuit. We also need to hook an LED to the pin that carries the Hwy3 output so we can visually observe if the circuit is operating correctly.

If you built or purchased the XS95 or XS40 Boards (Appendix A), you know that certain pins of the XC95108 or XC4005XL chip can be driven by the parallel port of a PC connected to the board. This lets us use the parallel port to apply test patterns to the chip. Another set of XC95108 or XC4005XL pins is connected to a 7-segment LED (see Figure 3.5 for the labeling of the LED segments I have used throughout this book). So the XS95 and XS40 Boards have what we need for the job.

However, the Foundation Series Implementation tools that compile the ABEL file, VHDL file, or schematic do not know anything about what FPLD pins are attached to the parallel port or the LED segments. Left on their own, the tools will assign the inputs and outputs of the Trip-Genie to any pins that are convenient. So we need a way to pass our required pin assignments to the Foundation Series tools. There are three ways to do this:

1. Embed the pin numbers in the PIN statements of the ABEL file,
2. Enter the pin numbers as attributes of the IBUF and OBUF buffers in the schematic,
3. Create a separate User Constraint File (UCF) that lists the pin number for each input and output.

Entering the pin numbers right into the ABEL file is very easy. Figure 2.26 shows the TRIPGENI.ABL file with the pin numbers added to allow for testing the design on the XS95 Board.

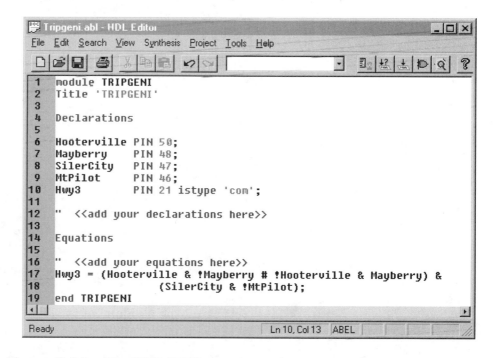

Figure 2.26   The TRIPGENI.ABL file with embedded pin assignments.

In the TRIPGENI.SCH schematic, we can double-click on any IBUF or OBUF symbol to set its attributes. For example, double-clicking on the IBUF attached to the Hooterville input terminal brings up the **Symbol Properties** window of

Figure 2.27. The following steps assign this input buffer to a pin 47 of the XC4005XL FPGA:

1. Click in the **Name** box and type LOC.

2. Click in the **Description** box and type p47.

3. Click the **Add** button.

4. Click **OK**.

You can repeat this procedure to assign the MtPilot, SilerCity, and Mayberry inputs to pins 44, 45, and 46, respectively, and to assign the Hwy3 output to pin 25.

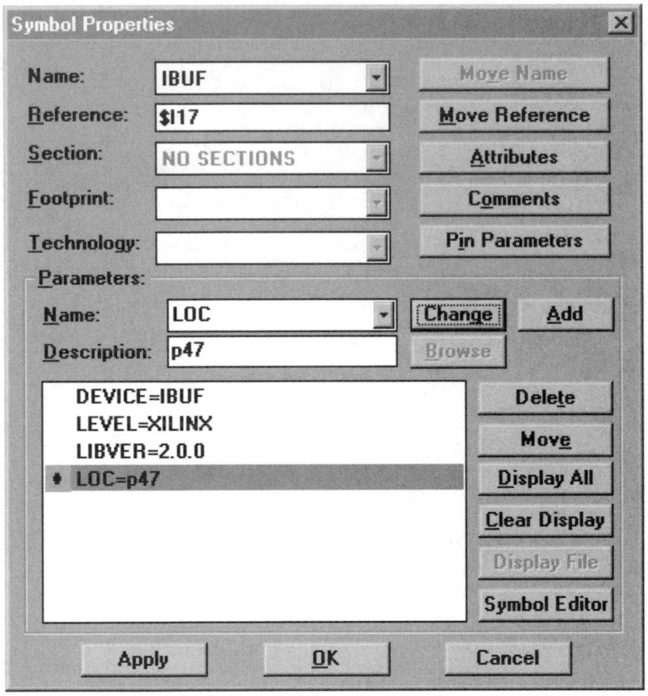

Figure 2.27 Setting the pin assignment attribute for the Hooterville IBUF in the Trip-Genie schematic.

Finally, we can use the HDL Editor to create a UCF (User Constraint File) that assigns the I/O terminals to specific FPLD I/O pins. In the **Flow** tab of the **Project Manager** window, click on the 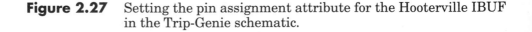 button and then click on the **create**

Programmable Logic Design Techniques   Chapter 2

empty radio button to bring up an empty editor window. Type the following lines into the window if you are using an XS95 Board:

```
NET Hooterville     LOC=p50;
NET Mayberry        LOC=p48;
NET SilerCity       LOC=p47;
NET MtPilot         LOC=p46;
NET Hwy3            LOC=p21;
```

or enter this if you are using an XS40 Board:

```
NET HOOTERVILLE     LOC=p47;
NET MAYBERRY        LOC=p46;
NET SILERCITY       LOC=p45;
NET MTPILOT         LOC=p44;
NET HWY3            LOC=p25;
```

(Note that we had to use upper-case letters for the names in the second UCF because the TRIPGENI.SCH schematic only allows upper-case letters for I/O terminal names.)

Then use the **File → Save As...** menu item to save the file as TRIPGENI.UCF. All this UCF does is tell the Foundation Series Implementation tools to assign the 4 city inputs to pins 46, 47, 48 and 50, and to assign the Hwy3 output to pin 21 of the XC95108. We already know from our knowledge of the XS95 Board that pins 46, 47, 48 and 50 can be driven from the PC parallel port while pin 21 drives the bottom-most, horizontal segment of the 7-segment LED. (You can fill in the equivalent pin numbers for the XC4005XL on the XS40 Board.)

Which method of entering pin assignments should you use? If you are only doing designs for one type of board (either the XS40 or XS95 Board), then it is probably easier to embed the pin assignments in the design file. However, if you plan to implement your designs on both boards, then placing the pin assignments in a separate UCF is a cleaner approach. Since the examples in this book are intended for both boards, I will use UCFs exclusively to assign pins.

Compiling the Logic Design

Once the pin-assignments are entered and we are back in the **Project Manager** window, click on the ▣ ▸ ▦ button or select the **Implementation → Implement Design...** menu item to begin the compilation process. The **Implement Design** window of Figure 2.28 will appear. The **Device** and **Speed** should mirror what you typed in when you started the project, and the **version** and **revision** names should be **ver1** and **rev1**, respectively. (You can change any of these if you want.)

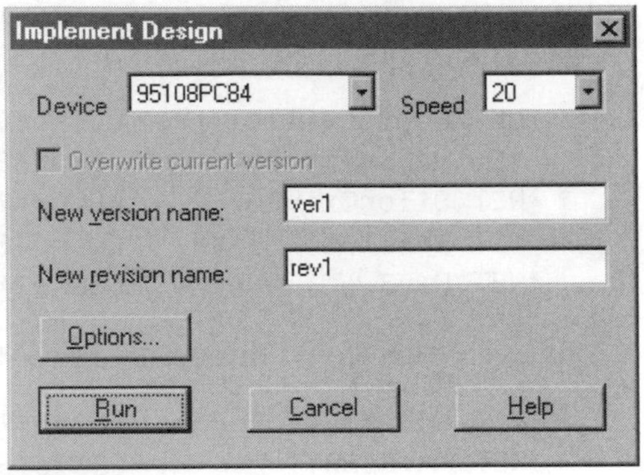

Figure 2.28 **Implement** window.

In the **Implement** window, click on the **Options...** button. This will bring up the window used to set the control parameters for the Foundation Series Implementation tools (Figure 2.29). The only thing we need to do is enter the location of the UCF into the **User Constraints** box. You can use the **Browse...** button to find the UCF, or you can type in its location. Then click on **OK** to return to the **Implement Design** window.

After returning to the **Implement Design** window, click on **Run**. The **Flow Engine** window appears which has a graphical rendition of the stages required to map a netlist into an FPLD (in this case, Figure 2.30 shows the steps needed to map into an XC9500 CPLD). The Foundation Series Implementation tools run through 4 stages:

Translate: The EDIF netlist is converted to an internal netlist format, design-rule checks are performed, and various optimizations of the logic circuit are tried to increase the speed of the circuit and/or decrease the number of gates used.

Fit: The gates in the netlist are assigned to specific macrocells and their interconnections are routed through the FastCONNECT matrix.

Timing: The propagation delays through the macrocells and the FastCONNECT matrix are computed and stored in a file for use during a timing simulation.

Bitstream: The stream of bits is generated which will be downloaded into the chip to make it carry out the logic functions described in the ABEL or schematic file.

If you are implementing the design for the XC4005XL FPGA and the XS40 Board, then the Flow Engine window shown in Figure 2.31 will appear. There are more stages in the implementation process when targeting an FPGA:

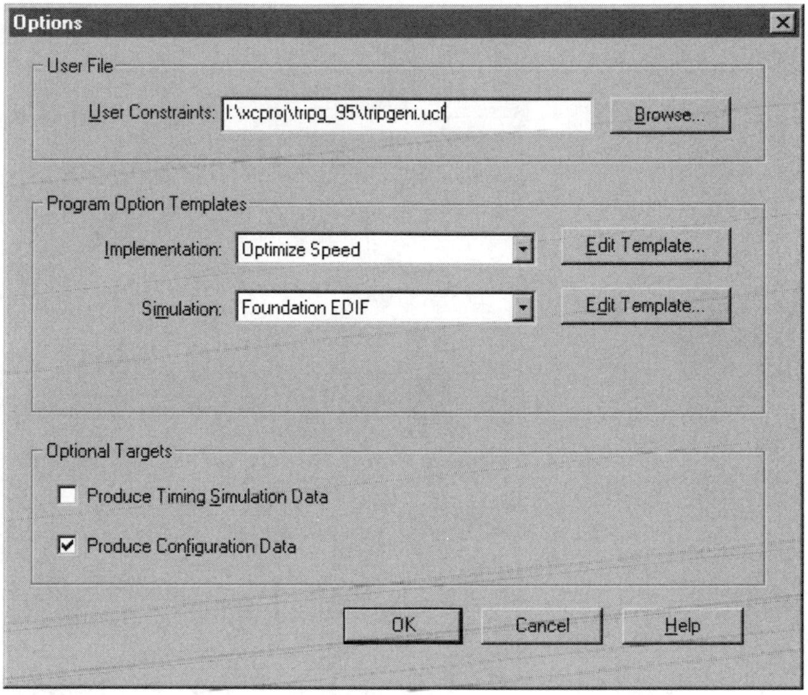

Figure 2.29 The window for setting control parameters for the Foundation Series Implementation tools.

Translate: The EDIF netlist is converted to an internal netlist format and design-rule checks are performed.

Map: Various optimizations of the logic circuit are tried to increase the speed of the circuit and/or decrease the number of gates used.

Place&Route: The gates in the netlist are assigned to specific CLBs and their interconnections are routed through the PSMs and other routing resources of the FPGA.

Timing: The propagation delays through the CLBs and the routing PSMs are computed and stored in a file for use during a timing simulation.

Configure: The stream of bits is generated which will be downloaded into the chip to configure it to carry out the logic functions described in the ABEL or schematic file.

The label below each stage will change from **Running** to **Completed** as it finishes. (If a problem occurs during a particular stage of processing, an **Aborted** label will appear instead and an indication of the type of error will appear in the lower dialog box.) After all the stages are finished, a pop-up window will

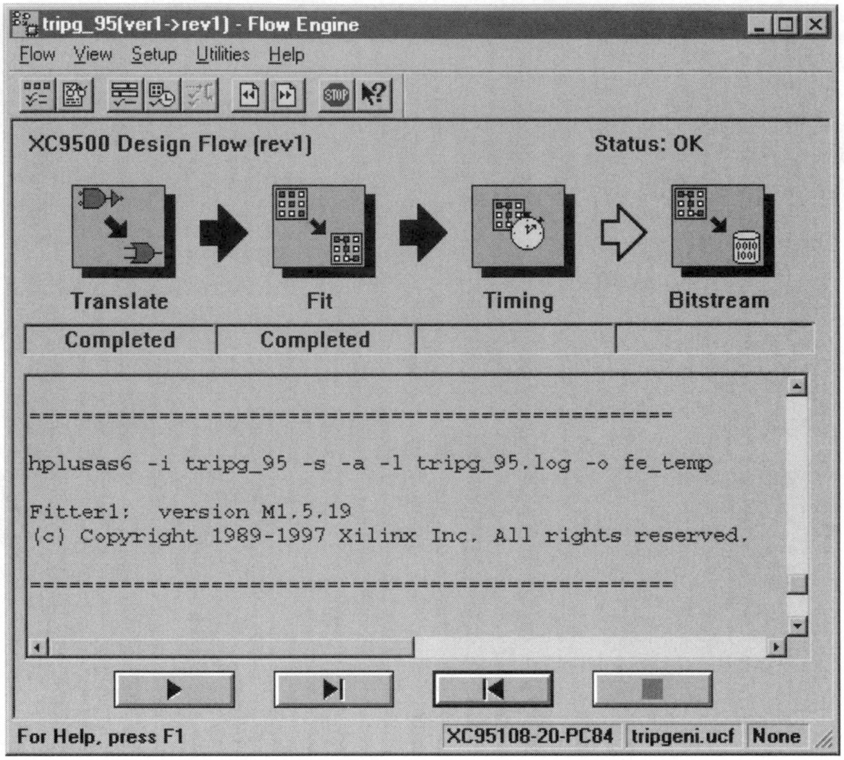

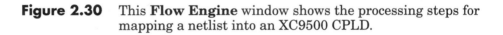

Figure 2.30   This **Flow Engine** window shows the processing steps for mapping a netlist into an XC9500 CPLD.

appear indicating the success or failure of the compilation process. Click **OK** in this window to return to the **Project Manager** window.

Performing a Timing Simulation

In this example we will not bother to perform a detailed timing simulation of the Hwy3 circuit. Chapter 5 will cover this subject.

Preparing Bitstreams for Downloading to the XS95

Now that the Hwy3 circuit bitstream has been generated, we need to do a little more processing before we can download it to an XC95108 CPLD on an XS95 Board. (The TRIPG_40.BIT bitstream file that is downloaded to an XC4005XL FPGA on an XS40 Board is OK as it is, so you can skip this section if that is what you are using.)

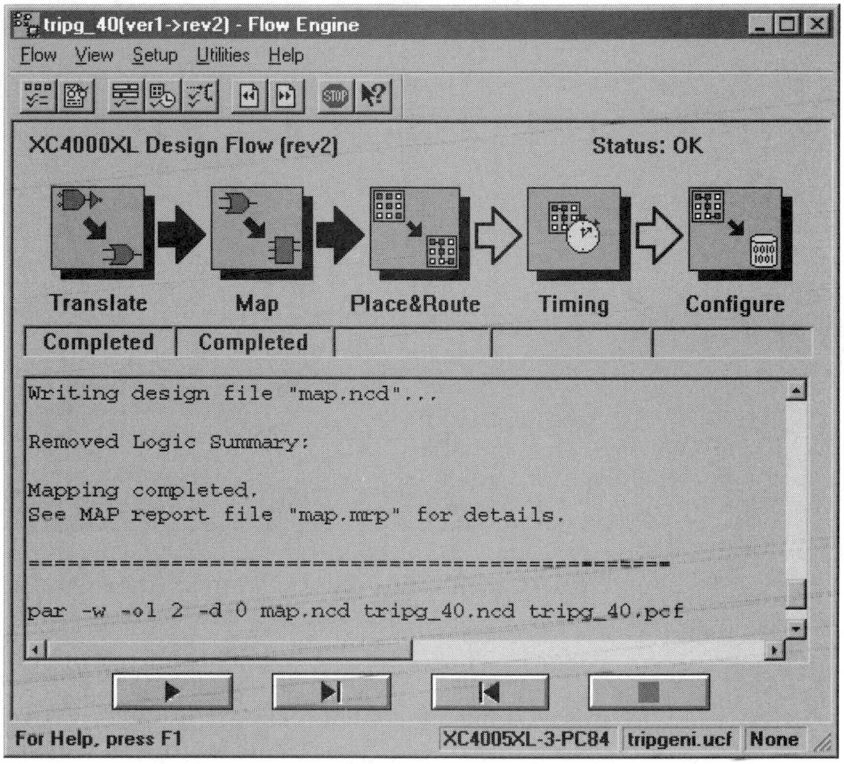

Figure 2.31 This **Flow Engine** window shows the processing steps for mapping a netlist into an XC4000XL CPLD.

To begin translating the bitstream file that was previously generated by the Flow Engine into a form we can download into an XC9500 CPLD, select the **Tools → Device Programming → JTAG Programmer** menu item or click on the ▣•◼ button in the **Design Manager** window. The **JTAG Programmer** window of Figure 2.32 will appear. This window lets us produce a stream of bits that is suitable for programming a set of one or more XC9500 chips which are attached end-to-end in a chain. Since we are only programming a single chip in this example, we do not need to add any more CPLDs to the chain shown in the window.

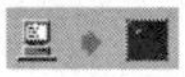

The XS95 Board downloading software requires that the bitstream be stored in an SVF file format. Select the **Output → Create SVF File...** menu item. Then an **SVF Options** pop-up window will appear. Select the radio button which makes the initial transition to Run-Test/Idle through Test-Logic-Reset (do not worry about what this means — it has to do with initialization of the JTAG circuitry in the XC9500 CPLD). Then click on **OK** and another pop-up window will appear that lets you specify the name of the SVF file (TRIPG_95.SVF in

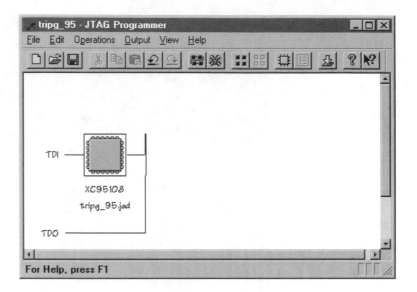

Figure 2.32 The **JTAG Programmer** window is used to translate bitstream files into a form that can be loaded into the XS95 Board.

this example) and what directory it will be placed in. Normally, the file will be placed in a directory associated with the version and revision of your design, but that makes it harder to get to when you need it. For that reason, I like to place it in the top directory for the project (C:\XCPROJ\TRIPG_95 in this example). After specifying the name and directory, click on the **Save** button.

Now select the **Operations → Program** menu item in the **JTAG Programmer** window. In the Options window that appears (Figure 2.33), there are several checkboxes. You do not need to change any of them from the default values. Just click on **OK** and the SVF will be generated. The intermediate results of the SVF generation are reported in a pop-up window (Figure 2.34). If all the status messages report success, then the SVF has been generated. Click on **OK** to return to the **JTAG Programmer** window. Then select **File → Exit** to return to the **Design Manager** window. You will be asked if you want to save the changes to tripg_95. You can click on the **No** button because this only saves the chain-description file (CDF) and has no effect on the SVF.

Downloading the Logic Design to the XS40 or XS95 Board

If you are targeting the XC95108 CPLD and the XS95 Board, at this point you should open a DOS window and type:

```
CD C:\XCPROJ\TRIPG_95
```

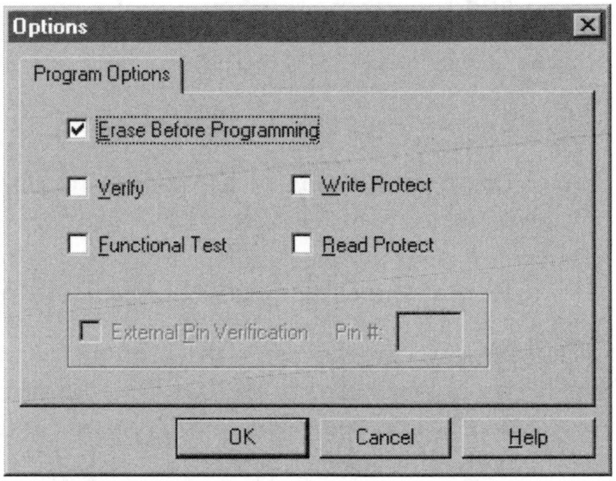

Figure 2.33 The **Options** window lets you initiate the generation of the SVF file.

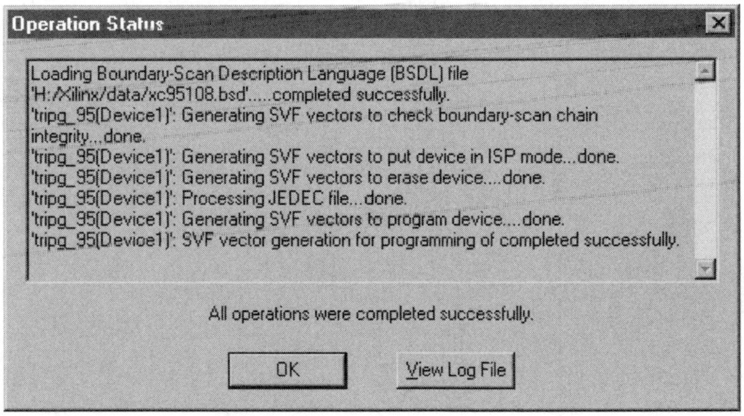

Figure 2.34 The **Operation Status** window reports the intermediate results that occur during the generation of the SVF.

to move to the project directory where the Trip-Genie design for the XS95 Board is stored. You should connect the XS95 to the parallel port of your PC and apply power to the board. Then you can download the Hwy3 circuit into the XC95108 CPLD on the XS95 board using the command:

```
XSLOAD TRIPG_95.SVF
```

On the other hand, if you are targeting the XC4005XL FPGA and the XS40 Board, you should connect the XS40 Board to the PC parallel port and type:

```
CD C:\XCPROJ\TRIPG_40
```

and then issue the command:

```
XSLOAD TRIPG_40.BIT
```

to download the Hwy3 circuit into the XC4005XL FPGA on the XS40 Board.

Upon completion of the XSLOAD command, the XS95 or XS40 Board will be programmed with the Hwy3 logic.

Testing the Downloaded Logic Design

Now we need to force the inputs to various logic levels to see if the Hwy3 circuit we downloaded to the XS95 or XS40 Board is working. The XSPORT program accepts a string of 1 and 0 bits and outputs it on the 8 data pins of the parallel port that are connected to pins of the XC95108 CPLD (XC4005XL FPGA) on the XS95 Board (XS40 Board). Given the command

XSPORT $b_7 b_6 b_5 b_4 b_3 b_2 b_1 b_0$

where b_7–b_0 represent binary bits, the correspondence between the bits in the string and the pins on the XC95108 or XC4005XL is as follows:

String Bit	XC95108 Pin	XC4005XL Pin
b_0	46	44
b_1	47	45
b_2	48	46
b_3	50	47
b_4	51	48
b_5	52	49
b_6	81	32
b_7	80	34

XSPORT sets the most significant bits to zero if it is given fewer than 8 bits. If XSPORT is given more than 8 bits, it keeps only the 8 least-significant bits.

The pin assignments we used in our UCF file leads to the following correspondence between the bitstring passed to XSPORT and the inputs to the Hwy3 circuit:

String Bit	Hwy3 Input
b_0	MtPilot
b_1	SilerCity
b_2	Mayberry
b_3	Hooterville
b_4	*** Not Used ***
b_5	*** Not Used ***
b_6	*** Not Used ***
b_7	*** Not Used ***

The UCF also assigned the Hwy3 output to pin 21 (pin 25) of the XC95108 CPLD (XC4005XL FPGA). This pin is attached to the lower-most, horizontal segment of the 7-segment LED digit (S0). When Hwy3 = 1, then S0 is bright. Otherwise, S0 is dark.

We can use XSPORT with all 16 combinations of inputs to completely test the Hwy3 circuit. Here are the commands you can issue and what the results should be:

Command	Hooterville	Mayberry	SilerCity	MtPilot	Hwy3
XSPORT 0000	0	0	0	0	0 (dark)
XSPORT 0001	0	0	0	1	0 (dark)
XSPORT 0010	0	0	1	0	0 (dark)
XSPORT 0011	0	0	1	1	0 (dark)
XSPORT 0100	0	1	0	0	0 (dark)
XSPORT 0101	0	1	0	1	0 (dark)
XSPORT 0110	0	1	1	0	1 (bright)
XSPORT 0111	0	1	1	1	0 (dark)
XSPORT 1000	1	0	0	0	0 (dark)
XSPORT 1001	1	0	0	1	0 (dark)
XSPORT 1010	1	0	1	0	1 (bright)
XSPORT 1011	1	0	1	1	0 (dark)
XSPORT 1100	1	1	0	0	0 (dark)
XSPORT 1101	1	1	0	1	0 (dark)
XSPORT 1110	1	1	1	0	0 (dark)
XSPORT 1111	1	1	1	1	0 (dark)

If you have done everything correctly, your tests of the downloaded design should match the simulation results and the truth-table.

Trip-Genie Project Design (HDL Mode)

In this section we will look at doing the Trip-Genie design using VHDL and the HDL design flow of the Xilinx Foundation Series software. The only real difference between the schematic and HDL design flows is the method by which the netlist is extracted from the design description. Therefore, this section will only discuss the steps of entering the VHDL code for the Trip-Genie design and the synthesis of the netlist from this VHDL code. The steps which follow (functional simulation, compiling into a bitstream, downloading the bitstream to an XS Board, and testing the downloaded design) are performed exactly as they were in the schematic design flow discussed in the previous section.

Starting a New Project

To begin, create a main directory for all your projects. (In my case, I place all HDL mode designs in C:\XCPROJ-V.) Then click on the **Foundation Project Manager** icon. This will bring up the window shown in Figure 2.35.

The **New Project** sub-window is displayed. (You can also bring up this sub-window by selecting the **File → New Project** menu item.) Then, enter the project name, project directory, and click on the HDL radio button to select the HDL design flow. Once you select the HDL mode, the chip family, chip part number, and device speed drop-down menus disappear. (You will set these values later during the synthesis phase of the design flow.) In Figure 2.35, I have created a project called TRIPG_95 in directory C:\XCPROJ-V. Then click **OK** to return to the **Project Manager** window.

Now your window should look just like Figure 2.36. There are three main panes in the **Project Manager** window. The left-hand pane displays the source files and libraries for your design in the **Files** tab, while the **Versions** tab shows design implementation files generated by the XILINX Foundation Software for all the versions of your design. The right-hand pane has a **Flow** tab for controlling the activation of tools used in various phases of the design flow, a **Contents** tab which shows the status of the files in your project, and a **Reports** tab which lists the report files generated by the tools as they do their work. The bottom pane lists diagnostic messages output by the Foundation software as it processes your design files. For now, you will spend most of your time working in the **Files** and **Flow** tabs of the left and right panes, respectively.

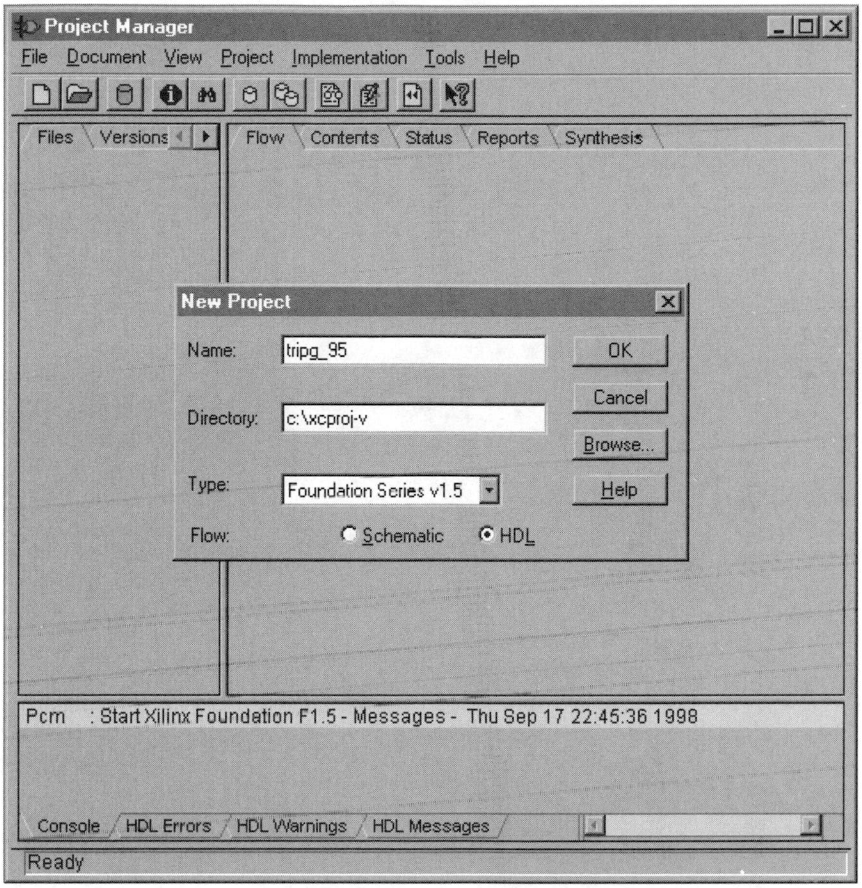

Figure 2.35   Foundation Series 1.5 Project Manager main window.

Entering the Design Using the VHDL Hardware Description Language

In this section, I will show you how to enter the design for the Hwy3 output of the Trip-Genie using VHDL. VHDL is a more complex language that ABEL and is more suitable for large designs intended for high-density FPLDs.

In the **Project Manager** window, select **Tools** → **Design Entry** → **HDL Editor** or click on the 📄 button in the Flow tab of the **Project Manager** window. In the window that appears, click on the radio-button labeled **Use HDL Design Wizard** and then click on **OK**. Then click on **Next** in the **Design Wizard** window that appears to reach the **Design Wizard-Language** window. Here, select the **VHDL** radio-button and click on **Next** to move to the **Design Wizard-Name** window. Enter TRIPGENI as the name and click **Next**.

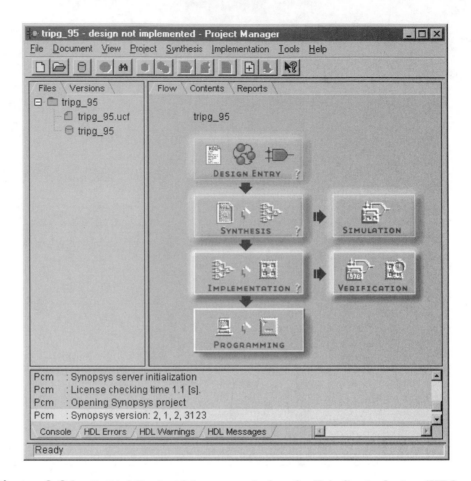

Figure 2.36 Initial Project Manager window for TripGenie design (HDL design flow).

Now we should be in the **Design Wizard-Ports** window where we specify the inputs and outputs for our design (Figure 2.37). To add an input, click on **New** and then type **Hooterville** (for example) in the **Name** field. Click on the **Input** radio-button to set the port to be an input. Repeat these steps for the **Mayberry**, **SilerCity**, and **MtPilot** inputs. To add the output of our circuit click on **New**, enter **Hwy3** in the **Name** field, and click on the **Output** radio-button. Now all the inputs and outputs are defined, so click on the **Finish** button.

(You can click on the **Advanced** button to bring up the **Advanced Port Settings** window. The drop-down menu in this window lets you set the type of any of the inputs or outputs. The default setting is **STD_LOGIC** and this is fine for this example.)

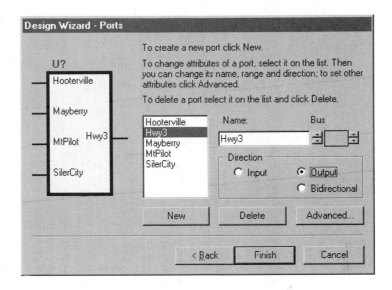

Figure 2.37 Adding inputs and outputs to the TRIPGENI design.

At this point, an **HDL Editor** window will appear (Figure 2.38) with the skeleton of the Trip-Genie logic. The first line of the TRIPGENI.VHD file uses the LIBRARY keyword followed by the names of the libraries we want to use in our design. Libraries are used to encapsulate functions that are generally useful in a wide variety of designs. In this example, we can access the macros, definitions, and functions from the IEEE library. A library can be subdivided into packages which further encapsulate features useful in a certain area or type of application. The USE keyword in line 2 indicates that our design will have access to ALL the features found in the STD_LOGIC_1164 package of the IEEE library. (The IEEE library and the STD_LOGIC_1164 package are standards which are supported by all VHDL tools.)

Following the library access control lines, lines 4 through 12 define the interface to the Trip-Genie circuit. The interface declares the inputs and outputs which an external circuit can use to gain access to the features and functions of this circuit. The interface definition begins with the ENTITY keyword followed by the name of the interface (TRIPGENI in this example). The PORT keyword on line 5 begins the list of input and output declarations which follows on lines 6 through 10. A declaration begins with the identifier of the input or output followed by a colon (:). Then the identifier is declared as an input or output using the IN and OUT keywords, respectively. This is followed by the type of the data going through the input or output. For most of the examples in this book we will use the STD_LOGIC type which allows logic signals to take on the standard 1 and 0 Boolean states as well as the undefined, high-impedance, and other states. Each input/output declaration is terminated with a

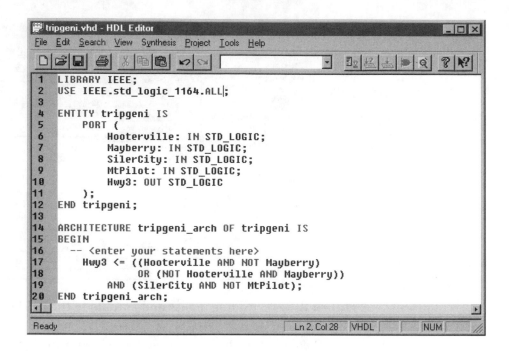

Figure 2.38 HDL Editor window with the filled-in Trip-Genie VHDL code skeleton.

semicolon (;) except the declaration on line 10 which is immediately before the parenthesis that ends the list (line 11). The entire interface definition is terminated by the END keyword followed by the name of the interface on line 12.

The interface definition is followed by an architecture definition on lines 14 through 20. The VHDL statements in the architecture section describe how the circuit actually carries out the operations on the input/output values passed through the interface. The ARCHITECTURE keyword begins this section, followed by the name of the architecture (TRIPGENI_ARCH) and the interface to which the architecture is linked (TRIPGENI).

The VHDL statements which describe the logic circuitry are bounded by the BEGIN and END keywords on lines 15 and 20, respectively. Lines 17 through 20 contain the VHDL statement for the Hwy3 Boolean equation we derived in the previous chapter:

```
Hwy3 <=    ((Hooterville AND NOT Mayberry) OR
            (NOT Hooterville AND Mayberry)) AND
            (SilerCity AND NOT MtPilot);
```

The Hwy3 output identifier is placed to the left of the VHDL signal assignment operator (<=). On the right-hand side, the primitive VHDL logical operators

(AND, OR, NOT) are used to combine the values of the 4 inputs to the Trip-Genie circuit. There is no operator precedence in VHDL, so parentheses must be used to group the AND and OR operations so the sum-of-products is computed correctly. The logic equation can be broken across multiple lines and is terminated by a semicolon.

Now we have to check to make sure we have not made any simple mistakes. Select the **Synthesis → Check Syntax** menu item. A small pop-up window will appear informing you that the VHDL code is being examined for errors. Within a few seconds, this window should be replaced by another window stating **Check Successful**. Click the **OK** button in this window.

(You are likely to get syntax errors when you first begin to use VHDL for your own designs. Each error will be highlighted in the **HDL Editor** window and an error message will appear at the bottom of the window. You can get error-free VHDL code examples by selecting the **Tools → Language Assistant** menu item. A **Language Assistant - VHDL** window will appear which lists topics on the left-hand side. Click on the + symbols to drill-down to a particular topic. Click on the topic and a VHDL code example for that topic will appear on the right side of the window. You can cut-and-paste the example into your own code and then edit it for your particular design.)

VHDL, unlike ABEL, is not synthesized within the HDL Editor window so the **Synthesis → Synthesize** menu item is blanked out. The synthesis of the netlist from the VHDL is done in the next section.

Now that the design entry is complete, select **File → Save** in the **HDL Editor** window. Then select **File → Exit**.

Upon returning to the **Project Manager** window, we must make the TRIP-GENI.VHD file a part of our project. Select the **Document → Add...** menu item and list items of type **HDL (*.VHD;*.VER;*.VE;*.V)** in the dialog window. Highlight TRIPGENI.VHD in the list and click **Open**. You should then see TRIPGENI.VHD in the list of files making up the TRIPGENI project (Figure 2.39).

Synthesizing the Netlist

Once the VHDL source file is complete, we next need to extract its netlist. The first thing I usually do is to select the **Synthesis → Force Analysis of All HDL Source Files** in the **Project Manager** window. This initiates a check of all the VHDL files to detect any errors. (The Foundation Series software usually does this without your intervention, but I like to make sure it is done before I initiate the synthesizer.)

Next, click on the ▒ ⬚ button in the right-hand pane of the **Project Manager** window. This brings up the Synthesis/Implementation window of Figure

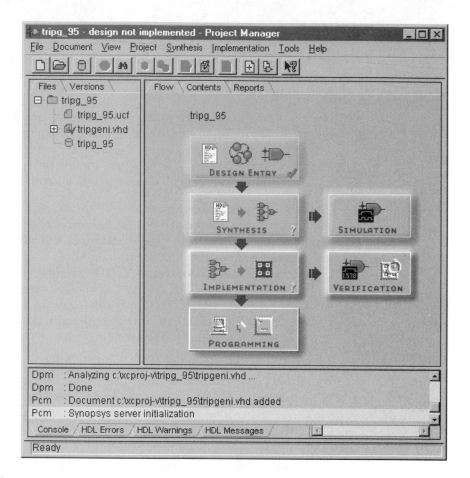

Figure 2.39 Project Manager window after the TRIPGENI.VHD file is added to the TRIPG_95 project.

2.40. When the window first appears, the name of the interface for the Trip-Genie design will be listed in the **Top Level** text box. The **Version Name** box shows **ver1** and this will be incremented each time you change the source code for the design.

In the **Target Device** area of the window, you will select the family, particular device type, and device speed in the drop-down menus (the XC95108 CPLD in the 84-pin package with the slowest speed is selected in Figure 2.40.) This lets the synthesis software know the type of chip architecture you are targeting so it can generate a netlist that takes advantage of the features of the chip.

You can also use the controls in the **Synthesis Settings** area to direct the synthesis tools to emphasize high-speed or area-efficient circuitry. There is also an **Insert I/O Pads** checkbox which controls whether input and output buffers will be placed on all I/O signals. This box should be checked since the TRIP-GENI.VHD code is at the top-level of this design. (This box would not be checked if the TRIPGENI.VHD circuitry was being included as a macro in a larger design.)

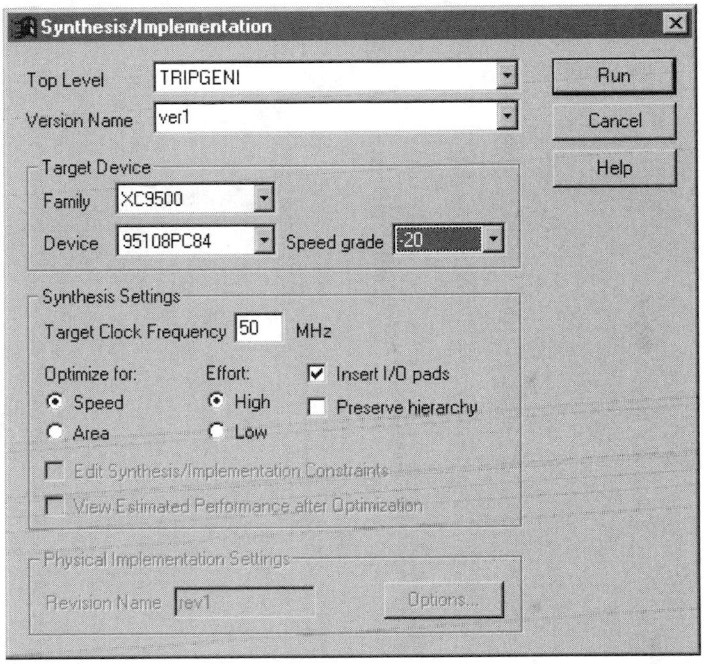

Figure 2.40   Window for controlling the synthesis process.

Clicking on the **Run** button starts the synthesis process. If the synthesis is successful, you should see a green check mark in the **Synthesis** box of the **Flow** tab in the **Project Manager** window. (You will see a red cross in the event of failure.)

At this point, you have extracted a netlist from the VHDL code that describes the Trip-Genie circuit. With this netlist you can do functional simulation, compile the netlist into a bitstream, and download and test the bitstream to an XS95 or XS40 Board in exactly the same way as was shown in the previous section on the schematic design flow. (The only difference you will encounter is that the HDL mode version of the Trip-Genie design is stored in the XCPROJ-V directory instead of the XCPROJ directory where the schematic mode design is kept.)

Projects

1. Design and implement the entire Trip-Genie discussed in the previous chapter. How difficult is it to implement this design with an FPLD as compared to using TTL SSI chips?

Combinational Logic

Objectives

- Describe combinational logic circuits.
- Show examples of common combinational circuits.
- Give further examples of the use of FPLDs in combinational logic circuits.
- Introduce the types of information included in the Foundation Implementation report files.

Discussion

Combinational logic is probably the easiest circuitry to design. The outputs from a combinational logic circuit depend only on the current inputs. (The circuit has no remembrance of what it did at any time in the past.) But combinational logic is the base on which all other logic design rests, so it pays to spend some time on it.

We already have experience in combinational circuit design from the previous two chapters. After specifying the problem and determining the number of inputs and outputs, we came up with a truth table that lists the logic values on each output for each combination of logic levels on the inputs. Then we manually reduced the truth table to a Boolean equation and implemented the equation by interconnecting logic gates. There are many techniques for minimizing and synthesizing a Boolean expression from a given truth table. And there are many textbooks describing these techniques. We will not be discussing minimization techniques here. The Foundation Series tools contain a complete set of tools for minimizing and synthesizing logic. Our job is to use these tools for designing logic and save our energy for tasks that have not been automated yet.

Much of logic design involves connecting simple, easily understood circuits to construct a larger circuit that performs a much more complicated function. In the next few subsections, I will describe several simple, often-used combinational logic circuits.

Multiplexers

A multiplexer circuit accepts N inputs and outputs the logic value of one of those inputs. The selection of which inputs goes out on the output is deter-

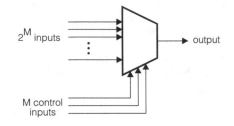

Figure 3.1   Inputs and output for a 2^M-to-1 multiplexer.

mined by a set of M control inputs. A multiplexer with M control inputs can steer up to 2^M inputs to a single output. The I/O for a multiplexer is shown in Figure 3.1. The truth table for a 2-to-1 multiplexer is shown in Table 3.1.

Table 3.1   Truth table for a 2-to-1 multiplexer.

Control	Input0	Input1	Output
0	0	0	0
0	0	1	0
0	1	0	1
0	1	1	1
1	0	0	0
1	0	1	1
1	1	0	0
1	1	1	1

It is often simpler to write these truth tables if we introduce the *don't-care* symbol X. This symbol is used in place of a logic 0 or 1 on an input when the input has no effect on the value of the output. With the multiplexer, for example, Input1 has no effect on the output when the Control input is 0. Similarly, Input0 has no effect when the Control is 1. Rewriting the multiplexer truth table with don't-cares results in Table 3.2.

Table 3.2   Truth table with "don't-cares" for a 2-to-1 multiplexer.

Control	Input0	Input1	Output
0	0	X	0
0	1	X	1
1	X	0	0
1	X	1	1

Combinational Logic   Chapter 3

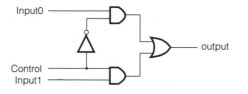

Figure 3.2   Logic circuitry for a 2-to-1 multiplexer.

The truth table in Table 3.2 has as much information as the original in Table 3.1. If we expanded each line with an X in it into two lines with the X replaced by a 0 or a 1, then we would be back to the original truth table. The schematic for the 2-to-1 multiplexer is shown in Figure 3.2.

Decoders

A decoder has M inputs and up to 2^M outputs. If the logic values on the M inputs are interpreted as a binary number of value P, then the Pth output will be at logic 1 while all the others are at logic 0. The I/O for a decoder is shown in Figure 3.3.

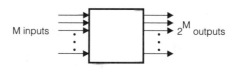

Figure 3.3   Inputs/outputs for an M-input, 2^M-output decoder.

The truth table for a 2-input, 4-output decoder is shown in Table 3.3. The corresponding circuitry for this decoder is shown in Figure 3.4.

Table 3.3   Truth table for a 2-4 decoder.

Inputs		Outputs			
Input1	**Input0**	**Out0**	**Out1**	**Out2**	**Out3**
0	0	1	0	0	0
0	1	0	1	0	0
1	0	0	0	1	0
1	1	0	0	0	1

Decoders can perform more complicated functions. For example, an LED decoder accepts a 4-bit binary number and outputs seven signals that will dis-

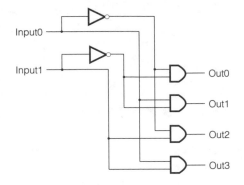

Figure 3.4   Logic circuitry for a 2-input, 4-output decoder.

play the corresponding numeral on a 7-segment LED digit (Figure 3.5). Table 3.4 shows the truth table for an LED decoder for the digits 0 through 9 (a logic 1 on an output indicates that the corresponding LED segment is lit).

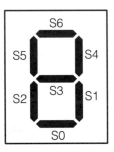

Figure 3.5   A 7-segment LED digit.

Priority Encoders

The opposite of a decoder is the priority encoder: It outputs the largest index of the input that is currently at logic 1. Defining this more exactly, we say that given $2^M - 1$ inputs in which input P is the input with the largest index that is also at logic 1, then the encoder output is an M-bit binary number representation of P. Of course, if no inputs are on, then the output is zero. Figure 3.6 shows the I/O for the priority encoder. Table 3.5 lists the truth table for a 3-input, 2-output priority encoder. The corresponding circuitry for this priority encoder is shown in Figure 3.7.

Table 3.4 Truth table for a 7-segment LED decoder.

4-bit Input				7-segment LED Driver Outputs							
D3	**D2**	**D1**	**D0**	**S6**	**S5**	**S4**	**S3**	**S2**	**S1**	**S0**	**Numeral**
0	0	0	0	1	1	1	0	1	1	1	"0"
0	0	0	1	0	0	1	0	0	1	0	"1"
0	0	1	0	1	0	1	1	1	0	1	"2"
0	0	1	1	1	0	1	1	0	1	1	"3"
0	1	0	0	0	1	1	1	0	1	0	"4"
0	1	0	1	1	1	0	1	0	1	1	"5"
0	1	1	0	1	1	0	1	1	1	1	"6"
0	1	1	1	1	0	1	0	0	1	0	"7"
1	0	0	0	1	1	1	1	1	1	1	"8"
1	0	0	1	1	1	1	1	0	1	1	"9"

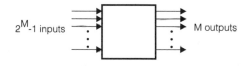

2^M-1 inputs M outputs

Figure 3.6   A 2^M-input, M-output encoder.

Table 3.5 Truth table for a 3-2 priority encoder.

Inputs			Outputs	
Input3	**Input2**	**Input1**	**Output1**	**Output0**
0	0	0	0	0
0	0	1	0	1
0	1	X	1	0
1	X	X	1	1

Parity Generators

An even-parity generator accepts M input bits and generates a single output such that the total number of logic ones is an even number. For example, if a three-input even-parity generator received 010 as an input, it would generate a 1 at its output so the total number of ones is 2—an even number. Parity can be useful for detecting errors if a computing system is designed so that numbers always have even parity. Then if a number is seen with an odd number of

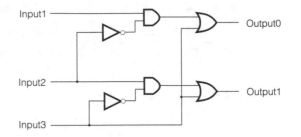

Figure 3.7   Logic circuitry for a 3-input, 2-output priority encoder.

1 bits (odd parity), it must have an error in it. (That is why some PC memories have 9 bits—8 to store the actual data and a ninth parity bit.)

The I/O for an even-parity generator is shown in Figure 3.8. Table 3.6 shows the truth table for a 3-input, even-parity generator, and Figure 3.9 depicts the circuit for it. This parity circuit makes use of an exclusive-OR (XOR) gate that operates according to the truth table in Table 3.7.

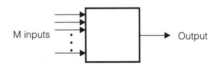

Figure 3.8   Inputs/outputs for an M-bit even-parity generator.

Table 3.6   Truth-table for a 3-bit parity generator.

Input2	Input1	Input0	Output
0	0	0	0
0	0	1	1
0	1	0	1
0	1	1	0
1	0	0	1
1	0	1	0
1	1	0	0
1	1	1	1

Adders

An M-bit adder accepts two M-bit binary numbers and outputs an M-bit sum. But it is possible the sum will not fit in M bits. For example, adding two 3-bit

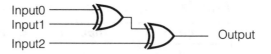

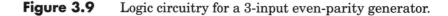

Figure 3.9   Logic circuitry for a 3-input even-parity generator.

Table 3.7   Truth-table for a 2-input XOR gate.

Input1	Input0	Output
0	0	0
0	1	1
1	0	1
1	1	0

numbers like 111 (7) and 110 (6), results in a sum of 1101 (13), which requires 4 bits to represent it. For this reason, an extra *carry* output bit is added to indicate when this overflow condition occurs. And if the adder can have a carry output, then it makes sense that it should also have a carry input. The symbol for this full adder is shown in Figure 3.10.

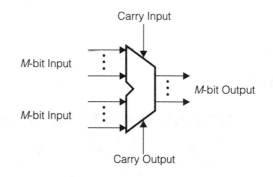

Figure 3.10   An *M*-bit full adder.

Later we will see how to build an $N \times M$-bit adder by chaining N of these M-bit adders through their carry inputs and outputs. For now, we will look at a simple 1-bit adder that obeys the truth table in Table 3.8. From this truth table, we can see that the sum output is the same as the output of a 3-input, even-parity generator. Also, the carry output goes high if two or more of the inputs are at logic 1. From this, we can draw the circuitry for the 1-bit adder, as shown in Figure 3.11.

Table 3.8   Truth-table for a 2-bit adder with carry input and output.

Input1	Input0	Carry Input	Sum Output	Carry Output
0	0	0	0	0
0	0	1	1	0
0	1	0	1	0
0	1	1	0	1
1	0	0	1	0
1	0	1	0	1
1	1	0	0	1
1	1	1	1	1

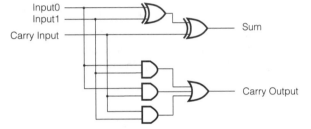

Figure 3.11   Logic circuitry for a 1-bit adder.

Experimental

1-Bit Adder for the XS40 Board

Let's build the 1-bit adder using schematics for both the XC4005XL and the XC95108. First, create a schematic mode project called ADD1_40 that will hold the design for the XC4005XL (Figure 3.12).

Start the schematic editor and connect the gates as shown in Figure 3.11. Figure 3.13 shows the 1-bit adder with the additional input and output buffers.

Once the schematic is complete, save it as ADD1.SCH. Then generate a netlist using the **Options → Create Netlist** menu item and export it in EDIF 200 format using the **Options → Export Netlist...** menu item. (You can also activate the **Options → Integrity Test** to make sure everything is OK before you move on.) Then exit the schematic editor.

Now let's do a simulation to make sure the adder is functioning correctly. Click on the ▩ button in the **Flow** tab of the **Project Manager** window. Within the simulator, use the **Signal → Add Signals...** menu item to add the input

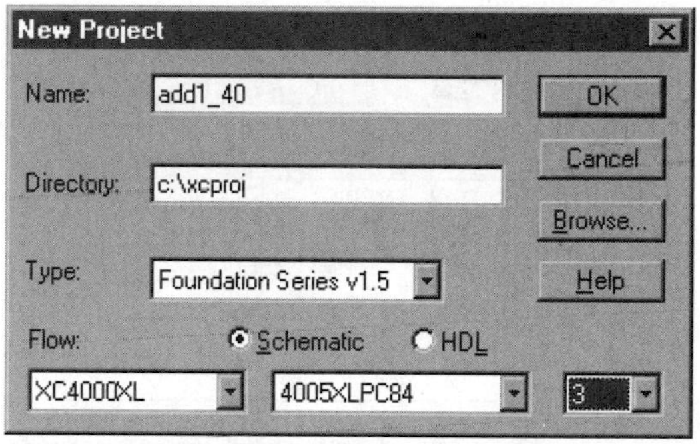

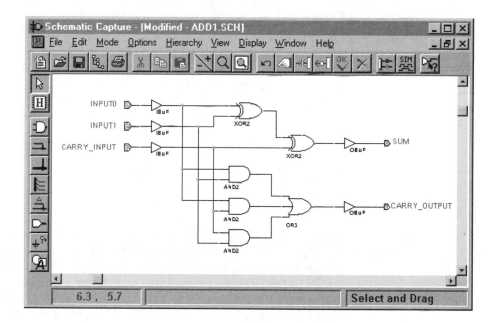

Figure 3.12 Initiating the 1-bit adder project for the XC4005XL FPGA.

Figure 3.13 Schematic for the 1-bit adder.

and output signals to the **Waveform Viewer**. Then use the **Signal → Add Stimulators...** menu item to connect the inputs to the lower 3 bits of the binary counter (Bc0, Bc1, Bc2). Finally, click on the ![button] **button** in the **Logic Simulator** toolbar and the waveforms shown in Figure 3.14 should appear. (You can use the ![button] button and the scrollbar in the **Waveform Viewer** window to expand and center the waveforms.) A careful check reveals that these waveform values match what is shown in the truth-table for the 1-bit adder.

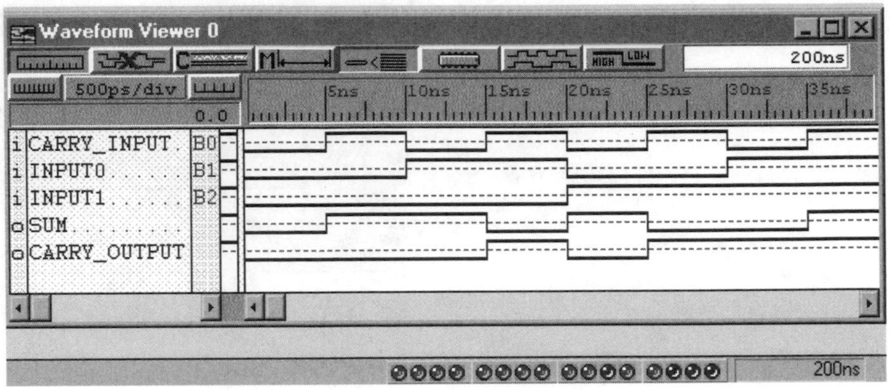

Figure 3.14 A simulation of the 1-bit adder.

After checking the 1-bit adder's operation, we can exit the simulator and begin compiling the design for the XC4005XL FPGA. First we need to assign the input and output terminals of the schematic to physical I/O pins of the XC4005XL. Use the HDL editor to edit the ADD1_40.UCF user-constraint file and add the following lines to it:

```
NET  INPUT1           LOC=P46;
NET  INPUT0           LOC=P45;
NET  CARRY_INPUT      LOC=P44;
NET  SUM              LOC=P25;
NET  CARRY_OUTPUT     LOC=P26;
```

Now click on the ▓▶▪▓ button to start compiling the design. Then click on the **Options...** button in the **Implement Design** window that appears. In the **Options** window, type the name of the UCF you generated, C:\XCPROJ\ ADD1_40\ADD1_40.UCF, into the **User Constraints** box (or use the **Browse...** button to select the UCF). Then click **OK** to return to the **Implement Design** window and click the **Run** button to start the compilation process. It should complete successfully and uneventfully.

At this point, we should have an ADD1_40.BIT file that we can download into the XS40 Board. But let's not rush off just yet. Instead, let's try to answer two questions:

1. How much of the XC4005XL FPGA was used to build the 1-bit adder?

2. Did the compiler assign the inputs and outputs to the pins we requested?

We can get answers to these questions from the report files generated by the Foundation Series Implementation tools. Select the **Reports** tab in the right pane of the **Project Manager** window. Then double-click on the icon labeled

Implementation Report Files. The **Report Browser** window of Figure 3.15 will appear. Double-click on any of the icons to view the contents of the associated report.

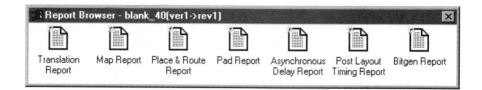

Figure 3.15   Various reports generated by the Foundation Series Implementation tools.

These reports summarize the following information:

Translation Report: Any problems detected when a netlist is being converted into the internal format used by the Foundation Implementation tools will be listed here. Design rule checks are also performed and the user-constraint file is checked for errors. The most likely error you will see in this file will result from an illegal constraint you have entered. These errors should not occur for the design examples in this book, but you may cause some when you begin experimenting on your own.

Map Report: In here, you will get information on what types of optimizations were attempted on the netlist. Logic gates are removed and added to optimize the circuit without changing its function. One example of logic removal occurs when you put a gate in a design but you do not use the output for anything. Another example is an AND gate with an input tied to logic 0 (GND) so that it's output is always low. The report also indicates how the logic gates were collected into groups and mapped into the LUTs of the FPGA CLBs.

Place&Route Report: This report records the various options we selected to affect the way the place-and-route is performed, lists any warnings and errors, and tallies the total processing time spent in the various place-and-route phases. It also reports various statistics on the timing performance of the routing. But most importantly (right now), it gives us the answer to our first question on how much of the XC4005XL device is used to build a 1-bit adder:

```
Device utilization summary:

        IO        5/112    4% used
                  5/61     8% bonded
        LOGIC     1/196    0% used
        IOB       5/112    4% used
        CLB       1/196    0% used
```

The 1-bit adder has 3 inputs and 2 outputs, so it uses 5 of the 112 I/O blocks (IOBs) available in the XC4005XL FPGA. However, only 61 of these IOBs are actually connected (or bonded) to physical pins on the 84-pin PLCC package. Thus, approximately 8% of the available I/O is used by the 1-bit adder as reported above.

Both the SUM and CARRY_OUTPUT circuits in our design have a single output and 3 inputs. Each circuit should easily fit in a single 4-input LUT, so we would expect to use a total of two LUTs for our 1-bit adder. And there are two 4-input LUTs in each CLB, so only a single CLB should be used by the 1-bit adder. This is indeed the case as shown above: only one of a total of 196 CLBs was used.

Pad Report: Our second question is answered in this report. It describes where our design's I/O terminals were placed with respect to the package pins. Here is some of the pertinent information for our design:

```
Pinout by Component Name:
+--------------------+--------------------+
|   Comp Name        |   Pin Number       |
+--------------------+--------------------+
| CARRY_INPUT        | P44                |
| CARRY_OUTPUT       | P26                |
| INPUT0             | P45                |
| INPUT1             | P46                |
| SUM                | P25                |
+--------------------+--------------------+
```

Notice that the actual pin assignments match what we put in our constraint file. This is just what we wanted and expected.

At the bottom of the report there is a list of the pin assignments in a format that can be used in a physical constraint file (PCF).

```
COMP "CARRY_INPUT" LOCATE = SITE "P44" ;
COMP "CARRY_OUTPUT" LOCATE = SITE "P26" ;
COMP "INPUT0" LOCATE = SITE "P45" ;
COMP "INPUT1" LOCATE = SITE "P46" ;
COMP "SUM" LOCATE = SITE "P25" ;
```

PCF constraints use a different syntax from those found in UCFs, but they are still useful if you ever get confused about the correct name to place in a constraint file for a particular terminal. Just run the Foundation Series Implementation tools without a constraint file and then use the pin names found in this report to generate a UCF. It is almost 100% certain that the pin assignments selected by the compiler will not be what you want, but you can edit the constraint file to fix that.

Asynchronous Delay Report: The propagation delays for each routed signal are listed in this report.

Post Layout Timing Report: This report lists any paths through the placed&routed logic which violate timing constraints. For example, a violation would be listed if you required the logic circuit to operate at 50 MHz (i.e., all logic operations must complete in 20 ns or less) and a path was found with a delay greater than 20 ns.

Bitgen Report: This report records all the options that were in effect when the bitstream file was generated. Any errors that occured during the bitstream generation are also listed here.

OK, we have answered our questions. Now we can start testing the actual design.

Open a DOS window and go to the C:\XCPROJ\ADD1_40 directory. Download the 1-bit adder circuit to the XS40 Board:

```
C:\XCPROJ\ADD1_40> XSLOAD ADD1_40.BIT
```

Now we can force values on the 3 inputs to the adder bit to see if its working. The mapping of the adder circuit input terminals, XC4005XL pins, and XSPORT arguments is shown below:

Adder Terminal	XC4005XL Pin	XSPORT
CARRY_INPUT	44	b_0
INPUT0	45	b_1
INPUT1	46	b_2
*** Not Used ***	47	b_3
*** Not Used ***	48	b_4
*** Not Used ***	49	b_5
*** Not Used ***	32	b_6
*** Not Used ***	34	b_7

The ADD1_40.UCF file also assigns the SUM and CARRY_OUT outputs to pins 25 and 26 of the XC4005XL FPGA, respectively. Pin 25 is attached to segment S0 of the 7-segment LED while pin 26 drives segment S1.

We can use XSPORT with all 8 combinations of inputs to completely test the 1-bit adder circuit. Here are the commands you can issue and what the results should be:

Command	Input1	Input0	CARRY_INPUT	SUM	CARRY_OUTPUT
XSPORT 000	0	0	0	0 (dark)	0 (dark)
XSPORT 001	0	0	1	1 (bright)	0 (dark)
XSPORT 010	0	1	0	1 (bright)	0 (dark)
XSPORT 011	0	1	1	0 (dark)	1 (bright)
XSPORT 100	1	0	0	1 (bright)	0 (dark)
XSPORT 101	1	0	1	0 (dark)	1 (bright)
XSPORT 110	1	1	0	0 (dark)	1 (bright)
XSPORT 111	1	1	1	1 (bright)	1 (bright)

If you have done everything correctly, your tests of your downloaded design should match the simulation results and the truth-table for the 1-bit adder.

1-Bit Adder for the XS95 Board

We can follow the same sequence of steps to design and test a 1-bit adder on the XS95 Board:

1. Create an ADD1_95 project targeted at the XC95108 CPLD in an 84-pin PLCC package.

2. Copy the ADD1.SCH schematic from the ADD1_40 project to this project as follows by selecting the **Document → Add...** menu item in the **Project Manager** window. The window of Figure 3.16 will appear. Select the ADD1.SCH file and click on the **Open** button to copy the schematic into the ADD1_95 directory.

3. Open the ADD1.SCH file and generate and export the netlist.

4. Do a functional simulation of the design.

5. Create an ADD1_95.UCF file with the following pin assignments:
   ```
   NET INPUT1          LOC=P48;
   NET INPUT0          LOC=P47;
   NET CARRY_INPUT     LOC=P46;
   NET SUM             LOC=P21;
   NET CARRY_OUTPUT    LOC=P23;
   ```

6. Compile the design and examine the report files. Remember to use the **Options** button in the **Implement Design** window to inform the Foundation Series Implementation tools of the location of your ADD1_95.UCF file.

7. After the implementation phase is done, double-click on the **Implementation Reports** files icon in the **Reports** tab. You will notice that the XC9500 fitter only generates the Translation, Fitting, and Post-Layout Timing reports. Double-clicking on the **Fitting Report** icon provides quite a bit of information. All we are interested in is the answers to our questions on pin assignments and device utilization. The pin assignments are as we expected:

```
********Resources Used by Successfully Mapped Logic**********
** LOGIC **
Signal         Total   Signals Loc     Pwr   Slew  Pin  Pin   Pin
Name           Pt      Used            Mode  Rate  #    Type  Use
CARRY_OUTPUT   3       3        FB3_12 STD   FAST  23   I/O   O
SUM            3       3        FB3_11 STD   FAST  21   I/O   O
** INPUTS **
Signal                         Loc             Pin  Pin   Pin
Name                                           #    Type  Use
CARRY_INPUT                    FB6_3           46   I/O   I
INPUT0                         FB6_5           47   I/O   I
INPUT1                         FB6_6           48   I/O   I
```

The device utilization is:

```
*******************  Resource Summary  *******************
Design    Device          Macrocells   Product Terms  Pins
Name      Used            Used         Used           Used
add1_95   XC95108-20-PC84 2/108 (1%)   6/540 (1%)     5/69 (7%)
```

Our circuit has two outputs, so we will need to use at least two macrocells (since each macrocell has only one output). This matches the figure reported in the Fitting Report. Both the CARRY_OUTPUT and SUM outputs use 3 product-terms each, for a total consumption of 6 product terms out of 540. As was the case with our XC4005XL-based design, we have used a negligible fraction of the full capabilities of our programmable logic blocks.

8. Now we have to translate the bitstream file that was generated by the Foundation Implementation tools into a form we can download into the XC95108 CPLD. To do this, select the **Tools → Device Programming → JTAG Programmer** menu item and bring up the **JTAG Programmer** window. Select the **Output → Create SVF File...** menu item and use the **Create a New SVF File** window to specify the file ADD1_95.SVF in the top-level C:\XCPROJ\ADD1_95 project directory. Click on **Save** to create the SVF file. Then select the **Operations → Program** menu item in the **JTAG Programmer** window and click on **OK** in the **Options** window that

appears. This will initiate the generation of the SVF file. Close the window after the SVF file is generated.

9. Open a DOS window, move to the C:\XCPROJ\ADD1_95 directory, and download the circuit into the XS95 Board:

```
C:\XCPROJ\ADD1_95> XSLOAD ADD1_95.SVF
```

10. Apply all 8 combinations of inputs using XSPORT and observe the results on LED segments S0 and S1. Once again, these should match the results of the truth-table and the simulation.

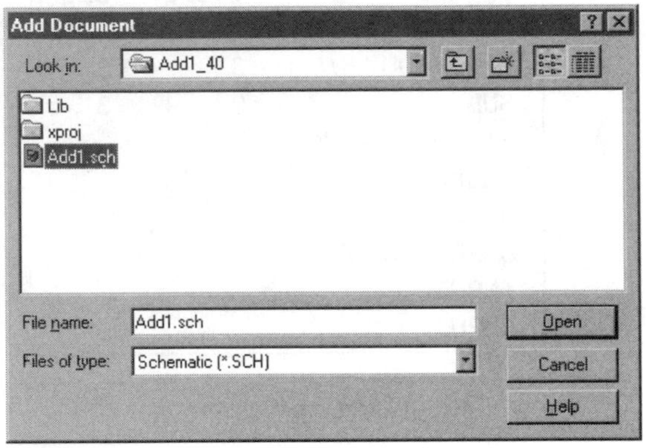

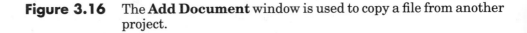

Figure 3.16 The **Add Document** window is used to copy a file from another project.

1-Bit Adder in VHDL

In this section we will create the 1-bit adder circuit using VHDL and the HDL design flow mode of the Xilinx Foundation Series software. We will target this to the XC4005XL FPGA, so start by creating a project called ADD1_40 in the XCPROJ-V directory (Figure 3.17). Make sure to click on the **HDL** radio button so the HDL design flow mode is enabled.

Start the HDL editor and elect to use the Design Wizard. Choose VHDL as the HDL and set the name of the VHDL design to **FULLADD**. Then use the **Design Wizard-Ports** window to set-up the inputs and outputs. From the circuit schematic, we know we need 3 inputs. Click the **New** button and type **CARRY_INPUT** into the **Name** field. Repeat these actions for the INPUT0 and INPUT1 inputs.

Click **New** again and enter **CARRY_OUTPUT** in the **Name** field. Click on the **Output** radio button to set this as an output bit. Repeat these actions for the SUM output. Figure 3.18 shows the assignment of I/O ports for the 1-bit adder.

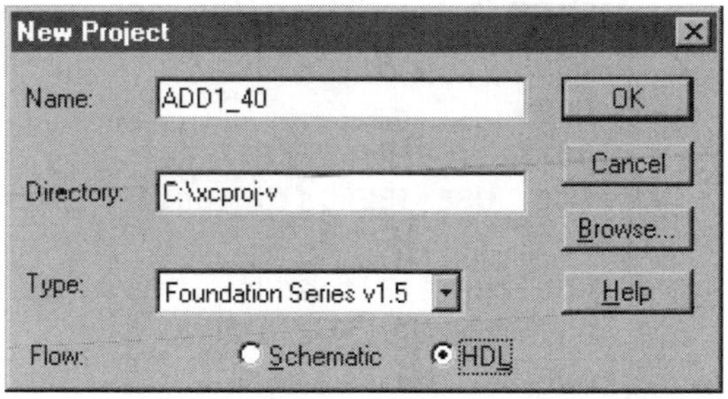

Figure 3.17   Initiating the HDL-mode 1-bit adder project for the XC4005XL FPGA.

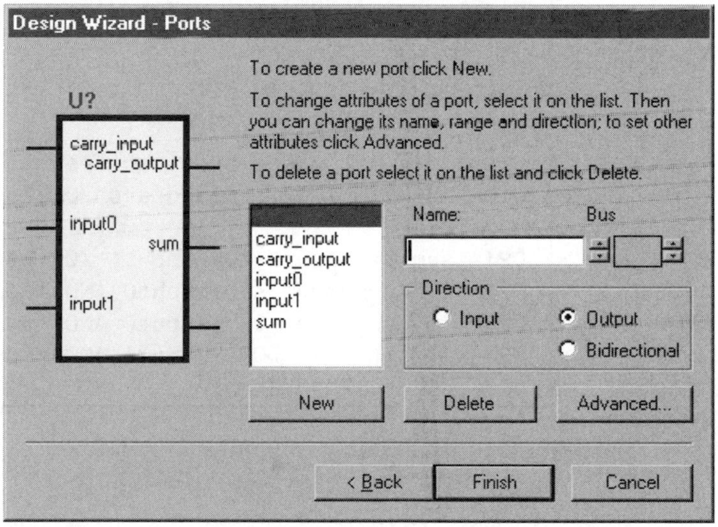

Figure 3.18   Input and output ports for the 1-bit adder.

Once the inputs and outputs are defined, click on the **Finish** button and an HDL Editor window appears with the skeleton of the 1-bit adder. Now we can enter the Boolean equations for the sum and carry outputs. Listing 3.1 shows the completed VHDL file with the truth-table (I added the line numbers for reference purposes; they are not part of the VHDL code).

Listing 3.1 starts by accessing the features of the STD_LOGIC_1164 package of the IEEE library (lines 3 and 4). The ENTITY block (lines 7 through 16) declares the 3 inputs and 2 outputs of the 1-bit adder. This is followed by the ARCHITECTURE block which holds the Boolean equations for the sum and carry outputs (line 21 and lines 22–24, respectively). The equation for the sum output is just the exclusive-or of the 3 inputs as shown in the schematic of

Listing 3.1   The VHDL description of the 1-bit adder.

```
001- -- 1-Bit Adder
002-
003- LIBRARY IEEE;
004- USE IEEE.std_logic_1164.ALL;
005-
006- -- 1-Bit Adder interface description
007- ENTITY fulladd IS
008-     PORT
009-     (
010-         input0: IN STD_LOGIC;        -- addend bit
011-         input1: IN STD_LOGIC;        -- addend bit
012-         carry_input: IN STD_LOGIC;   -- carry input bit
013-         sum: OUT STD_LOGIC;          -- sum output bit
014-         carry_output: OUT STD_LOGIC  -- carry output bit
015-     );
016- END fulladd;
017-
018- -- 1-Bit Adder architecture
019- ARCHITECTURE fulladd_arch OF fulladd IS
020- BEGIN
021- sum <= input0 XOR input1 XOR carry_input;
022- carry_output <= (input0 AND input1)
023-               OR (input0 AND carry_input)
024-               OR (input1 AND carry_input);
025- END fulladd_arch;
```

Figure 3.11. The carry output is defined by a sum-of-products equation that also follows the gate circuitry of Figure 3.11.

Once the VHDL code is entered, do a syntax check using the **Synthesis →** **Check Syntax** menu item. This checks to make sure you have got all the commas, semi-colons, braces, and brackets in the right place. If there are no errors, then save the code as FULLADD.VHD. Then close the **HDL Editor** window and use the **Document → Add** menu item in the **Project Manager** window to add the FULLADD.VHD file to the ADD1_40 project.

Now we need to synthesize the netlist from the VHDL code. Click on the ▣ ▶ ▥ button in the **Flow** tab of the **Project Manager** window. This will bring up the **Synthesis Implementation** window shown in Figure 3.19. Set the options as shown to target the 1-bit adder at the XC4005XL FPGA. (You can also select the XC95108 CPLD at this point if you are going to test your design with an XS95 Board.) Then click on the **Run** button and the Foundation Series synthesis software will build the netlist.

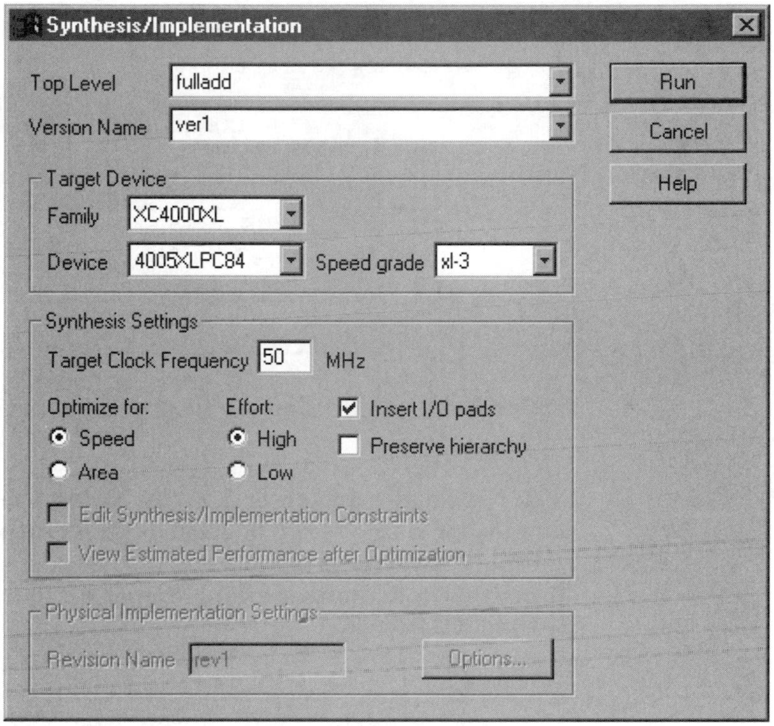

Figure 3.19 Targeting the synthesis of the 1-bit adder to the XC4005XL FPGA.

Once the netlist for the 1-bit adder has been created, you can simulate its operations, compile a bitstream, download the bitstream to an XS Board, and test it with XSPORT commands just as was shown in the previous two sections.

LED Decoder for the XS95 Board

Now we can try a more complicated design like the LED decoder. Let's build it using ABEL for the XC95108 CPLD first, and then move it to the XC4005XL FPGA. Start by creating a project called LEDDCD95 for the XC95108 (Figure 3.20).

Start the HDL editor and choose ABEL as the HDL. Then use the Design Wizard to set-up the inputs and outputs. From the truth-table, we know we need 4 inputs and 7 outputs. We could type in all these names, but there is a smarter way. Click on the **New** button and type **D** into the **Name** field. Then click on the upper-left ▣ button under the **Bus** label. You will see the D in the name field change to D[1:0]. Two more clicks will change it to D[3:0]. This specifies our 4 inputs (D0, D1, D2, and D3) as a 4-bit input vector.

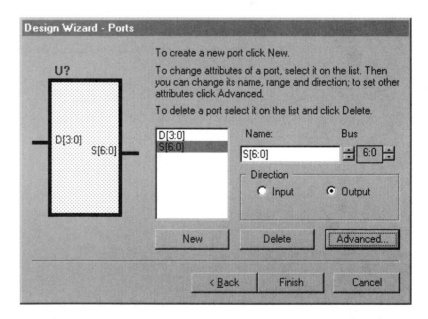

Figure 3.20 Initiating the LED decoder project for the XC95108 CPLD.

Click **New** again and enter **S** in the **Name Field**. Then click the ▣ button to advance the name to S[6:0]. Click on the **Output** radio button to set this as a 7-bit output vector. Remember to also click the **Advanced** button and set the output type to **Combinatorial**. Figure 3.21 shows the assignment of I/O ports for the LED decoder.

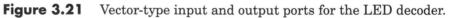

Figure 3.21 Vector-type input and output ports for the LED decoder.

Combinational Logic   Chapter 3

Once the inputs and outputs are defined, click on the **Finish** button and an HDL Editor window appears with the skeleton of the LED decoder. We could try to enter the Boolean equations for the LED decoder into the ABEL file, but this is too much work. The Foundation Series tools can derive these equations for us if we just enter the truth-table. Listing 3.2 shows the completed ABEL file with the truth-table (I added the line numbers for reference purposes; they are not part of the ABEL code).

Listing 3.2   The ABEL HDL description of the LED decoder.

```
001- module LEDDCD
002- Title 'LEDDCD'
003-
004- Declarations
005-
006- D3..D0 PIN;
007- D = [D3..D0];
008- S6..S0 PIN istype 'com';
009- S = [S6..S0];
010-
011- Equations
012-
013- TRUTH_TABLE
014-    (D -> [S6, S5, S4, S3, S2, S1, S0])
015-      0 -> [1,   1,   1,   0,   1,   1,   1 ];
016-      1 -> [0,   0,   1,   0,   0,   1,   0 ];
017-      2 -> [1,   0,   1,   1,   1,   0,   1 ];
018-      3 -> [1,   0,   1,   1,   0,   1,   1 ];
019-      4 -> [0,   1,   1,   1,   0,   1,   0 ];
020-      5 -> [1,   1,   0,   1,   0,   1,   1 ];
021-      6 -> [1,   1,   0,   1,   1,   1,   1 ];
022-      7 -> [1,   0,   1,   0,   0,   1,   0 ];
023-      8 -> [1,   1,   1,   1,   1,   1,   1 ];
024-      9 -> [1,   1,   1,   1,   0,   1,   1 ];
025-
026- end LEDDCD
```

Listing 3.2 has a bunch of new things we have not seen before, so let's take some time to review it:

Line 6: The 4 inputs D0, D1, D2 and D3 are declared here. The '..' notation lets you enter an entire vector range by giving only the start and end points.

Line 7: This is just an equate statement that lets us use D as short-hand for the [D3..D0] input vector.

Lines 8–9: These declare the 7-element output vector and its short-hand name.

Line 13: TRUTH_TABLE is the ABEL keyword that signals the beginning of a truth-table (obviously).

Line 14: The inputs and outputs for the truth-table are listed on this line. The '->' separates the list of inputs (on the left-hand side) from the list of outputs (on the right-hand side). The short-hand name for the inputs is used, but the entire set of output names is written out.

Lines 15–24: Each combination of inputs and the resulting outputs are listed on these lines. The input combinations are written as decimal digits instead of Boolean values. This is allowable because we have used a vector field so the ABEL complier has enough information to determine what we mean. (For example, on line 21 we set $D = 6$ which ABEL expands into the assignments $D3 = 0$, $D2 = 1$, $D1 = 1$, and $D0 = 0$.) We could have also used the vector field for the outputs on line 14, but it makes more sense to list the outputs for the LED segments individually. Then if we make a mistake on a particular segment, it is easier to go back and modify the truth-table without inadvertently affecting another output.

Once the ABEL code is entered, do a syntax check using the **Synthesis** → **Check Syntax** menu item. This checks to make sure you have got all the commas, semi-colons, braces, and brackets in the right place. If there are no errors, then save the code as LEDDCD.ABL. Then select the **Synthesis** → **Options** menu item and click the **Chip** radio-button in the **ABEL6** window that appears (since this is meant to be a self-contained design). Selecting the **Synthesis** → **Synthesize** menu item will generate the netlist. Then close the **HDL Editor** window and use the **Document** → **Add** menu item in the **Project Manager** window to add the LEDDCD.ABL file to the LEDDCD95 project.

We could simulate the design now, but that is a waste of time. The function of the LED decoder is to give us a visual presentation of a number. The simulator will only show us the waveforms, so we will have to mentally construct which LED segments are lit by these waveforms. This is an error-prone task which we can avoid just by downloading the design into the CPLD and observing the 7-segment LED for each input combination.

We begin the process of compiling the LED decoder for the XC95108 CPLD by creating a user-constraint file for the XS95 Board. Open an **HDL Editor** window and enter the following lines:

```
NET D0    LOC=P46;
NET D1    LOC=P47;
NET D2    LOC=P48;
```

```
NET D3    LOC=P50;
NET S0    LOC=P21;
NET S1    LOC=P23;
NET S2    LOC=P19;
NET S3    LOC=P17;
NET S4    LOC=P18;
NET S5    LOC=P14;
NET S6    LOC=P15;
```

Then save the file as LEDDCD95.UCF and exit the editor.

Now click on the ▒▒ ◆ ▒▒ button to start compiling the design. Then click on the **Options...** button in the **Implement Design** window that appears. In the **Options** window, type the name of the UCF you generated, C:\XCPROJ\ LEDDCD95\LEDDCD95.UCF, into the **User Constraints** box (or use the **Browse...** button to select the UCF). Then click **OK** to return to the **Implement Design** window and click the **Run** button to start the compilation process. It should complete successfully and uneventfully.

After the compiling is done, select the **Reports** tab in the right pane of the **Project Manager** window. Then double-click on the **Fitting Report** so we can check how much of the XC95108 CPLD was used to build the LED decoder. The following lines tell the story:

```
*******************  Resource Summary  *******************
Design     Device          Macrocells  Product Terms  Pins
Name       Used            Used        Used           Used
leddcd95   XC95108-20-PC84  7/108  (6%)  26/540  (4%)  11/69  (15%)
```

The LED decoder uses 7 of the 108 available macrocells. Looking further into the report, we can get an idea of which parts of the circuit consume the most resources:

```
*********Resources Used by Successfully Mapped Logic**********
** LOGIC **
Signal      Total  Signals Loc      Pwr   Slew Pin  Pin   Pin
Name        Pt     Used             Mode  Rate #    Type  Use
S0          5      4       FB3_11   STD   FAST 21   I/O   O
S1          3      4       FB3_12   STD   FAST 23   I/O   O
S2          2      4       FB3_8    STD   FAST 19   I/O   O
S3          4      4       FB3_5    STD   FAST 17   I/O   O
S4          4      4       FB3_6    STD   FAST 18   I/O   O
S5          4      4       FB3_2    STD   FAST 14   I/O   O
S6          4      4       FB3_3    STD   FAST 15   I/O   O
```

The circuit which drives the S0 LED uses 5 product terms, so it is the most complex. Conversely, the S2 circuitry is simpler because it only sums together two product terms. The rest of the circuitry is somewhere in the middle. A total of twenty-six of the 540 available product terms are used, so even this circuit leaves us quite a bit of room for more logic.

The report file also lists the Boolean equations it derived for each decoder output. Here is the equation for S0:

```
S0  = /D0 * D1 */D3
    + /D0 */D1 */D2
    + /D1 */D2 * D3
    +  D1 */D2 */D3
    +  D0 */D1 * D2 */D3
```

Normally, we would not care about this. But if for some reason you need to build the LED decoder using a schematic editor and simple gates, this report file is a good place to look to get the equations.

Now we have to translate the bitstream file that was generated by the Flow Engine into a form we can download into the XC95108 CPLD. To do this, select the **Tools → Device Programming → JTAG Programmer** menu item and bring up the **JTAG Programmer** window. Select the **Output → Create SVF File...** menu item and use the **Create a New SVF File** window to specify the file LEDDCD95.SVF in the top-level C:\XCPROJ\LEDDCD95 project directory. Click **Save** to create the SVF file. Then select the **Operations → Program** menu item in the **JTAG Programmer** window and click on **OK** in the **Options** window that appears. This will initiate the generation of the SVF file. Close the window after the SVF file is generated.

At this point, we should have a LEDDCD95.SVF file that we can download into the XS95 Board. Open a DOS window and go to the C:\XCPROJ\LEDDCD95 directory. Download the circuit to the XS95 Board as follows:

```
C:\XCPROJ\LEDDCD95> XSLOAD LEDDCD95.SVF
```

Once the design is loaded into the CPLD, we can force values on the 4 inputs and watch the 7-segment LED to see if its working. The mapping of the LED decoder circuit input terminals, XC95108 pins, and XSPORT arguments is shown below:

LED Decoder Terminal	XC95108 Pin	XSPORT Argument
D0	46	b_0
D1	47	b_1
D2	48	b_2
D3	50	b_3
*** Not Used ***	51	b_4
*** Not Used ***	52	b_5
*** Not Used ***	81	b_6
*** Not Used ***	80	b_7

The LEDDCD95.UCF file also assigned the [S0..S6] outputs to pins 21, 23, 19, 17, 18, 14, and 15 of the XC95108 CPLD, respectively. These pins are attached to the S0, S1, S2, S3, S4, S5, and S6 segments of the 7-segment LED, respectively.

We can use XSPORT with all 10 combinations of inputs to test the LED decoder circuit. Here are the commands you can issue and what digit should be displayed:

Command	D3	D2	D1	D0	7-seg. LED
XSPORT 0000	0	0	0	0	"0"
XSPORT 0001	0	0	0	1	"1"
XSPORT 0010	0	0	1	0	"2"
XSPORT 0011	0	0	1	1	"3"
XSPORT 0100	0	1	0	0	"4"
XSPORT 0101	0	1	0	1	"5"
XSPORT 0110	0	1	1	0	"6"
XSPORT 0111	0	1	1	1	"7"
XSPORT 1000	1	0	0	0	"8"
XSPORT 1001	1	0	0	1	"9"

LED Decoder for the XS40 Board

We can follow the same sequence of steps to design and test LED decoder for the XS40 Board:

1. Create an LEDDCD40 project targeted at the XC4005XL FPGA in an 84-pin PLCC package.

2. Copy the LEDDCD.ABL schematic from the LEDDCD40 project to this project as follows by selecting the **Document → Add...** menu item in the **Project Manager** window. Select the LEDDCD.ABL file in the **Add Document** window and click on the **Open** button to copy the ABEL file into the LEDDCD40 directory.

3. Open the LEDDCD.ABL file and synthesize the netlist.

4. Create an LEDDCD40.UCF file with the following pin assignments:

    ```
    NET D0      LOC=P44;
    NET D1      LOC=P45;
    NET D2      LOC=P46;
    NET D3      LOC=P47;
    NET S0      LOC=P25;
    NET S1      LOC=P26;
    NET S2      LOC=P24;
    NET S3      LOC=P20;
    NET S4      LOC=P23;
    NET S5      LOC=P18;
    NET S6      LOC=P19;
    ```

5. Compile the design and examine the report files. Remember to use the **Options** button in the **Implement Design** window to inform the Foundation Series Implementation tools of the location of your LEDDCD40.UCF file.

6. After the implementation phase is done, double-click on the **Implementation Reports** files icon in the **Reports** tab. Double-clicking on the **Place&Route Report** icon lets us examine the device utilization:

```
Device utilization summary:
   IO            11/112      9% used
                 11/61      18% bonded
   LOGIC          4/196      2% used
   IOB           11/112      9% used
   CLB            4/196      2% used
```

The LED decoder has 7 outputs and each output is a function of 4 inputs. So each output of the decoder can be computed by a single 4-input LUT.

Seven LUTs can be packed into 4 CLBs, which explains the reported logic usage.

7. Open a DOS window, move to the C:\XCPROJ\LEDDCD40 directory, and download the circuit into the XS40 Board:

```
C:\XCPROJ\LEDDCD40> XSLOAD LEDDCD40.BIT
```

8. Apply all 10 combinations of inputs using XSPORT and observe the numeral displayed on the 7-segment digit.

LED Decoder in VHDL

As our final example, we will describe the LED decoder using VHDL and the HDL design flow mode of the Xilinx Foundation Series software. We will target this to the XC95108 CPLD, so start by creating a project called LEDDCD95 in the XCPROJ-V directory (Figure 3.22). Make sure to click on the **HDL** radio button so the HDL design flow mode is enabled.

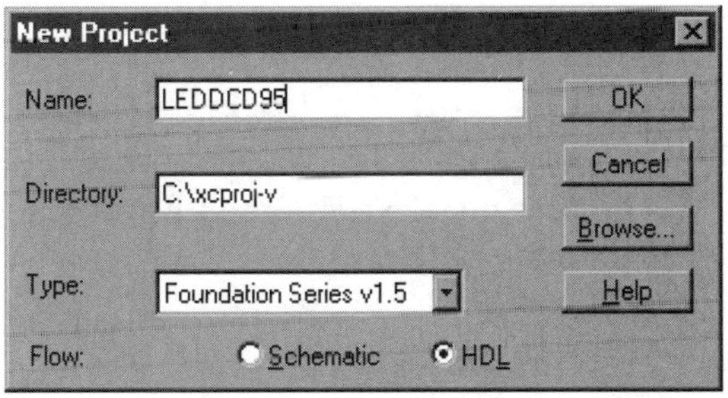

Figure 3.22   Initiating the HDL-mode LED decoder project for the XC95108 CPLD.

Start the HDL editor and elect to use the Design Wizard. Choose VHDL as the HDL and set the name of the VHDL design to **LEDDCD**. Then use the **Design Wizard-Ports** window to set-up the inputs and outputs. From the truth-table, we know we need 4 inputs and 7 outputs. Click on the **New** button and type **D** into the **Name** field. Then click on the upper-left ▦ button under the **Bus** label. You will see the D in the name field change to D[1:0]. Two more clicks will change it to D[3:0]. This specifies our 4 inputs (D0, D1, D2, and D3) as a 4-bit input vector.

Click **New** again and enter **S** in the **Name Field**. Then click the ▦ button to advance the name to S[6:0]. Click on the **Output** radio button to set this as a 7-bit output vector. Figure 3.23 shows the assignment of I/O ports for the LED decoder.

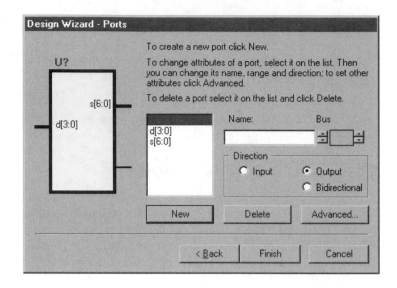

Figure 3.23 Input and output ports for the 1-bit adder.

Once the inputs and outputs are defined, click on the **Finish** button and an HDL Editor window appears with the skeleton of the LED decoder. Now all we have to do is express the LED decoder truth table using VHDL statements and the Foundation Series synthesis software will take care of the rest. Listing 3.3 shows the completed VHDL file.

As we have seen in our other VHDL examples, Listing 3.3 starts by accessing the features of the STD_LOGIC_1164 package of the IEEE library (lines 3 and 4). The ENTITY block (lines 7 through 13) declares the 4 inputs and 7 outputs that form the interface to the LED decoder. Note that only two lines of VHDL source code are required to specify all the inputs and outputs. Instead of listing each input and output separately as an IN or OUT of type STD_LOGIC, we have created arrays d and s of inputs and outputs, respectively, with the type set to STD_LOGIC_VECTOR. Following the type is a parenthesized list of the array bounds separated with a keyword that specifies the ordering of the array elements. For example, the d input array consists of the 4 inputs [d3,d2,d1,d0] in that order. (If we had set the d array bounds as (0 TO 3), then the array elements would be ordered as [d0,d1,d2,d3]).

Following the ENTITY block is the ARCHITECTURE block which holds the truth table for the LED decoder. Unlike ABEL, VHDL has no specific truth table construct. Instead, VHDL uses a WITH-SELECT-WHEN statement (lines 20 through 31). Line 20 specifies that the value of the 4 elements in the d array (the selector) will be used to select a value that will be loaded into the 7-element s array (the receptor). For example, line 21 states that when the

Listing 3.3   The VHDL description of the LED decoder.

```
001- -- LED Decoder for 7-segment LED
002-
003- LIBRARY IEEE;
004- USE IEEE.std_logic_1164.ALL;
005-
006- -- LED Decoder interface description
007- ENTITY leddcd IS
008- PORT
009- (
010- d: IN std_logic_vector (3 DOWNTO 0); -- 4-bit hex input
011- s: OUT std_logic_vector (6 DOWNTO 0) -- 7 LED drivers
012- );
013- END leddcd;
014-
015- -- LED Decoder architecture description
016- ARCHITECTURE leddcd_arch OF leddcd IS
017- BEGIN
018- -- for all decimal values 0-9, drive the appropriate segments
019- -- on the 7-segment LED to display the numeral
020- WITH d SELECT
021- s <= "1110111" WHEN "0000",-- display numeral '0'
022-       "0010010" WHEN "0001",-- display numeral '1'
023-       "1011101" WHEN "0010",-- display numeral '2'
024-       "1011011" WHEN "0011",-- display numeral '3'
025-       "0111010" WHEN "0100",-- display numeral '4'
026-       "1101011" WHEN "0101",-- display numeral '5'
027-       "1101111" WHEN "0110",-- display numeral '6'
028-       "1010010" WHEN "0111",-- display numeral '7'
029-       "1111111" WHEN "1000",-- display numeral '8'
030-       "1111011" WHEN "1001",-- display numeral '9'
031-       "0000000" WHEN OTHERS;-- blank display for non-decimals
032- END leddcd_arch;
033-
```

input array [d3,d2,d1,d0] = [0,0,0,0], then the output array is set such that [s6,s5,s4,s3,s2,s1,s0] = [1,1,1,0,1,1,1].

The values of d following the WHEN keyword have to be mutually exclusive (i.e., no value can be repeated). But it is not necessary to list every possible value for d. Line 31 uses the keyword OTHERS to set [s6,s5,s4,s3,s2,s1,s0] = [0,0,0,0,0,0,0] whenever d does not match one of the values on lines 21 through 30. In effect, OTHERS matches any unspecified value of the selector.

Experimental

Once the VHDL code is entered, do a syntax check using the **Synthesis →
Check Syntax** menu item. If there are no errors, then save the code as
LEDDCD.VHD. Then close the **HDL Editor** window and use the **Document
→ Add** menu item in the **Project Manager** window to add the LEDDCD.VHD
file to the LEDDCD95 project.

Now we need to synthesize the netlist from the VHDL code. Click on the 
█ ◆ ▷ button in the **Flow** tab to bring up the **Synthesis/Implementation**
window shown in Figure 3.24. Set the options as shown to target the LED
decoder at the XC95108 CPLD. (You can also select the XC4005XL FPGA at
this point if you are going to test your design with an XS40 Board.) Then click
on the **Run** button and the Foundation Series synthesis software will build the
netlist.

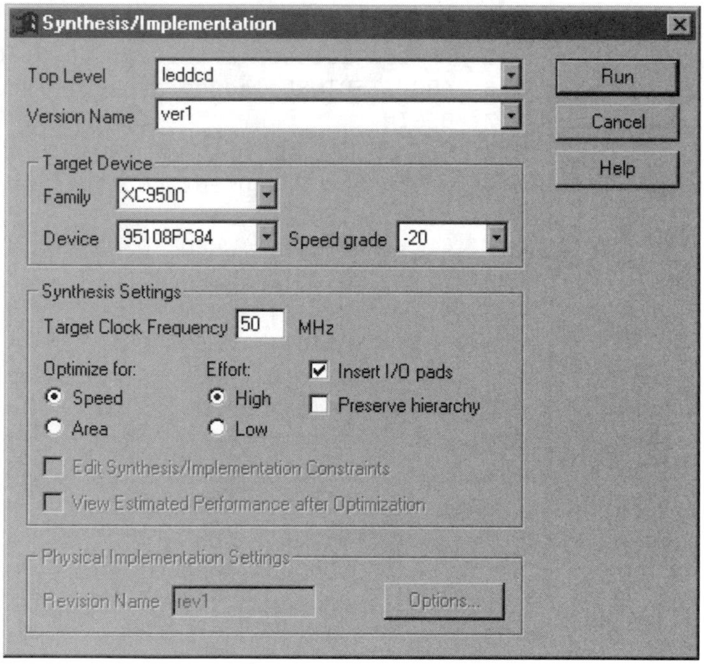

Figure 3.24   Targeting the synthesis of the LED decoder at the XC95108
CPLD.

Once the netlist for the 1-bit adder has been created, you can simulate its operations, compile a bitstream, download the bitstream to an XS Board, and test
it with XSPORT commands just as was shown in the previous two sections.

Projects

1. Design an 8-to-1 multiplexer. Then design a 4-bit-wide, 2-to-1 multiplexer that accepts two 4-bit inputs and places one of them on its 4-bit output. How are these two circuits different? How much of the XC95108 and XC4005XL does each circuit consume?

2. Modify an LED decoder so that it puts out a valid numeral for the inputs of 10, 11, 12, 13, 14, and 15. The numerals for these inputs should look like "A", "b", "C", "d", "E", and "F" (as close as you can get with the seven segments available). How much of the XC95108 and XC4005XL does the new design consume?

3. Create a new design project called BARDCD. This new design should accept a 3-bit input and have 7 outputs. The number of outputs that are on should be the same as the binary representation of the input. Thus, if all the inputs are 0 (i.e., 000), no outputs should be on. If the input is 001, only one output should be on. If the input is 010 (2), two outputs should be on. Such a design would be useful for an amplitude meter on a stereo receiver: Depending on the strength of the signal, more LEDs will light up. Arranging the LEDs in a stack would give you a good visual indication of the amplitude of the signal. How much of the XC95108 and XC4005XL does the new design consume?

4. Create a new design file called ADD2. This new design should accept a pair of 2-bit inputs. The output of the design should be the binary sum of the two 2-bit inputs. How many output bits will be needed? How much of the XC95108 or XC4005XL does the new design consume? How many input patterns are needed to test completely the new design to make sure it all works? Extend your design to build an 8-bit adder. How many steps would be needed to test such a design?

Modular Designs and Hierarchy

Objectives

- Describe the advantages of logic encapsulation and hierarchical design.
- Show how to describe logic modules and interconnect them in a hierarchical fashion using a HDL.
- Show how to describe logic modules and interconnect them in a hierarchical fashion using a schematic editor.

Discussion

By now, you have built combinatorial logic circuits and tested them by using the simulator and by downloading them into the XS40 or XS95 Board. It has all been so simple! So simple, in fact, that you might be asking the question, "Since the computer can figure out the logic design from just the truth table, what do you need me for?" Well, maybe we don't! On the other hand, we have calculators but we still have accountants. The important fact is that the tedious calculations have been automated, leaving us with the more challenging tasks to perform that computers do not yet do so well. For example, accountants now spend less time adding numbers and more time figuring out neat ways to evade the tax regulations. Likewise, you can spend more time thinking up neat circuits and less time mechanically grinding out the details.

However, the computer-aided design programs will always have limitations, so you cannot completely ignore the details of how the gates are connected to build a design. Having knowledge of how logic is transformed and minimized allows you to check the output from the programs and determine if some terrible mistake is being made. At other times, the CAD programs will fail to find a solution even though you know one exists. In these cases, you will have to go in by hand and give a more detailed specification to help the CAD program figure it all out.

More detailed specifications are also needed if you are pushing the limits of your devices. For example, you may be trying to build a circuit that performs some operation very quickly, so you need to reduce the number of gates the signals must pass through on their way from the inputs to the outputs. The CAD programs you are using may not handle this situation, so you may be required to do more work. So, for now, you have not been completely automated out of the picture.

There is another case where you have to be more involved in the design of the gate-level logic. This occurs when a truth table representation for a design is theoretically possible but impractical. For example, an adder that adds two 8-bit numbers could be specified by a truth table having $2^8 \times 2^8 = 65,536$ entries, but you would not want to try to write it out! And while the CAD program could derive the circuit (if it had enough memory), the result would be a circuit that is large and impractical. The solution to building such circuits lies in the human ability to decompose a problem into a set of simpler problems and then recombine the smaller solutions into a solution for the original problem. In the case of the 8-bit adder, you might decompose the problem into a set of 1-bit addition problems. Then you would design a circuit module that adds 1-bit numbers using some combination of AND, OR, and NOT gates. Finally, you combine all these 1-bit adder modules to create a single 8-bit adder. The result is a design with a hierarchy of levels (see Figure 4.1). At the bottom-most level are the simple logic gates you are familiar with. At the next level these gates have been combined to create the 1-bit adder modules. At the top of the hierarchy is the 8-bit adder. But it does not have to stop there because the 8-bit adder could also be used as a module to build a 32-bit adder which could then be used in a microprocessor that is just a module in a multiprocessor… So a hierarchy can extend over many levels.

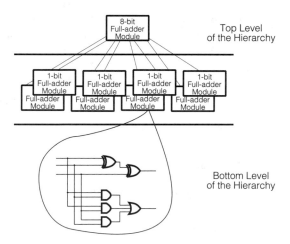

Figure 4.1   The design hierarchy of an 8-bit adder.

Some of the advantages of modular design and hierarchy are as follows:

Design reuse: A well-designed module can be used in multiple designs, which saves you work later on.

Information hiding: Encapsulating a design into a module lets you ignore the details of how a circuit operates and concentrate on the relationship between the inputs and outputs of the module.

Replication: Building a large circuit by duplicating a small group of modules is much easier than building a large circuit by stitching together a large number of primitive gates.

Testing: It is difficult to find the source of problems when you are testing a single large circuit. But a modular design lets you completely test each small module so that any errors found in the final system are probably located in the interfaces between the modules.

Team-based design: Groups of designers can work in parallel on a circuit that is divisible into a set of modules with well-defined interfaces, thus leading to a faster completion.

Not surprisingly, these are the same advantages that modular programming languages provide. The next section will show you how to create modular, hierarchical logic designs using the ABEL, VHDL, and the schematic editor.

Experimental

A Parity Generator for the XS95 Board

Let's expand on the parity generator from the previous chapter. We could specify a truth-table for an 8-bit even-parity generator, but it would have $2^8 = 256$ possible combinations of inputs. Not impossible to type in, but not pleasant. Instead, let's use a hierarchical approach and break the problem apart into simpler problems.

Suppose we knew of a circuit that could generate the parity for a 4-bit number. Could we use this circuit to calculate the parity of an 8-bit number? Of course! (Why else would I bother to mention it?) We can just compute the parity of bits

0..3 and bits 4..7 using two of the 4-bit parity circuits. Then, compute the total parity of the 8-bit number using the following table:

Parity 0..3	Parity 4..7	Parity 0..7
Even	Even	Even
Odd	Even	Odd
Even	Odd	Odd
Odd	Odd	Even

Assuming that parity circuits output a logic 1 whenever the parity is odd and a 0 when parity is even, the above table can be re-written as follows:

Parity 0..3	Parity 4..7	Parity 0..7
0	0	0
1	0	1
0	1	1
1	1	0

This is just the truth table for an XOR gate. So the circuit for an 8-bit parity detector now looks like the one of Figure 4.2(a).

Now the question naturally arises: "So how do I build the 4-bit parity circuits that I just used in the 8-bit circuit?" Well that is easy: build a couple of 2-bit parity generators and connect their outputs to an XOR gate (see Figure 4.2(b)). Carrying the reasoning further shows that the 2-bit parity generators are just single XOR gates. The total 8-bit even-parity generator is shown in Figure 4.2(c).

The problem decomposition has given us a hierarchy: at the bottom level are 2-bit parity generators (a.k.a. XOR gates), followed by 4-bit parity generators at the next hierarchical level, topped off by an 8-bit parity generator at the top level.

How would we describe this hierarchy and these modules in ABEL? That is all shown in Listings 4.1, 4.2, and 4.3. The module names, module-instantiation names, and module input/output names used in the ABEL code are listed in Figure 4.2.

2-bit even-parity generator (Listing 4.1): The MODULE keyword (line 1) starts the definition of the 2-bit parity generator module, followed by the name of the module (evpar2). Line 2 uses the INTERFACE keyword to define a parenthesized list of inputs (b0 and b1) and a single output (par). In this case, the INTERFACE keyword is used to define the external interface of this module for those modules that will use it. The two inputs and the output are

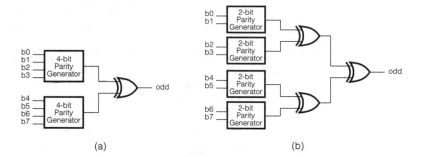

(a)   (b)

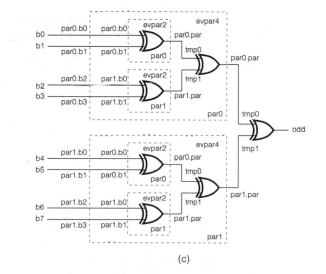

(c)

Figure 4.2   Construction of a parity generator.

Listing 4.1   ABEL description of a 2-bit even-parity generator.

```
001- MODULE evpar2
002- INTERFACE (b0,b1->par);"define interface
003- TITLE '2-bit even-parity generator'
004- DECLARATIONS
005-    b1..b0 PIN;"declare interface pins
006-    par PIN ISTYPE 'COM';
007- EQUATIONS
008-    par = b1 $ b0;        "just a simple XOR
009- END evpar2       "end 2-bit parity module
```

actually declared in lines 5 and 6 following the DECLARATIONS keyword. The Boolean equation on line 8 XORs the b0 and b1 inputs and outputs the parity on the odd output. (The output is called odd because it goes to a logic one if the inputs have odd parity. The total combination of input bits and the output bit has even parity, however.) Then the END keyword ends the 2-bit parity generator module definition. This ABEL code is stored in the EVPAR2.ABL file.

4-bit even-parity generator (Listing 4.2): The interface for the 4-bit parity-generator module, evpar4, is defined on line 2. It receives 4 input bits (b3..b0) and generates a single parity output (par). These inputs and output are declared on lines 5 and 6. The interface to the 2-bit parity generator module is specified on line 8. This replicates the interface information originally declared on the INTERFACE line of the evpar2 module. Then two instances of the 2-bit module are declared on line 10 using the FUNCTIONAL_BLOCK keyword. The two instances of the evpar2 module are named par0 and par1. Two new nodes (tmp0 and tmp1) are declared on line 12 for the sole purpose of holding the outputs from the par0 and par1 modules (this is somewhat like using temporary variables). Lines 15 and 16 specify that the lower 2 bits of input to the 4-bit parity generator (b0 and b1) are connected to the inputs of one of the 2-input parity generators (par0.b0 and par0.b1). (Note how the inputs and outputs of a module are specified by giving the module name and then appending the name of the input as a dot-extension.) The output of the module (par0.par) is connected to the tmp0 node on line 17. The equivalent connections for handling the upper 2 input bits with the second 2-bit parity generator are listed on lines 19–21. Finally, line 23 XORs the outputs of the two modules (now on nodes tmp0 and tmp1) and passes the final parity onto the output of the current module (odd). This ABEL code is stored in the EVPAR4.ABL file.

8-bit even-parity generator (Listing 4.3): The construction of the 8-bit parity-generator module from two 4-bit parity detectors is similar to the construction of the 4-bit parity detector described above. Note that this is the top-level of our design so no INTERFACE definition is needed to define its inputs and outputs since no other module will use it. Note how the use of vector notation on lines 15 and 18 reduces the number of lines needed to specify connections to the lower-level modules. This ABEL code is stored in the EVPAR8.ABL file.

You can create a schematic mode project called EVPAR_95 for the XS95 Board or an EVPAR_40 project for the XS40 Board. In either case, add the EVPAR8.ABL file to the project and then open it with the HDL Editor. Create a netlist for the design by selecting the **Synthesis → Synthesize** menu item in the **HDL Editor** window. The synthesizer will automatically look in the EVPAR2.ABL and EVPAR4.ABL files to get the descriptions of the lower-level modules.

After synthesizing the parity generator's netlist, you can run a functional simulation to see how it operates. Add the b0-b7 inputs and the odd output to **Waveform Viewer** window using the **Signal → Add Signals...** menu item. The 8 inputs should then be attached to the 8 lowest bits of the binary counter in the **Stimulator Selection** window. Then click on the ![button] button in the **Logic Simulator** window toolbar a few times. This will exercise the parity generator for all possible input combinations. A partial waveform for the parity

Listing 4.2   ABEL description of a 4-bit even-parity generator.

```
001- MODULE evpar4
002- INTERFACE ([b3..b0]->par); "4 inputs, 1 output
003- TITLE '4-bit even-parity generator'
004- DECLARATIONS
005-     b3..b0 PIN;
006-     par PIN ISTYPE 'COM';
007- "declare the interface to the 2-bit parity module
008-     evpar2 INTERFACE (b0,b1->par);
009- "declare 2 instances of the 2-bit parity module
010-     par1,par0 FUNCTIONAL_BLOCK evpar2;
011- "these hold the outputs of the 2-bit parity modules
012-     tmp0,tmp1 NODE ISTYPE 'COM';
013- EQUATIONS
014- "compute parity for lower 2 bits
015-     par0.b0 = b0;     "connect inputs of the 4-bit module
016-     par0.b1 = b1;     " to the inputs of the 2-bit modules
017-     tmp0 = par0.par; "connect module output to tmp0 node
018- "compute parity for upper 2 bits
019-     par1.b0 = b2;     "now connect the upper two inputs
020-     par1.b1 = b3;     " to the second 2-bit module
021-     tmp1 = par1.par; "and connect module output to tmp1
022- "combine outputs of the 2-bit parity modules
023-     par = tmp1 $ tmp0;
024- END evpar4
```

generator is shown in Figure 4.3. Note that the odd output goes high only when the total number of 1 bits in the b0-b7 inputs is 1, 3, 5, or 7. Thus, adding the odd output makes the total number of 1 bits even.

When using the XS95 Board, the appropriate XC95108 CPLD pin assignments for the 8 inputs of the parity generator are shown below. The output of the circuit drives the S0 segment of the 7-segment LED.

```
NET b0        LOC=P46;
NET b1        LOC=P47;
NET b2        LOC=P48;
NET b3        LOC=P50;
NET b4        LOC=P51;
NET b5        LOC=P52;
NET b6        LOC=P81;
NET b7        LOC=P80;
NET odd       LOC=P21;# LED segment S0
```

After compiling the even-parity generator into a bitstream and converting the bitstream into the EVPAR_95.SVF file, you can download it to the XS95 Board

Listing 4.3   ABEL description of an 8-bit even-parity generator.

```
001- MODULE evpar8 "top-level module has same name as file
002- "no interface definition in top-level module
003- TITLE '8-bit even-parity generator'
004- DECLARATIONS
005-    b7..b0 PIN;"8 inputs
006-    odd PIN ISTYPE 'COM'; "even-parity output
007- "declare the interface to the 4-bit parity module
008-    evpar4 INTERFACE ([b3..b0]->par);
009- "declare 2 instances of the 4-bit parity module
010-    par1,par0 FUNCTIONAL_BLOCK evpar4;
011- "these hold the outputs of the 4-bit parity modules
012-    tmp0,tmp1 NODE ISTYPE 'COM';
013- EQUATIONS
014- "connect lower 4 inputs to one 4-bit module
015-    par0.[b3..b0] = [b3..b0];
016-    tmp0 = par0.par; "connect module output to tmp0 node
017- "connect upper 4 inputs to other 4-bit module
018-    par1.[b3..b0] = [b7..b4];
019-    tmp1 = par1.par; "connect module output to tmp1 node
020- "compute final parity for 8 bits
021-    odd = tmp1 $ tmp0;
022- END evpar8
```

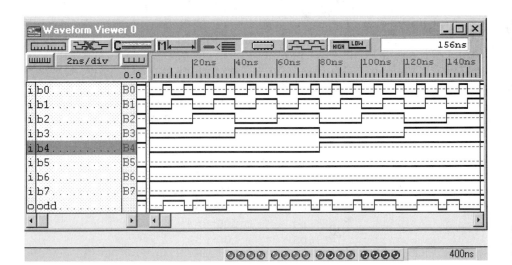

Figure 4.3   Simulated waveforms for the 8-bit even-parity generator.

(remember to store the SVF into the top-level directory of your EVPAR_95 project):

```
C:\XCPROJ\EVPAR_95> XSLOAD EVPAR_95.SVF
```

Now you can do actual tests on the parity generator. For example, the command

```
C:\XCPROJ\EVPAR_95> XSPORT 01001001
```

should light up the S0 segment of the 7-segment LED because the input pattern contains an odd number of 1 bits. On the other hand, the command

```
C:\XCPROJ\EVPAR_95> XCPORT 01011001
```

will should extinguish the S0 segment because the input pattern contains an even number of 1 bits.

You may also notice LED segments besides S0 digit turning on and off as you test the parity detector. If this occurs, it is because the Foundation Series Implementation tools assigned the outputs of the lower-level parity detector modules to pins which are attached to other segments of the LED digit. (Foundation did not knowingly do this - it was just looking for some pins to use and happened to pick some that were attached to the LED.) While unintended, this could be useful for debugging your circuit if you knew which segment was being affected by each parity detector module. You can find this information in the pin assignment report file.

A Parity Generator in VHDL for the XS95 Board

Let's rebuild the 8-bit parity generator using VHDL. Create an HDL mode project called EVPAR_95 in the XCPROJ-V directory. Then create the EVPAR.VHD file containing the VHDL code of Listing 4.4. (This is probably most easily done using an empty HDL Editor file rather than trying to use the HDL Design Wizard.)

The EVPAR.VHD file contains three modules: a 2-bit parity generator (lines 3–17), a 4-bit parity generator (lines 20–44), and an 8-bit parity generator (lines 47–81). Each module begins by including the STD_LOGIC_1164 package of the IEEE library, followed by an ENTITY block which describes the module interface and then an ARCHITECTURE section.

The evpar2 module has two inputs (b0 and b1) and a single output (par) declared on lines 9 and 10, respectively. As shown on line 16 of the ARCHITECTURE section, evpar2 exclusive-ORs the two inputs and places the result on the par output.

Listing 4.4 Hierarchical VHDL description of an 8-bit even-parity generator.

```
001- -- Even-Parity Generator (hierarchical description)
002-
003- -- 2-bit even-parity generator
004- LIBRARY IEEE;
005- USE IEEE.std_logic_1164.ALL;
006-
007- ENTITY evpar2 IS
008- PORT (
009- b0,b1: IN STD_LOGIC;
010- par: OUT STD_LOGIC
011- );
012- END evpar2;
013-
014- ARCHITECTURE evpar2_arch OF evpar2 IS
015- BEGIN
016- par <= b0 XOR b1;
017- END evpar2_arch;
018-
019-
020- -- 4-bit even-parity generator
021- LIBRARY IEEE;
022- USE IEEE.std_logic_1164.ALL;
023-
024- ENTITY evpar4 IS
025- PORT (
026- b0,b1,b2,b3: IN STD_LOGIC;
027- par: OUT STD_LOGIC
028- );
029- END evpar4;
030-
031- ARCHITECTURE evpar4_arch OF evpar4 IS
032- COMPONENT evpar2
033- PORT
034- (
035- b0,b1: IN STD_LOGIC;
036- par: OUT STD_LOGIC
037- );
038- END COMPONENT;
039- SIGNAL tmp: STD_LOGIC_VECTOR (1 DOWNTO 0);
040- BEGIN
041- par0: evpar2 PORT MAP(b0=>b0,b1=>b1,par=>tmp(0));
042- par1: evpar2 PORT MAP(b0=>b2,b1=>b3,par=>tmp(1));
```

Listing 4.4 Hierarchical VHDL description of an 8-bit even-parity generator. (Cont'd.)

```
043- top: evpar2 PORT MAP(b0=>tmp(0),b1=>tmp(1),par=>par);
044- END evpar4_arch;
045-
046-
047- -- 8-bit even-parity generator
048- LIBRARY IEEE;
049- USE IEEE.std_logic_1164.ALL;
050-
051- ENTITY evpar8 IS
052- PORT
053- (
054- b: IN STD_LOGIC_VECTOR (7 DOWNTO 0);
055- odd: OUT STD_LOGIC
056- );
057- END evpar8;
058-
059- ARCHITECTURE evpar8_arch OF evpar8 IS
060- COMPONENT evpar2
061- PORT
062- (
063- b0,b1: IN STD_LOGIC;
064- par: OUT STD_LOGIC
065- );
066- END COMPONENT;
067- COMPONENT evpar4
068- PORT
069- (
070- b0,b1,b2,b3: IN STD_LOGIC;
071- par: OUT STD_LOGIC
072- );
073- END COMPONENT;
074- SIGNAL tmp: STD_LOGIC_VECTOR (1 DOWNTO 0);
075- BEGIN
076- par0: evpar4 PORT MAP(b0=>b(0),b1=>b(1),b2=>b(2),
077-b3=>b(3),par=>tmp(0));
078- par1: evpar4 PORT MAP(b0=>b(4),b1=>b(5),b2=>b(6),
079-b3=>b(7),par=>tmp(1));
080- top: evpar2 PORT MAP(b0=>tmp(0),b1=>tmp(1),par=>odd);
081- END evpar8_arch;
```

The evpar4 module uses the evpar2 module to build a 4-bit parity generator. In order to use the other module as a component, the evpar4 module needs to know about the interface to the evpar2 module. This is done using the COM-

PONENT statement of VHDL. The COMPONENT statement is placed between the opening statement of the ARCHITECTURE section (line 31) and the BEGIN keyword (line 40). Following the COMPONENT keyword is the name of the interface of the module that is going to be used (evpar2). This is followed by the PORT keyword and then a copy of the I/O declarations found on lines 9 and 10 is placed on lines 35 and 36. Then the COMPONENT block is ended with the END COMPONENT keywords.

Once the interface to the evpar2 module is known, it can be instantiated in the ARCHITECTURE section to build the 4-bit parity generator circuitry (lines 41–43). Each of these lines starts with a distinct identifier that names the module being instantiated. Each identifier ends with a colon followed by the type of module that is being instantiated (evpar2 in all three statements). This is followed by the PORT MAP keywords and a parenthesized list of connections between the instantiated module and signals in the top-level module (evpar4). For example, on line 42 the b0 input of the par1 module is connected to the b2 input of the evpar4 top-level module. And b1 of the par1 module is connected to the b3 input of evpar4. Finally, the par output of the par1 module is connected to the tmp(1) signal. (The name of the I/O terminal for the lower-level module is always placed to the left of the => connection operator, while the name of the I/O or signal in the top-level module is placed to the right.)

Where did the tmp(1) signal come from? This is an element of the 2-element tmp array that was declared on line 39 (the other element is tmp(0)). The SIGNAL keyword indicates that the tmp array contains the values for signals, and the type STD_LOGIC_VECTOR indicates these signals carry logic values. The placement of the SIGNAL statement between the ARCHITECTURE and BEGIN statements means that the tmp array elements can only be accessed within the scope of the evpar4 module (i.e., the values in tmp are private and cannot be seen by any other modules). The tmp array is used to hold the outputs from the par0 and par1 modules (lines 41 and 42). Then the tmp array elements are used as inputs to the top module (line 43). The top module combines the parity bits from the two 2-bit fields and places the final parity bit for all 4 bits onto the par output.

The evpar8 module is very similar to the evpar4 module. The main difference is that the evpar8 module uses both the 2-bit and 4-bit parity generator modules to do its calculations. Therefore it must use COMPONENT statements within its ARCHITECTURE section to declare the interfaces for the evpar2 module (lines 60–66) and the evpar4 module (lines 67–73). The 4-bit parity generators are used to compute the parity for the lower 4 bits of the input vector (lines 76–77) and the upper 4 bits (lines 78–79). The outputs of the 4-bit parity generators are combined by a 2-bit parity generator to create the final output (line 80).

Once you have completed entering the VHDL code into the **HDL Editor** window, you can use the **Synthesis** → **Check Syntax** menu item to look for errors. Close the **HDL Editor** window after correcting any errors. Select the **Document** → **Add** menu item in the **Project Manager** window to add the EVPAR.VHD file to the project. Then click on the ▣ ▸ ▣ button in the **Flow** tab of the **Project Manager** window. The **Synthesis/Implementation** window will appear as usual, but there is a difference. You now have a hierarchical description of a logic design and you need to tell the Foundation Series synthesis software which module is at the top of the hierarchy. To do this, click on the **Top Level** drop-down menu in the **Synthesis/Implementation** window and a list of all the modules in the EVPAR.VHD file will appear (Figure 4.4). Just select evpar8 as the top level module. You must also target the project to an XC95108 CPLD. Then click on **Run** to start the synthesis of a netlist for the 8-bit parity generator. Once the netlist is available, you can simulate, compile, download, and test the parity generator on the XS95 Board exactly as I discussed in the previous section. Because of the way the Foundation Series synthesis software formats VHDL I/O array names, you will have to change the format of the EVPAR_95.UCF file to look as follows:

```
NET  b<0>          LOC=P46;
NET  b<1>          LOC=P47;
NET  b<2>          LOC=P48;
NET  b<3>          LOC=P50;
NET  b<4>          LOC=P51;
NET  b<5>          LOC=P52;
NET  b<6>          LOC=P81;
NET  b<7>          LOC=P80;
NET  odd           LOC=P21;# LED segment S0
```

A Parity Generator for the XS40 Board

The ABEL or VHDL code for the even-parity modules should also work for the XC4005XL FPGA on the XS40 Board. However, when you try to compile the even-parity generator for the XS40 Board using the following user-constraints:

```
NET  b0            LOC=P44;
NET  b1            LOC=P45;
NET  b2            LOC=P46;
NET  b3            LOC=P47;
NET  b4            LOC=P48;
NET  b5            LOC=P49;
NET  b6            LOC=P32;
NET  b7            LOC=P34;
NET  odd           LOC=P25;
```

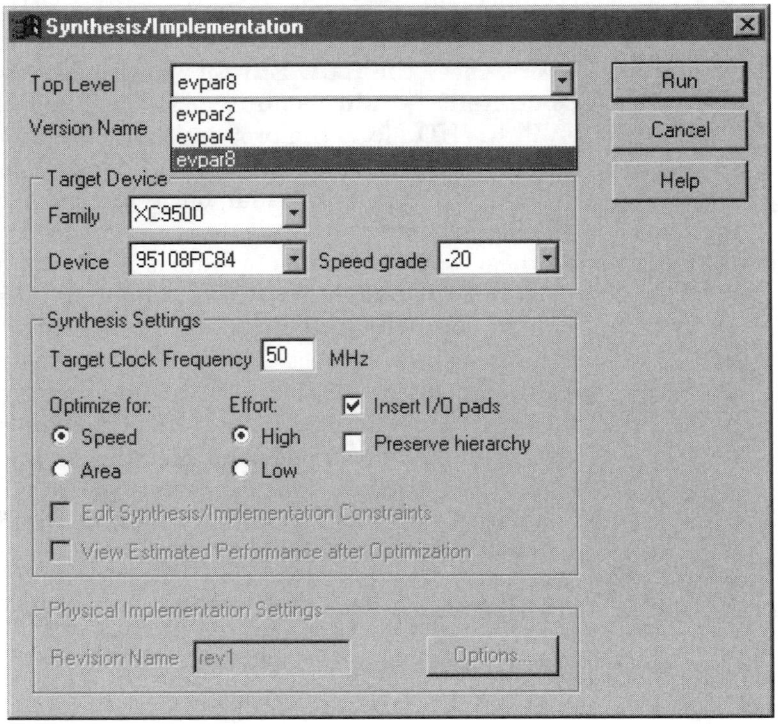

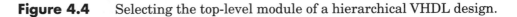

Figure 4.4   Selecting the top-level module of a hierarchical VHDL design.

you will get an error message like this from the Foundation Series Implementation tools:

ERROR:baste:262 - Bad format for LOC constraint P34 on symbol 'b7.PAD' (pad signal=b7). No such site for this device. This may also indicate that a non-constrainable site (such as a VCC, GND, mode, configuration, or otherspecial-purpose pin) has been used as a site name.

ERROR:baste:262 - Bad format for LOC constraint P32 on symbol 'b6.PAD' (pad signal=b6). No such site for this device. This may also indicate that a non-constrainable site (such as a VCC, GND, mode, configuration, or other special-purpose pin) has been used as a site name.

The XS40 Board uses M0 and M2 of the XC4000 FPGA (pins 32 and 34, respectively) as dedicated inputs that are driven by the PC parallel port. But M0 and M2 are also special-purpose pins that set the mode of the XC4005XL FPGA when it powers up. The Foundation Series software will not let us specify user-constraints that assign nets to these pins. One way to access these pins is to use special-purpose symbols in the schematic editor. Therefore, we need a way to get our ABEL or VHDL version of the parity generator into a schematic so we can use these pins.

First, create an EVPAR_40 project targeted toward the XC4005XL FPGA. Once the project directory exists, use the standard file copying tools of your operating system to bring the EVPAR2.ABL, EVPAR4.ABL, and EVPAR8.ABL files from the C:\XCPROJ\EVPAR_95 directory to the C:\XCPROJ\ EVPAR_40 directory. If you are using the VHDL version of the parity generator, then copy the EVPAR.VHD file from C:\XCPROJ-V\EVPAR_95 into the C:\XCPROJ\EVPAR_40 directory. (Do not use the **Document → Add...** menu item to do this because we do not want these ABEL or VHDL files actually added to the project.)

Once the ABEL or VHDL files have been copied, open the EVPAR8.ABL or EVPAR.VHD file in an **HDL Editor** window. Select the **Synthesis → Options** menu item. Click on the **Macro** radio-button in the window that appears. This directs the synthesizer to create a netlist for a module that will be included within a larger design. Click **OK** to remove the window. Then select **Synthesis → Create Macro** to create the netlist for the even-parity macro and add it to the EVPAR_40 project library.

Now, how do we actually use the parity generator macro once it is available? We do it the same way that we added simple logic gates. Just select the **Mode → Symbols** menu item to bring up the **SC Symbols** window. Scroll through the list of symbols until you find the EVPAR8 part. Highlight this part, move your cursor into the drawing area of the window, and then click to drop the parity generator into the schematic. After doing these steps, your **Schematic Capture** window should look like Figure 4.5.

Now we can connect the inputs and outputs of the parity generator's EVPAR8 symbol to I/O buffers and terminals. Use input terminals and IBUF symbols for the B0, B1, B2, B3, B4, and B5 inputs. Use an OBUF and an output terminal for the ODD output. These connections are shown in Figure 4.6. (You do not have to assign the terminals the same name as the input on the EVPAR8 symbol. I just did that for simplicity.)

The B6 and B7 inputs of the EVPAR8 need to be connected to special symbols associated with the M0 and M2 pins of the XC4005XL FPGA. The special symbols can be found in the **SC Symbols** window under the names **MD0** and **MD2**. Just select one of each type and drop them into the schematic. Then connect them to the B6 and B7 inputs of the EVPAR8 symbol through IBUFs (see Figure 4.6).

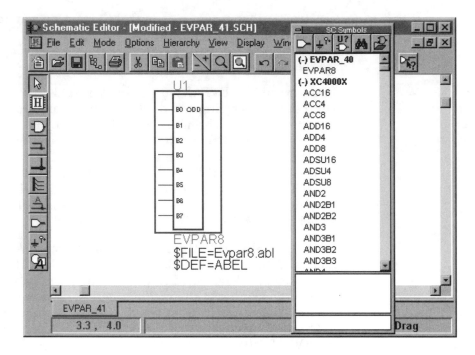

Figure 4.5   Adding the even-parity generator symbol to a schematic.

After you have gotten the inputs and outputs connected, just use the **Options** → **Create Netlist**, **Options** → **Integrity Test**, and **Options** → **Export Netlist...** menu items to create an EDIF netlist.

With the netlist generated, we can do a functional simulation. But when we open the **Component Selection** window to begin selecting signals for the **Waveform Viewer** window, the B6 and B7 inputs are not found in the **Signal Selection** pane (see Figure 4.7). They are absent because these inputs are attached to the special-purpose pins. But we need to drive values onto these 2 inputs to provide a full set of 8 input values to the parity generator.

The solution to our problem is to double-click on the **U1 - EVPAR8** item in the **Chip Selection** pane. This places a list of the inputs and outputs for the parity generator module in the right-most pane of the **Component Selection** window. We can select the B6 and B7 input pins for the module and click on the **Add** button. These inputs appear as U1.B6 and U1.B7 in the **Waveform Viewer** window.

Once the inputs and outputs are added, the 8 inputs should be attached to the 8 lowest bits of the binary counter in the **Stimulator Selection** window. Then click on the ![button] button in the **Simulator** window a few times. This will exercise

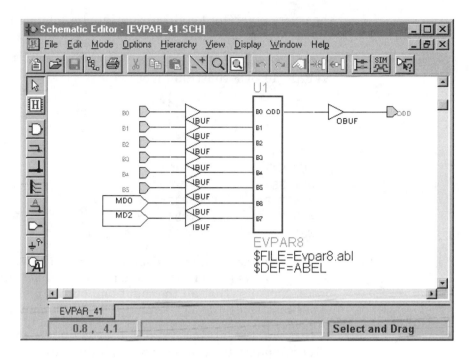

Figure 4.6 Connecting the even-parity generator symbol to the I/O pins.

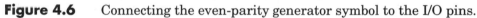

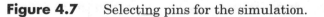

Figure 4.7 Selecting pins for the simulation.

the parity generator for all possible input combinations. The waveforms should resemble those of Figure 4.3.

Before compiling the design, place the following constraints in the EVPAR_40.UCF constraint file:

```
NET B0        LOC=P44;
NET B1        LOC=P45;
NET B2        LOC=P46;
NET B3        LOC=P47;
NET B4        LOC=P48;
NET B5        LOC=P49;
# NET B6      LOC=P32;
# NET B7      LOC=P34;
NET ODD       LOC=P25;# LED segment S0
```

The constraints for inputs B6 andB7 are commented-out so they have no effect (I just left them in there so I would not forget that these inputs were still being used). Note also that all the input and output names have been converted to upper-case since the I/O terminals in a schematic always have upper-case names.

After compiling the even-parity generator with the user-constraint file shown above, we can open the Place&Route report file and find the following information:

```
Device utilization summary:
    IO           7/112      6% used
                 7/61       11% bonded
    LOGIC        1/196      0% used
    SPECIAL      2/1167     0% used

    IOB          7/112      6% used

    CLB          1/196      0% used

    MODE0        1/1        100% used
    MODE2        1/1        100% used
```

The last two lines indicate we were successful in our attempt to use the M0 and M2 special purpose pins as inputs.

The configuration bitstream for the even-parity generator can be downloaded into the XS40 Board like so:

```
C:\XCPROJ\EVPAR_40> XSLOAD EVPAR_40.BIT
```

Now you can do actual tests on the parity generator. For example, the command

```
C:\XCPROJ\EVPAR_40> XSPORT 00101100
```

will light up the S0 segment of the 7-segment LED because the input pattern contains an odd number of 1 bits. On the other hand, the command

```
C:\XCPROJ\EVPAR_40> XCPORT 10110001
```

will extinguish the S0 segment because the input pattern contains an even number of 1 bits.

An 8-Bit Adder for the XS95 Board

Addition is a frequently-used operation, so it makes sense to define an adder module that can be used in other designs. We can use the 1-bit adder module from the previous chapter to build larger adders. The connections needed to build an 8-bit adder from 1-bit adder modules are shown in Figure 4.8. The only question that remains is procedural: How do we use the Foundation Series schematic editor to make a 1-bit adder module and connect them together?

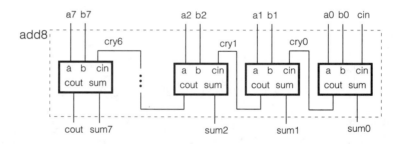

Figure 4.8   The construction of an 8-bit adder module using 8 1-bit adder modules.

Let's start by creating the ADD8_95 schematic mode project for the XC95108 CPLD. Then, use the **Document → Add** menu item to add the C:\XCPROJ\ ADD1_95\ADD1.SCH schematic file (see Figure 3.13) to the current project.. Double-click on the schematic name to open it. Once the 1-bit adder schematic is displayed, edit it as follows:

1. Double-click on the **INPUT0** terminal. When the **I/O Terminal** window appears, change the name of the terminal to **A**. Follow the same procedure to change **INPUT1** to **B**, **CARRY_INPUT** to **CIN**, and **CARRY_OUTPUT**

to **COUT**. The shorter names will be easier to display when we use these modules to build larger adders.

2. Bypass each IBUF and OBUF buffer with a wire that connects its input to its output.

3. Remove the IBUF and OBUF buffers. Since we are going to encapsulate the 1-bit adder into a module, we do not necessarily want the module's I/O to be attached to pins of the CPLD. Notice that the I/O terminal will disappear along with the buffer if you did not wire the buffer's input to its output. That is the reason for the previous step.

Once you have completed all these steps, your schematic should look like Figure 4.9. Next, we have to create a symbol to represent the 1-bit adder for when we use it in other schematics. To do this, select the **Hierarchy → Create Macro Symbol from Current Sheet** menu item. The **Create Symbol** window shown in Figure 4.10 will appear. Type the name by which you want to refer to the 1-bit adder module into the **Symbol Name** field (I chose MY_ADD1, but you might want to be more creative). You can also enter a phrase into the **Comment** field to reference the function of this module. Then click on **OK**. Then you will be informed that MY_ADD1 has been added to the list of symbols and asked if you want to modify the symbol for MY_ADD1. Just click on **No**.

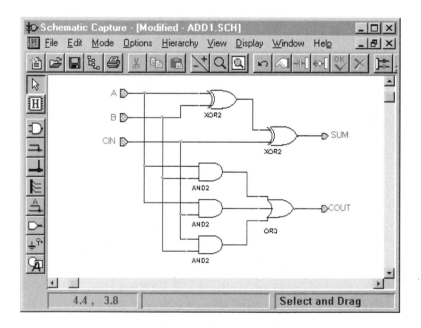

Figure 4.9   ADD1.SCH modified for encapsulation into a module (compare with Figure 3.13).

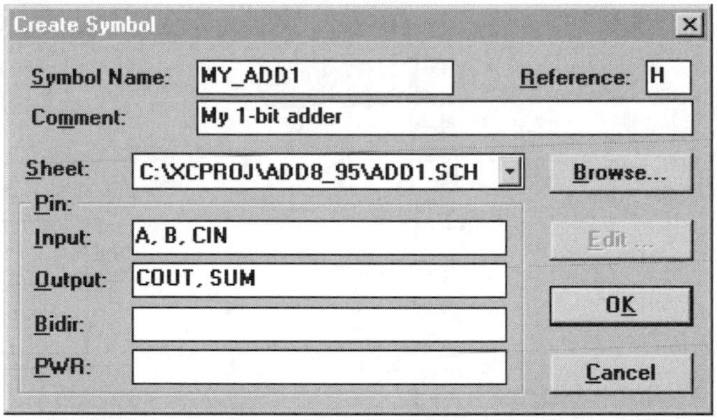

Figure 4.10 Creating a symbol for the 1-bit adder module.

At this point, you should be in a blank schematic editor screen. So select the **File → New Sheet** menu item. An empty schematic sheet labeled ADD8_951.SCH will appear. Select the **Mode → Symbols** menu item to display the list of parts you can enter into your schematic. Scroll through the parts list into the M section and you should find the **MY_ADD1** module (if that is what you called it). Click on MY_ADD1 and move your cursor into the schematic drawing area. The symbol for a 1-bit adder should be attached to your cursor. Drop 8 of them into the drawing and arrange them as a column.

Along with the 1-bit adders, we need IBUF and OBUF modules. Select IBUF8 from the list of modules and place two of them near the top of the column of adder bits. The IBUF8 module is a group of 8 IBUFs which we will use to bring in the 8-bit A and B operands for the adder. Also add a single IBUF module for the CIN carry input. At the bottom of the column, add an OBUF8 module for the 8-bit SUM output, and a single OBUF for the COUT carry output. At this point, the top portion of your schematic should look something like the one in Figure 4.11.

With the parts in place, we can now connect them. First, connect the COUT output of each 1-bit adder module to the CIN input of the 1-bit adder beneath it. This creates the carry-chain. Then connect the CIN of the top 1-bit adder to the output of the IBUF module and connect the COUT of the bottom 1-bit adder to the input of the OBUF module.

Next, select the **Mode → Draw Buses** menu item. Click on the output of one of the IBUF8 modules and draw a bus downward to the bottom of the column of 1-bit adders. At the bottom, double-click to end the bus. An **Edit Bus** window will appear (Figure 4.12). Type **AIB** in the **Bus Name** field, since this bus will carry the buffered A inputs into the adder. Then set the upper and lower bounds of the bus range to 7 and 0, respectively, since this is supposed to

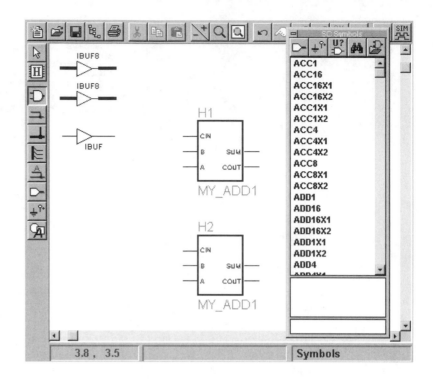

Figure 4.11   Placing the components for the 8-bit adder.

be an 8-bit bus. Select **None** as the **I/O Marker**. Then click **OK**. Repeat this procedure for the other IBUF8 module, but name the bus **BIB**. Then create a bus called SUMOB that attaches to the input of the OBUF8 module. This bus will carry the SUM output to the buffer. Now your schematic should look something like the one in Figure 4.13.

With the buses in place, we next need to connect the 1-bit adders to them. Select the **Mode → Draw Bus Taps** menu item to start the process. Then click your cursor on the AIB bus. This selects the bus which you will be tapping into. Now click your cursor on the A input of the top 1-bit adder module. A wire will extend from the A input and tap into the AIB bus. The wire will be labeled AIB0, indicating an attachment was made to the least-significant bit of the AIB bus. Clicking on the A input of the next 1-bit adder will create another tap labeled AIB1. Move downward through the adder bits to create the AIB2-AIB7 taps. Then press the ESC key to end the bus-tapping process. Repeat these steps for connecting the B inputs of each 1-bit adder to the BIB bus, and for connecting the SUM outputs to the SUMOB bus. If you make a mistake when tapping into a bus, just double-click on the name of the tap and you can edit it to be what you want.

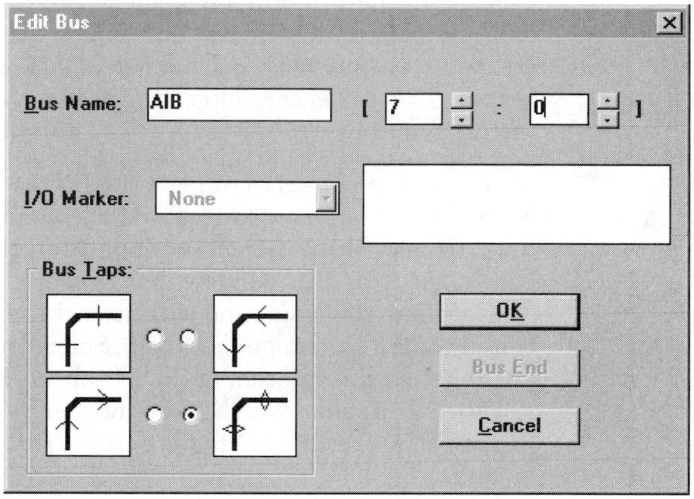

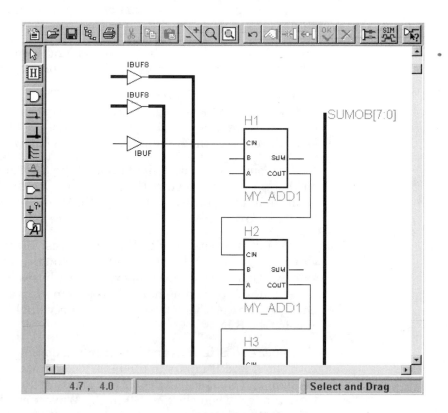

Figure 4.12 Window for setting bus names and types.

Figure 4.13 Partial wiring for the 8-bit adder.

That is all the internal wiring. Now we can connect the I/O terminals. First, create a CIN input terminal and wire it to the input of the IBUF module. Then create an output terminal named COUT and wire it to the OBUF module's output. This takes care of getting the carry input into the circuit and getting the carry output out.

We need to use buses to connect the IBUF8 and OBUF8 modules to terminals. Select **Mode → Draw Buses** and click on the input to the IBUF8 that drives the AIB bus. Move the cursor slightly to the left and double-click to end the bus. In the **Edit Bus** window that appears, type **A** in the **Bus Name** field since this is where the A operand will enter the circuit. Then set the upper and lower bounds of the bus range to 7 and 0. Select **Input** as the **I/O Marker**. Then click **OK** to finish the creation of the I/O bus. A bold input terminal symbol labeled **A[7:0]** will appear which is connected to the input of the IBUF8 module. Repeat this process to create a **B[7:0]** bus connected to the second IBUF8 module.

For the output of the adder, click on the output of the OBUF8 module, move the cursor slightly to the right, and double-click. Set the bus name to be SUM and select **Output** as the **I/O Marker**. Then click **OK** and the **SUM[7:0]** set of output terminals will appear.

Figure 4.14 and Figure 4.15 are the top and bottom portions of the final schematic for the 8-bit adder after all the bus connections are completed.

Save the schematic once it is finished. Then use the **Options → Create Netlist**, **Options → Integrity Test**, and **Options → Export Netlist...** menu items to create, check, and export the netlist (in EDIF 200 format) for the 8-bit adder. Then close the **Schematic Editor** window.

At this point, we can do a simulation of the 8-bit adder to see if it works. Click on the ⬛ button to start the simulator. Use the **Signal → Add Signals...** menu item to add the A bus, B bus, SUM bus, and CIN and COUT signals to the **Waveform Viewer** window (see Figure 4.16).

The A and B inputs are displayed as buses, but we will need to assign individual input lines to bits of the binary counter in the **Stimulator Selection** window. Therefore, we need to expand the buses into their individual signals. Click on the A bus name in the **Waveform viewer** and then select **Signal → Bus → Flatten**. This lets you see the individual bus signals. Repeat the process for the B bus.

Now open the **Stimulator Selection** window. We will attach the lower bits of both the A and B inputs to the binary counter. Click on **A0** in the **Waveform Viewer** and then on bit **Bc0** of the binary counter in the **Stimulator Selection** window. Then click on **B0** in the **Waveform Viewer** and on **Bc1** in the **Stimulator Selection** window. Repeat this process for A1, B1, A2, and B2

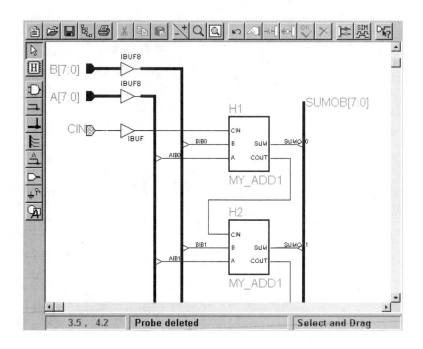

Figure 4.14 The top portion of the complete 8-bit adder.

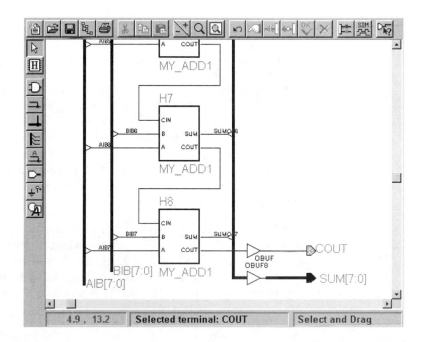

Figure 4.15 The bottom portion of the complete 8-bit adder.

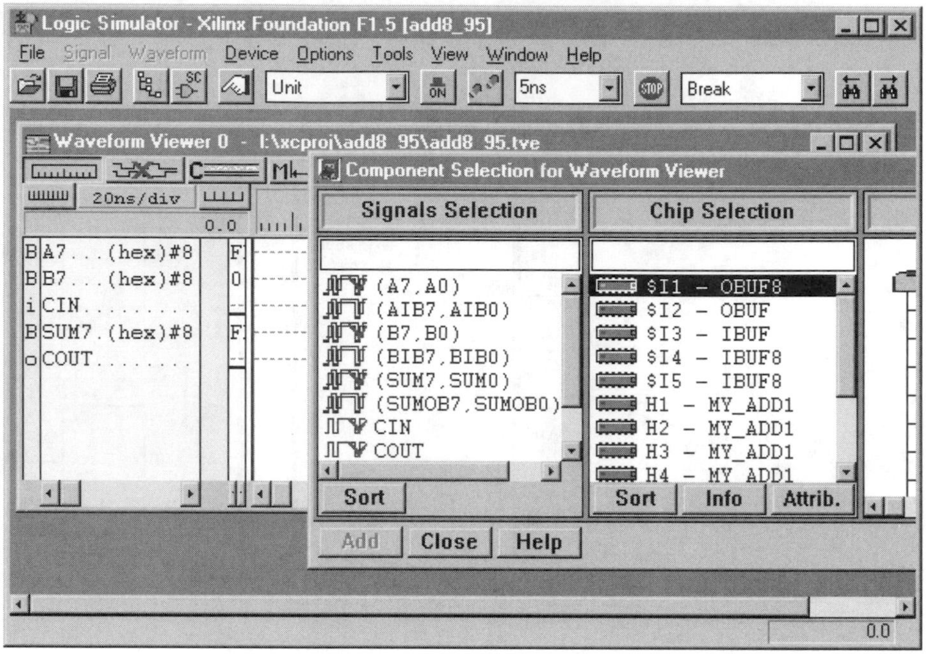

Figure 4.16 Adding buses of the 8-bit adder to the **Waveform Viewer** window.

with successive bits of the binary counter. We have now attached varying signals from the binary counter to both the A and B buses.

We now have varying signals from the binary counter driving the lower 3 bits of the A and B buses. But we also need to attach valid logic levels to the upper bits of these buses and to the CIN input. To do this, highlight an unattached input in the **Waveform Viewer** window and then click on the **0** button in the **Keyboard** area of the **Stimulator Selection** window. This will tie the input to a logic 0. After connecting the rest of the inputs to 0, close the **Stimulator Selection** window. Your **Waveform Viewer** should look like the one shown in Figure 4.17.

Now that the inputs have the appropriate stimulation, we can re-assemble the buses. Click on **A7** in the **Waveform Viewer** window, and then hold the Shift-key down while clicking on **A0**. All the A bus inputs will be highlighted. Then select **Signal** → **Bus** → **Combine**. This will assemble the A7-A0 signals back into an 8-bit bus. Repeat these steps to rebuild the B bus.

Now we can run a simulation of the 8-bit adder. Click the **Long** button in the floating **Simulator** window. A portion of the resulting waveforms are shown in figure 4.18. Note that at all times the value displayed on the SUM bus equals

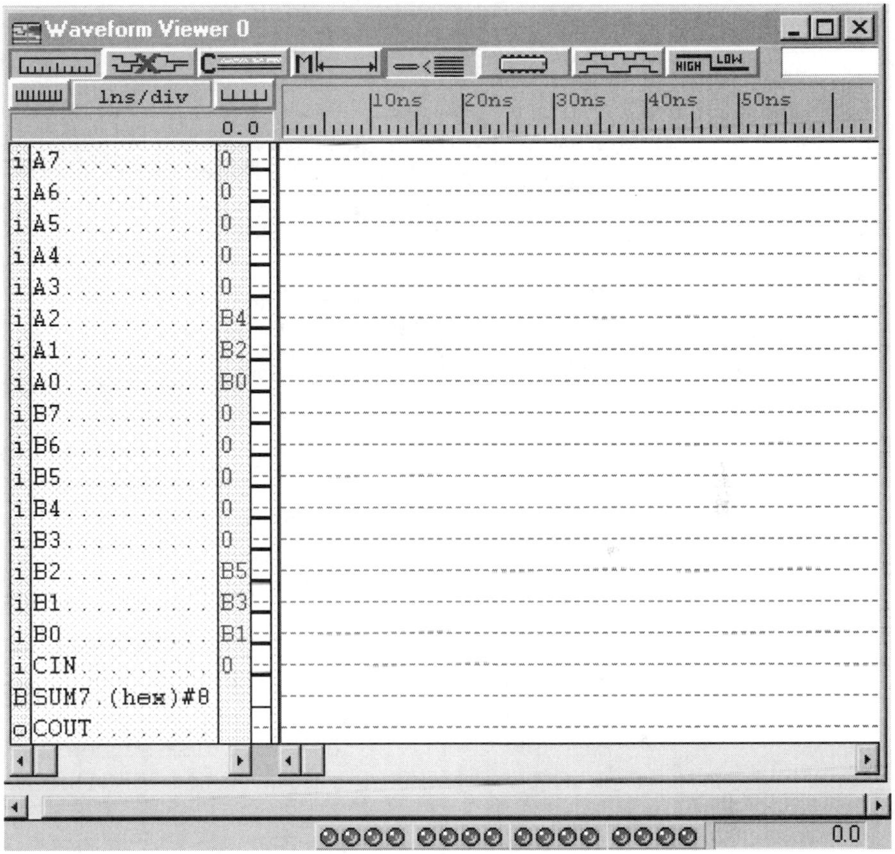

Figure 4.17 The **Waveform Viewer** window after stimulators have been added to the 8-bit adder inputs.

the sum of the two values displayed on the A and B buses. This was not an exhaustive test, but it does give us some hope that the 8-bit adder is functioning correctly. (If your results do not agree with those shown, you may have to change the direction of your buses. Click on a bus name and then select **Signal → Bus → Direction**. The hexadecimal value represented by the bus bits will be recomputed by reversing the bit-order of the bus lines. Do this for each bus that appears to have a problem.)

The adder appears to be working correctly. Place the following constraints in the ADD8_95.UCF file before compiling the 3-bit adder for the XS95 Board (note the addition of the <...> surrounding each index for the individual inputs making up each bus):

```
NET CIN          LOC=P46;
NET A<0>         LOC=P47;
```

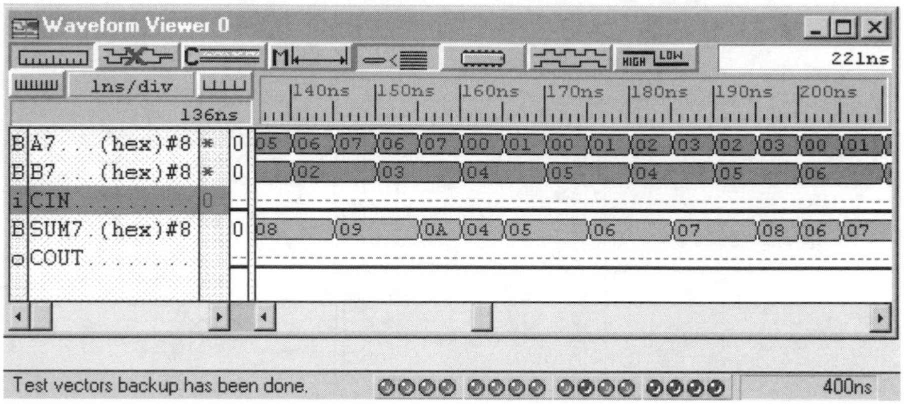

Figure 4.18 Waveforms from the simulation of the 8-bit adder.

```
NET A<1>          LOC=P48;
NET A<2>          LOC=P50;
NET B<0>          LOC=P51;
NET B<1>          LOC=P52;
NET B<2>          LOC=P81;
NET SUM<0>        LOC=P21;# LED segment S0
NET SUM<1>        LOC=P23;# LED segment S1
NET SUM<2>        LOC=P19;# LED segment S2
# the following constraints assign unused ADD8 terminals to out-of-the-way pins
NET COUT          LOC=P7;
NET A<3>          LOC=P75;
NET A<4>          LOC=P79;
NET A<5>          LOC=P82;
NET A<6>          LOC=P84;
NET A<7>          LOC=P1;
NET B<3>          LOC=P61;
NET B<4>          LOC=P55;
NET B<5>          LOC=P57;
NET B<6>          LOC=P58;
NET B<7>          LOC=P83;
NET SUM<3>        LOC=P2;
NET SUM<4>        LOC=P53;
NET SUM<5>        LOC=P56;
NET SUM<6>        LOC=P54;
NET SUM<7>        LOC=P6;
```

The PC's parallel port is limited to driving 8 of the 8-bit adder inputs. I allocated one bit to the carry input, and 3 bits to each of the A and B buses. The remainder of the A and B input bits were assigned to pins on the XS95 Board that would not interfere with anything. (You can tie these pins to ground manually, but it is not necessary since they only affect the upper bits of the SUM output.) In addition, the LED digit segments are driven by only the lower 3 output bits of the SUM bus. The rest of the SUM bits and the carry output are assigned to other pins that are out of the way.

After the bitstream file is generated, convert it to an SVF stored in the top directory of the ADD8_95 project using the **Tools → Device Programming → JTAG Programmer** menu item. Once the ADD8_95.SVF file is built, download it to the XS95 Board. Then you can test the design. The bit assignments for the XSPORT command are as follows:

XSPORT bit	Adder Input
b0	CIN
b1	A0
b2	A1
b3	A2
b4	B0
b5	B1
b6	B2
b7	*** Not Used ***

The assignment of LED segments to bits of the sum is as follows:

LED Segment	Sum Bit
S0	SUM0
S1	SUM1
S2	SUM2
S3–S6	*** Not Used ***

As an example, to add 2 (010) and 3 (011) with no carry input (0), you would issue the command:

```
C:\XCPROJ\ADD8_95\> XSPORT 0100110
```

Then you would expect a result of 5 (101). Therefore, the lower-left and the bottom segments of the 7-segment LED should be on.

At this point, you have used the 1-bit adder module to build an 8-bit adder. You could take the 8-bit adder, remove the IBUFs and OBUFs, and save it as the MY_ADD8 8-bit adder module. The MY_ADD8 module could be used wherever you need an 8-bit adder in a design. It is a lot less work than re-coding the adder each time you need it. Eventually, you will build a set of modules you often use in your designs. These modules play a role analogous to that of a "parts cabinet". In fact, this is what the list of parts shown in the **SC Symbol** window of the schematic editor is.

An 8-Bit Adder for the XS40 Board

You can build the 8-bit adder module for the XS40 Board in much the same way that we did for the XS95 Board: create the MY_ADD1 1-bit adder module and then replicate it and add connections to build an 8-bit adder.

But there is one modification that is needed. As we saw in the even-parity generator design, 2 of the 8 outputs from the PC parallel port drive special-purpose inputs on the XC4005XL FPGA of the XS40 Board. That means we have to use special symbols in the schematic because it is illegal to write constraints that assign nets or terminals to the special-purpose pins.

The 8-bit adder is driven by 7 outputs of the parallel port: one for the carry input, and three for the lower bits of each of the two numbers that are being added. That means we need to use at least one special-purpose pin. In this example, I will use the M0 pin. Figure 4.19 shows the circuit that results from the following modifications in the wiring of the 8-bit adder:

1. The BIB2 connection to the BIB bus has been removed.

2. A new input buffer has been connected to the B input of the H3 instance of the MY_ADD1 1-bit adder module.

3. The input of the new IBUF module is driven by the MD0 symbol which corresponds to an input from the special-purpose M0 pin.

The result of these modifications is that the third bit of the B input to the 8-bit adder is now driven by the value on the M0 pin and not from the B2 bus bit. So the B2 bit is assigned to an out-of-the-way pin in the following ADD8_40.UCF user-constraint file:

```
NET CIN          LOC=P44;
NET A<0>         LOC=P45;
NET A<1>         LOC=P46;
NET A<2>         LOC=P47;
NET B<0>         LOC=P48;
NET B<1>         LOC=P49;
NET B<2>         LOC=P6;     # assign to some out-of-the-way pin
NET SUM<0>       LOC=P25;    # LED segment S0
NET SUM<1>       LOC=P26;    # LED segment S1
NET SUM<2>       LOC=P24;    # LED segment S2
# the following constraints assign unused ADD8 terminals to out-of-the-way pins
NET COUT         LOC=P7;
```

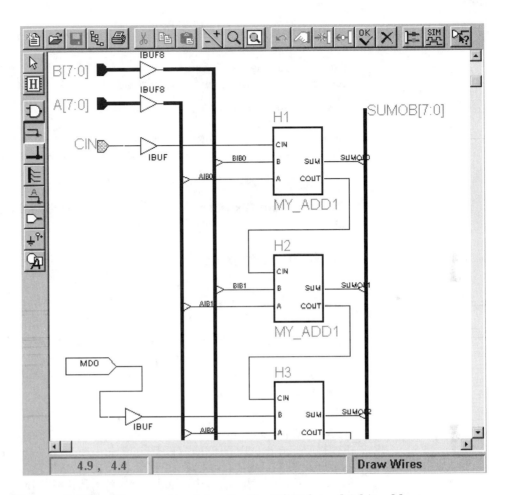

Figure 4.19 Changes in wiring for the XS40-based 8-bit adder.

```
NET  A<3>            LOC=P78;
NET  A<4>            LOC=P79;
NET  A<5>            LOC=P82;
NET  A<6>            LOC=P84;
NET  A<7>            LOC=P3;
NET  B<3>            LOC=P5;
NET  B<4>            LOC=P60;
NET  B<5>            LOC=P56;
NET  B<6>            LOC=P58;
NET  B<7>            LOC=P59;
NET  SUM<3>          LOC=P83;
NET  SUM<4>          LOC=P4;
NET  SUM<5>          LOC=P50;
```

```
NET  SUM<6>          LOC=P57;
NET  SUM<7>          LOC=P51;
```

After compiling the design for the XC405XL FPGA, the configuration bit-stream for the adder can be downloaded into the XS40 Board like so:

```
C:\XCPROJ\ADD8_40> XSLOAD ADD8_40.BIT
```

Now you can do actual tests with the adder. For example, to add 3 (011) and 3 (011) with the carry input set (1), you would issue the command:

```
C:\XCPROJ\ADD8_95\> XSPORT 0110111
```

Then you would expect a result of 7 (111). Therefore, the lower-left, lower-right, and the bottom segments of the 7-segment LED should be on.

An 8-Bit Adder in VHDL

In this section we will build an 8-bit adder for both the XS95 Board and then the XS40 Board using VHDL and the HDL design flow mode of the Foundation Series software. We will also see how to create our own parts library in VHDL.

Start by creating an HDL mode project called ADD8_95 in the XCPROJ-V directory. Then open an **HDL Editor** window and create a VHDL file called XSE.VHD which will hold a library of components for use in other designs. For now, XSE.VHD will contain only a single component that is a single-bit full adder (Listing 4.5).

The bottom portion of the XSE.VHD file (lines 21 through 42) is just the ENTITY and ARCHITECTURE sections of the fulladd design we did in chapter 3. The only new feature in this file is the PACKAGE declaration section on lines 8 through 19. Following the PACKAGE keyword is the package name we have chosen (adder) followed by a COMPONENT statement declaring the interface to the fulladd module. This COMPONENT statement has exactly the same form as we used in the EVPAR.VHD file when all the modules were declared in the same file. If there were multiple modules we wanted to place in the adder package, then we would have multiple COMPONENT statements in the PACKAGE section.

Save the XSE.VHD file and close the **HDL Editor** window. Now we have to make the library known to the Foundation Series software. Start by clicking on the **Synthesis → New Library...** menu item in the **Project Manager** window. Type a library name into the **New Library** window that appears. I chose "XSE", but you can pick any name. After clicking on the **OK** button, the **Project Manager** window will resemble Figure 4.20. The library symbol (▤) and **xse** appears in the **Files** tab on the left-hand side of the **Project Manager** window.

Listing 4.5   VHDL library file containing a single-bit full adder component.

```
001- -- Xilinx Student Edition Library
002- -- Contains some components used in design examples.
003-
004- -- package declaration for 1-bit full-adder component
005- LIBRARY IEEE;
006- USE IEEE.std_logic_1164.ALL;
007-
008- PACKAGE adder IS
009- COMPONENT fulladd
010- PORT
011- (
012-        input0: IN STD_LOGIC;
013-        input1: IN STD_LOGIC;
014-        carry_input: IN STD_LOGIC;
015-        sum: OUT STD_LOGIC;
016-        carry_output: OUT STD_LOGIC
017- );
018- END COMPONENT;
019- END adder;
020-
021- -- interface and architecture definitions for the 1-bit full-adder
022- LIBRARY IEEE;
023- USE IEEE.std_logic_1164.ALL;
024-
025- ENTITY fulladd IS
026-     PORT (
027-        input0: IN STD_LOGIC;
028-        input1: IN STD_LOGIC;
029-        carry_input: IN STD_LOGIC;
030-        sum: OUT STD_LOGIC;
031-        carry_output: OUT STD_LOGIC
032-     );
033- END fulladd;
034-
035- ARCHITECTURE fulladd_arch OF fulladd IS
036- BEGIN
037- sum <= input0 XOR input1 XOR carry_input;
038- carry_output <= (input0 AND input1)
039- OR (input0 AND carry_input)
040- OR (input1 AND carry_input);
041- END fulladd_arch;
042-
```

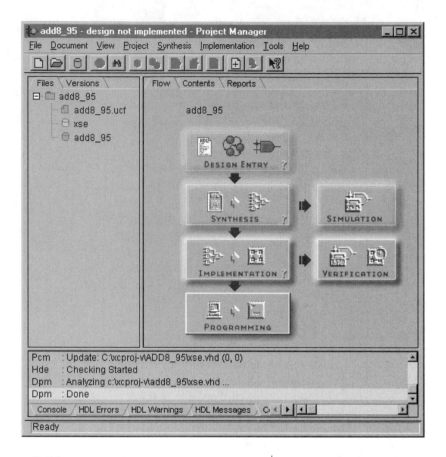

Figure 4.20   The xse library has been created in the ADD8_95 HDL mode project.

Now we have to add VHDL source files to the xse library, so right-click on ⊟xse and select **Add Source Files to "xse"** from the pop-up menu that appears. This will cause an **Add Document** window to come into view. Select the XSE.VHD file and click on **Open**. The XSE.VHD file will be added to the library and the **Project Manager** window will appear as in figure 4.21.

Now we can create the top-level VHDL file that describes an 8-bit adder using the fulladd module from the xse library. Listing 4.6 shows this VHDL code which is stored in the BYTEADD.VHD file.

The first major difference in this file is that the xse library is appended to the LIBRARY statement on line 4. Then the USE statement on line 6 gives the rest of the file access to all the components of the adder package of the xse library.

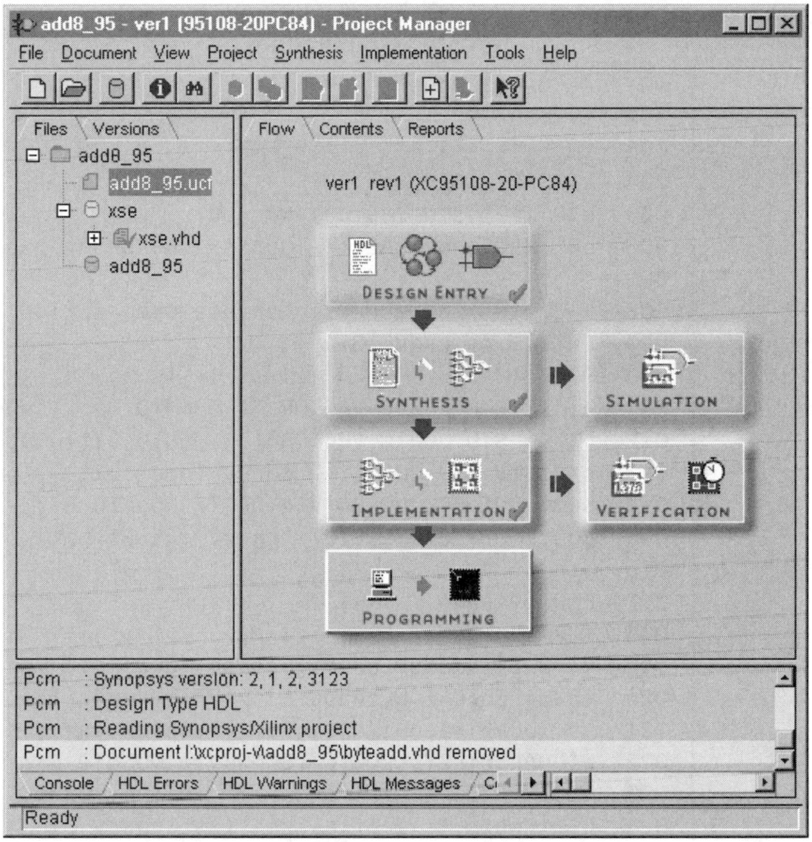

Figure 4.21 The XSE.VHD file has been added to the xse library.

The ENTITY section of the 8-bit adder is named byteadd on line 9. The PORT statement declares two 8-bit addend inputs (a and b) and a carry input (cin) on lines 11 through 13. These are followed by the declaration of an 8-bit summation output (sum) and a carry output bit (cout) on lines 14 and 15.

The ARCHITECTURE section starts on line 20. Note that there is no need for the byteadd module to use a COMPONENT statement to gain access to the full-adder bit. This has already been taken care of by accessing the xse library on lines 4 through 6.

The byteadd module could be written using 8 separate instantiations of the fulladd module to create an 8-bit adder. Instead, we use a FOR loop on line 26 to iteratively instantiate 8 copies of the fulladd module. The GENERATE and END GENERATE statements on lines 27 and 36 surround the instantiation of a fulladd module. The loop counter, i, is used to index the input and output arrays as the looping proceeds. For example, the first time through the loop the

Listing 4.6   VHDL code for iteratively-instantiating a full adder component to create an 8-bit adder.

```
001- -- 8-bit Adder
002- -- page 98
003-
004- LIBRARY IEEE,xse;
005- USE IEEE.std_logic_1164.ALL;
006- USE xse.adder.ALL;-- access the adder package in the XSE library
007-
008- -- byte-wide adder interface description
009- ENTITY byteadd IS
010- PORT (
011- a: IN STD_LOGIC_VECTOR (7 DOWNTO 0);-- addend byte
012- b: IN STD_LOGIC_VECTOR (7 DOWNTO 0);-- addend byte
013- cin: IN STD_LOGIC;-- carry input bit
014- sum: OUT STD_LOGIC_VECTOR (7 DOWNTO 0);-- sum byte
015- cout: OUT STD_LOGIC-- carry output bit
016- );
017- END byteadd;
018-
019- -- 8-bit adder architecture (generated structural)
020- ARCHITECTURE byteadd_arch OF byteadd IS
021- -- local signal array to transfer carry bits between adder stages
022- SIGNAL c: STD_LOGIC_VECTOR (8 DOWNTO 0);
023- BEGIN
024- c(0)<=cin;-- initialize first carry signal to carry input value
025- -- for-loop will generate 8 full-adder bits and hook them together
026- lbl0: FOR i IN 0 TO 7
027-    GENERATE
028-    lbl1: fulladd PORT MAP
029- ( -- connect the adder bit I/O to I/O of the 8-bit adder
030- input0=>a(i), -- addend inputs connected to i'th bit of the
031- input1=>b(i), --   a and b addend input arrays
032- carry_input=>c(i), -- carry_input feeds i'th carry signal
033- sum=>sum(i),-- sum connects to the i'th sum output
034- carry_output=>c(i+1)-- carry output becomes the input
035- );-- to the next stage of the adder
036-    END GENERATE;
037- cout<=c(8);-- last bit of carry array drives carry output
038- END byteadd_arch;
```

inputs to the fulladd module are a(0), b(0), c(0) (which was connected to the carry input on line 24), and the module's sum and carry outputs are connected to sum(0) and c(1), respectively. The c(1) array element connects the carry output from the first adder bit to the carry input of the next adder bit on the next loop iteration.

After the BYTEADD.VHD file is created, use **Document** → **Add...** to add it to the ADD8_95 project. Then click on the ▯ ◦ ▯ button in the **Flow** tab of the **Project Manager** window. The **Synthesis/Implementation** window will appear. Because of the hierarchical nature of this design, you need to tell the Foundation Series synthesis software which module is at the top of the hierarchy. To do this, click on the **Top Level** drop-down menu in the **Synthesis/Implementation** window and a list of all the modules in the project will appear. Just select **byteadd** as the top level module. You must also target the design toward an XC95108 CPLD. Then click on **Run** to start the synthesis of a netlist for the 8-bit adder. Once the netlist is available, you can simulate, compile, download, and test the adder on the XS95 Board just as was done in a preceding section.

We can retarget this design to the XS40 Board and the XC4005XL FPGA, but we need to figure out how to use the MD0 pin as an input. We could create a macro from the BYTEADD.VHD file, include it in a schematic, and then connect an MD0 symbol one of its inputs. But this would require us to create a schematic mode project, and such projects do not allow us to add our private xse library. So we would have to re-write the BYTEADD.VHD code so it does not need the xse library. That is too much work. It would be easier if we could access the MD0 and MD2 pins directly from the VHDL file. Listing 4.7 shows the 8-bit adder VHDL code with modifications to do just that.

The COMPONENT statements on lines 24 and 25 declare the interfaces to modules for the MD0 input pin and an input buffer, respectively. The MD0 module has a single output (MD0) while the IBUF input buffer has an input (I) and an output (O).

On line 30, the MD0 output is connected to the tmp signal. Then the tmp signal is connected to the input of an IBUF on line 31. The output of the IBUF drives the bib(2) element of the bib signal array. The effect of these VHDL statements is to drive the MD0 input onto the intermediate bus bit b(2). The other bits of the intermediate bus are being driven by bits from the b input on lines 28 and 29. This replicates the input arrangement for the 8-bit adder we created using schematics as shown in Figure 4.19. The intermediate bus is connected to the full adder modules on line 39. Other than that, Listing 4.7 is no different from Listing 4.6.

Now where did the IBUF and MD0 modules come from? These modules are actually described by Xilinx Netlist Format (XNF) files. Listing 4.8 shows the contents of the MD0.XNF file. Line 5 defines the single output (O) pin with the net name of MD0. For the most part, you have to find the input and output pins in an XNF file and then place the associated net names in a COMPONENT statement before you can use the module in a VHDL design. I always copy the

Listing 4.7   8-bit adder VHDL code modified to get an input through the MD0 pin of the XC4005XL FPGA.

```
001- -- 8-bit Adder
002-
003- LIBRARY IEEE,xse;
004- USE IEEE.std_logic_1164.ALL;
005- USE xse.adder.ALL;
006-
007- -- byte-wide adder interface description
008- ENTITY byteadd IS
009- PORT (
010- a: IN STD_LOGIC_VECTOR (7 DOWNTO 0);-- addend byte
011- b: IN STD_LOGIC_VECTOR (7 DOWNTO 0);-- addend byte
012- cin: IN STD_LOGIC;-- carry input
013- sum: OUT STD_LOGIC_VECTOR (7 DOWNTO 0);-- sum byte
014- cout: OUT STD_LOGIC-- carry output
015- );
016- END byteadd;
017-
018- -- 8-bit adder architecture (generated structural)
019- ARCHITECTURE byteadd_arch OF byteadd IS
020- -- local signal arrays
021- SIGNAL c: STD_LOGIC_VECTOR (8 DOWNTO 0);
022- SIGNAL bib: STD_LOGIC_VECTOR (7 DOWNTO 0); -- intermediate bus
023- SIGNAL tmp: STD_LOGIC;
024- COMPONENT MD0 PORT(MD0: OUT std_logic); END COMPONENT;
025- COMPONENT IBUF PORT(I: IN std_logic; O: OUT std_logic);
026-                                     END COMPONENT;
027- BEGIN
028- bib(7 DOWNTO 3) <= b(7 DOWNTO 3);
029- bib(1 DOWNTO 0) <= b(1 DOWNTO 0);
030- u0: MD0 PORT MAP(MD0=>tmp);-- connect MD0 input to tmp
031- u1: IBUF PORT MAP(I=>tmp, O=>bib(2)); -- tmp to intermediate bus
032- c(0)<=cin;
033- -- for-loop will generate 8 full-adder bits and hook them together
034- lbl0: FOR i IN 0 TO 7
035- GENERATE
036- lbl1: fulladd PORT MAP
037- (
```

Listing 4.7   8-bit adder VHDL code modified to get an input through the MD0 pin of the XC4005XL FPGA. (Cont'd.)

```
038- input0=>a(i),
039- input1=>bib(i), -- connect to intermediate bus bit
040- carry_input=>c(i),
041- sum=>sum(i),
042- carry_output=>c(i+1)
043- );
044- END GENERATE;
045- cout<=c(8);
046- END byteadd_arch;
```

MD0.XNF and MD2.XNF files into the directory of any project that requires the use of these pins. (But do not use the **Document → Add...** menu item to actually add them to your HDL mode project because this is not allowed.) The XNF for the other modules is usually stored in the Foundation Series software directory hierarchy in a place like Xilinx\Synth\xilinx\macros\Xc4000e\v6_xnf. You can also include schematic-based circuits into a VHDL design by exporting their XNF netlists from the Schematic Editor tool and then placing their inputs and outputs into COMPONENT statements.

Listing 4.8   XNF file for the MD0 pin of the XC4005XL FPGA.

```
001- LCANET, 5
002- PROG, ACTIVE-CAD-2-XNF, 2.5.5.50, Sun Sep 13 11:53:37 1998
003- PART, 4000E
004- SYM, MD0, MD0, SCHNM=MD0, LIBVER=2.0.0
005- PIN, I, O, MD0
006- END
007- EOF
```

Once the BYTEADD.VHD file is complete and added to the ADD8_40 project in the XCPROJ-V directory, you can synthesize it for an XC4005XL FPGA. (Do not forget to select **byteadd** as the top-level module in the **Top Level** drop-down menu of the **Sythesis/Implementation** window.) Since the b(2) input is not used in Listing 4.7, the Foundation Series software does not synthesize this input into the netlist file. Therefore, you have to slightly modify the ADD8_40.UCF file so that the constraint on B<2> is removed or commented-out as follows:

```
NET CIN          LOC=P44;
NET A<0>         LOC=P45;
NET A<1>         LOC=P46;
NET A<2>         LOC=P47;
NET B<0>         LOC=P48;
NET B<1>         LOC=P49;
```

```
# NET  B<2>          LOC=P6;      # DON'T LIST B<2>!!
  NET  SUM<0>        LOC=P25;     # LED segment S0
  NET  SUM<1>        LOC=P26;     # LED segment S1
  NET  SUM<2>        LOC=P24;     # LED segment S2
# the following constraints assign unused ADD8 terminals to out-of-the-way pins
  NET  COUT          LOC=P7;
  NET  A<3>          LOC=P78;
  NET  A<4>          LOC=P79;
  NET  A<5>          LOC=P82;
  NET  A<6>          LOC=P84;
  NET  A<7>          LOC=P3;
  NET  B<3>          LOC=P5;
  NET  B<4>          LOC=P60;
  NET  B<5>          LOC=P56;
  NET  B<6>          LOC=P58;
  NET  B<7>          LOC=P59;
  NET  SUM<3>        LOC=P83;
  NET  SUM<4>        LOC=P4;
  NET  SUM<5>        LOC=P50;
  NET  SUM<6>        LOC=P57;
  NET  SUM<7>        LOC=P51;
```

At this point, you can compile, download, and test the 8-bit adder in the XS40 Board.

Projects

1. Create the MY_ADD8 module from the 8-bit adder design.

2. Create an LED decoder module that takes a 4-bit input and outputs the hexadecimal numeral on the LED digit.

3. Attach the LED decoder module to the output of the 8-bit adder so that the sum is displayed in a more easily understood form.

4. A 2's-complement circuit outputs the negation of its input. This operation is performed by inverting each bit of the input and then adding 1 to it. Use the MY_ADD8 module to make an 8-bit 2's-complement circuit.

5. Design a circuit that will take an 8-bit input and display the number of 1's in the input on an LED digit.

Electrical Characteristics

Objectives

- Discuss some of the electrical characteristics of various types of digital inputs and outputs.

- Introduce the concept of gate delay and its effect on the speed of a design.

Discussion

So far, you have written ABEL and VHDL and drawn schematics that describe logic circuits, compiled them, simulated them, downloaded them to your XS40 or XS95 Board, and—magically—they seem to do what you expected. With all the emphasis placed on entering and simulating the design files, it is easy to forget that there is actually something happening electrically. This chapter will discuss and examine some simple electrical characteristics of digital circuits that you will need to know so you can figure out why things sometimes do not work as expected.

I/O Characteristics

A simple inverter gate with a standard output is shown in Figure 5.1a. The inverter has two transistors: T_h which is connected between the output pin of the inverter and the +5 V supply, and T_l which is connected between the output pin and the ground. When the input of the inverter is pulled to ground (Figure 5.1(b)), transistor T_l is turned off and T_h is turned on. This means that T_l acts like an open switch while T_h acts like a closed switch. Thus, there is now a low-resistance path between the output pin and the +5 V supply. This pulls the output pin up to +5 V. When the input is tied to +5 V (Figure 5.1c), the opposite set of actions occurs: T_h turns off and T_l turns on, thus creating a low-resistance path from the output pin to ground. This drags the output to the ground potential.

What happens if we apply a voltage to the input that is between +5 V and ground? In this case, both T_l and T_h could turn on and "fight" each other. T_l tries to pull the output to ground while T_h tries to pull the output to +5 V. The output will settle to a voltage between +5 V and ground and large amounts of current will flow from the +5 V supply through T_h and T_l to ground. The question is,

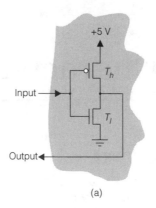

(a)

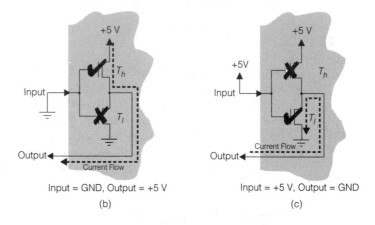

Input = GND, Output = +5 V

(b)

Input = +5 V, Output = GND

(c)

Figure 5.1 Operation of a simple inverter.

how close must an input be to ground or +5 V in order to prevent this situation? The answer depends on the detailed electrical characteristics of T_l and T_h, but for TTL the answer is that a valid low logic level is any voltage between 0 V and 0.8 V, and a valid high logic level is any voltage between 2.0 V and 5.0 V. For CMOS logic, the voltage ranges are slightly different: 0 V–1.0 V and 3.5 V–5.0 V, respectively.

What happens if we do not connect the input to anything? In that case, the state of the circuit is unclear. The inputs to the transistors are not driven to any solid voltage and the transistors may be on or off. Thus, the output voltage could be anywhere between +5 V and ground. To prevent unconnected inputs from causing logic circuits to enter an indeterminate state, some ICs have pull-up resistors from every input to the +5 V supply (see Figure 5.2a). When you leave the input unconnected, the pull-up resistor, R_p, brings the inputs of the transistors up to +5 V. The pull-up resistor typically has a large resistance value (over 10K) so that when you want to pull the input to ground, you do not have to supply much power to do it. (You would not want to use a low-resistance pull-up because pulling the input low would cause a lot of current to flow from +5 V through the pull-up resistor to ground.)

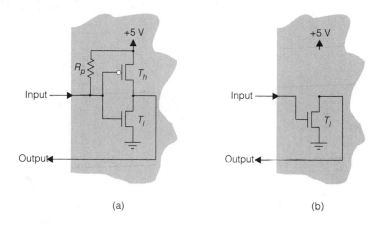

(a)   (b)

Figure 5.2   FPLD I/O options.

The output circuitry can also be modified to have different electrical characteristics. The standard output, as we have seen, can be pulled to +5 V or ground through the low resistance of a transistor that is turned on. There are some cases, however, when you only want the output to be pulled to ground but never to +5 V. Such an open-drain output is shown in Figure 5.2b. When the input is at a high logic level, the output is pulled to ground through transistor T_l. But when the input is pulled to a low logic level, the output is not pulled to +5 V (as it is with the standard output) because T_h is missing. Since T_h is missing and T_l is turned off, the output is not connected to either +5 V or ground, so it floats. Of course, you could attach an external resistor from the output to +5 V to pull it high in this case.

One of the uses of open-drain outputs is performing a wired-AND operation in combination with a pull-up resistor (see Figure 5.3). If two open-drain outputs are tied directly together, the result will be a low output level if either output is low. If both outputs are floating, then the wired-AND output is pulled to +5 V by the pull-up resistor. You do not want to try building a wired-AND with gates having standard outputs. That is because one gate output may be trying to pull the wired-AND node low through its T_l while the other gate tries to pull the node high through its T_h. The resulting voltage on the wired-AND node is indeterminate, a large amount of current flows, and you might burn up one or both gates.

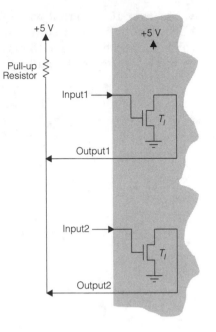

Figure 5.3   A wired-AND using open-drain outputs.

Another type of output is the tristate output. A tristate output can drive to +5 V or ground, but it can also be made to float. That is different than an open-drain output, which can only drive to ground or float. The circuitry for a tristate output is shown in Figure 5.4. Two new T_g transistors are added that are controlled by an enable input. When the enable input is pulled to +5 V, the T_g transistors act like closed switches and the gate is enabled (i.e., the output can be driven to +5 V or ground). When the enable line is pulled to ground, however, both T_g transistors open and completely disconnect the output from both the +5 V supply and ground. So the output is disabled and it floats. What use is this? Multiple tristate outputs can be connected to the same input of another gate and the enable lines can be used to prevent the outputs from interfering with each other. When one output is driving the input, the other output is disabled so it floats and cannot interfere with the first output. Then, by disabling the first output and enabling the second, the second output can transmit a signal without interference. Thus, the input pin can receive a signal from either of two outputs by intelligently controlling the enable lines of the outputs. This principle will be used when you create multidrop buses. For example, the memory bus in your PC allows the microprocessor to receive information from any one of a large number of memory chips. All the memory chip outputs can be hooked to the same bus because they all have tristate outputs that are enabled only when a particular chip is supposed to transmit some

Electrical Characteristics   Chapter 5

data to the microprocessor. They will not interfere with each other as long as only one output is enabled at a time.

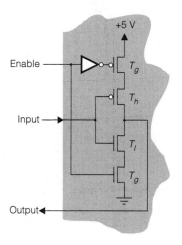

Figure 5.4   Circuitry for a tristate output.

Where can all this get you in trouble? Obviously, if you use an open-drain output and do not provide a pull-up resistor, you cannot expect to drive the output to +5 V. Or if you use a tristate output and do not enable it, you won't be able to drive the output to either +5 V or ground. But there can be other, more subtle, problems. Suppose you have a single logic gate output that is driving a lot of inputs of other logic gates (see Figure 5.5). Further, suppose that these other inputs all have 10K pull-up resistors. Now, your output can be pulled to ground through T_l, but T_l has some finite resistance (i.e., it is not a perfect conductor). Suppose that the ON-resistance of T_l is about 100 Ω. The 100 Ω resistance will be trying to pull node A to ground while the pull-up resistors try to pull A to +5 V. Thus, a voltage divider is formed. The question is, How many inputs can be connected to the output before the combined pull-up resistors bring the voltage on node A above 0.8 V (the upper limit for a TTL low logic level)? The answer is given by solving

$$0.8 \text{ V} = 5.0 \text{ V} \times \frac{100 \ \Omega}{100 \ \Omega + \dfrac{10{,}000 \ \Omega}{N}}$$

and it turns out that $N = 19$. So you can drive 19 other gate inputs from the single output. This is the maximum allowable fanout of your gate output. Trying to drive more gates could lead to problems caused by violating the limits on allowable logic levels. (Note that driving the inputs high is no problem because now the pull-up resistors help T_h pull the output to +5 V.)

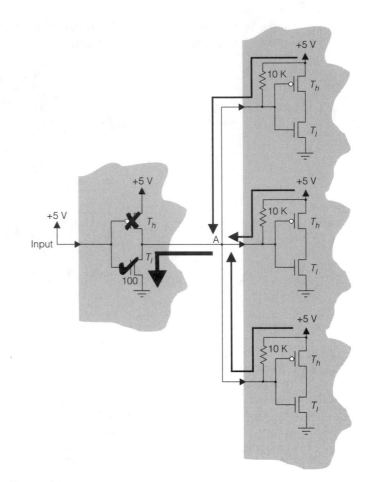

Figure 5.5   Logic gate fanout.

Circuit Delays

Driving 19 other gates may seem like quite a bit, so you might think that worrying about fanout is not a high priority. However, there are other factors you must consider. One of these is the delay in transmitting a signal. On every circuit node there is capacitance. Capacitance acts like a charge bucket that has to be filled or emptied before the node can reach +5 V or ground, respectively. There is capacitance you want, such as the power filtering capacitors on your XS40 or XS95 Boards. There is also parasitic capacitance, which arises just from the geometry of your circuit and the fundamental laws of electromagnetism. The parasitic capacitance causes most of your problems. Suppose that you are trying to pull 19 inputs to ground through a single output. Suppose also that each input has 10 pF of capacitance (1 pF = 10^{-12} F). Thus, there is a total of 19×10 pF = 190 pF of capacitance being discharged through the 100 Ω

resistance of T_l. How long will it take to discharge all this capacitance and pull all the inputs to ground? The time constant of this resistance-capacitance combination is 100 Ω × 190 pF = 19 ns. It takes approximately two or three times the time constant for the voltage to decay below the threshold, so that is between 38 and 57 nanoseconds.

Assume the worst case and let the time delay, t_d, be 57 ns. That is 57 *billionths* of a second, so you may say it is no big deal. However, consider a 200-MHz PC. It computes a new answer every $1/(200 \times 10^6)$ s = 5 ns. So in the time it takes the output to drive all 19 inputs to ground, the PC is able to perform almost 12 complete operations! The lesson is clear: Fast circuits need to limit their fanout.

The delays caused by fanout are not the only ones you have to worry about. The logic gates themselves have a built-in delay (see Figure 5.6). When the input to a gate changes, the output does not change instantaneously. This delay is primarily caused by parasitic capacitances inside the IC. The time between the change in the input to the change in the output is called propagation delay or gate delay. There are two types of gate delay:

t_{PHL}: This is the time delay between the change in the input and the time the output drops from +5 V to ground. (Propagation high → low)

t_{PLH}: This is the time delay between the change in the input and the time the output rises from ground to +5 V. (Propagation low → high)

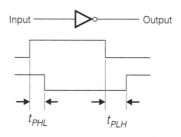

Figure 5.6   Gate propagation delay.

Usually, t_{PHL} and t_{PLH} are not too different so we lump them together into a single gate delay parameter called t_{pd}. The t_{pd} of the XC95108 CPLD macrocells ranges from 7.5 to 20 ns, depending upon how much you are willing to pay XILINX (faster is more expensive). The situation is more complicated for the XC4005XL FPGA because time delays are dependent on the way logic gates are mapped into the LUTs within a CLB. For instance, if the logic can be mapped into a single F or G LUT, the result is available on the CLB output just 1.6 ns after the inputs arrive (based on the timing for an XC4005XL FPGA with −3 speed rating). But if the logic function is more complicated and requires the use of the H LUT to combine the outputs from the F and G LUTs, then the propagation delay rises to 2.7 ns. Both of these delays are much lower than the one for the XC95108 because they ignore the delays of getting the inputs and outputs through the I/O pins.

Why worry about gate delay? Because the more gates your signals have to go through, the longer it will take for the final output to be computed. Figure 5.7 shows one potential path for the carry signal in a 3-bit ripple-carry adder. Note that the carry signal ripples from the least-significant bit to the most-significant bit, picking up two gate delays through each stage of the adder. Assuming each AND-OR circuit of the carry chain is placed in a single macrocell (CLB), the maximum delay through the 3-bit adder is 3×20 ns = 60 ns (3×3.2 ns = 9.6 ns) for the slowest XC95108 CPLD (XC4005XL FPGA). By extension, the maximum delay through a 32-bit ripple-carry adder is 32×20 ns = 640 ns (32×4.7 ns = 102.4 ns). So a 32-bit ripple-carry adder could do about 1,500,000 adds per second (6,650,000 adds per second). This is much slower than the microprocessor in a PC can add 32-bit numbers. Why does the adder in the PC work so much faster?

One answer is the carry-lookahead circuit. Carry-lookahead circuits compute several carry bits in parallel, so the cascading delays in the ripple-carry adder are eliminated. Assume that two binary numbers, $a_{N-1}a_{N-2}...a_2a_1a_0$ and $b_{N-1}b_{N-2}...b_2b_1b_0$, are being added. A carry, c_i, is generated out of bit position i if both $a_i = 1$ and $b_i = 1$. A carry can be propagated through bit position i if either a_i or b_i is 1 and the carry input from the previous bit, c_{i-1}, is 1. Thus, the carry generation and carry propagation signals are

$$G_i = a_i b_i$$

$$P_i = a_i \oplus b_i$$

The following equations show how the G_i and P_i signals are used in a 4-bit carry-lookahead adder:

$$c_0 = G_0 + P_0\, c_{in}$$

$$c_1 = G_1 + P_1\, c_0$$
$$c_1 = G_1 + P_1\, (G_0 + P_0\, c_{in})$$
$$c_1 = G_1 + P_1\, G_0 + P_1\, P_0\, c_{in}$$

$$c_2 = G_2 + P_2\, c_1$$
$$c_2 = G_2 + P_2\, (G_1 + P_1\, G_0 + P_1\, P_0\, c_{in})$$
$$c_2 = G_2 + P_2\, G_1 + P_2\, P_1\, G_0 + P_2\, P_1\, P_0\, c_{in}$$

$$c_3 = G_3 + P_3\, c_2$$
$$c_3 = G_3 + P_3\, (G_2 + P_2\, G_1 + P_2\, P_1\, G_0 + P_2\, P_1\, P_0\, c_{in})$$
$$c_3 = G_3 + P_3\, G_2 + P_3\, P_2\, G_1 + P_3\, P_2\, P_1\, G_0 + P_3\, P_2\, P_1\, P_0\, c_{in}$$

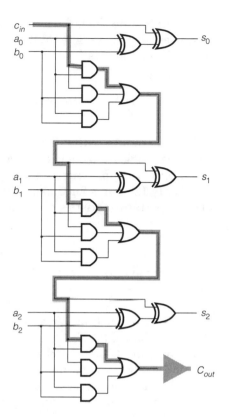

Figure 5.7   Cumulative gate delays in the carry path of a ripple-carry adder.

What is the maximum delay through the carry-lookahead adder? It takes one gate delay to generate the G_i and P_i signals simultaneously for all the bits. Then the sum of products equation for each c_i can be computed in two gate delays (one for the AND gates and one for the OR gate). So the final carry-out of the adder is computed after three gate delays. However, the total addition is not complete until the carry bits have been used to compute the sum bits, s_i. This takes another gate delay, so the sum is finally ready after four gate delays. The actual delay through an XC95108 or XC4005XL will be determined by how these gates are fitted into the macrocells and LUTs, but the basic analysis remains valid.

Suppose you kept repeating the preceding equations until you had all the equations necessary to build a 32-bit adder. What is the maximum delay through a 32-bit adder with full carry-lookahead? The answer is four gate delays! How much logic would it take to build such an adder? Quite a bit! For carry c_i, one XOR gate is needed to compute P_i, and the outputs of $i + 2$ AND gates must be OR'ed together (assuming that you can find AND and OR gates that can handle the large number of inputs involved here). Thus, each carry bit requires $i + 4$ gates. The total number of gates used just for handling the carry in an N-bit adder is

$$\# \text{ of gates} = \sum_{i=0}^{N-1} i + 4 = \frac{N(N-1)}{2} + 4N$$

So a 32-bit adder with full carry-lookahead requires 624 gates just for handling the carry logic (and some of these are large 32-input gates). The carry logic for each bit of a ripple-carry adder uses four gates, so a 32-bit ripple-carry adder uses only 128 gates for the total carry logic (and these are small 2-input gates). The important statistics are as follows:

	Gates	Gate Delay
32-bit ripple-carry circuit	128	64
32-bit carry-lookahead circuit	624	4

To go fast, a much larger and more complicated circuit is needed. Full-carry-lookahead for a 32-bit adder uses a lot of logic so carry-lookahead is typically done only on 4- or 8-bit chunks (see Figure 5.8). The carry output from each section feeds into the next section, thus giving a combination of carry-lookahead and ripple-carry techniques. If 8-bit lookahead is used in a 32-bit adder, the maximum gate-delay through the adder is

$$(32\text{bits} / 8\text{bits}) \times 3 + 1 = 13 \text{ gate delays}$$

and the number of gates used in the carry circuitry is

$$(32\text{bits}/8\text{bits}) \times [(8)(8 - 1)/2 + 4(8)] \text{ gates} = 240 \text{ gates.}$$

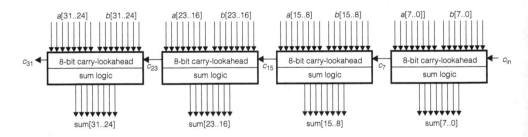

Figure 5.8   A 32-bit adder made from 8-bit carry-lookahead adder sections.

Experimental

The following sections will let you experiment with different types of I/O options and investigate the effects of gate delays.

Using FPLD I/O Options

Using Tristate Outputs

The circuit of Figure 5.9 is a simple use of a buffer with a tristate output. We have a 2-line to 4-line decoder (D2_4E) whose outputs are passed through a tristate buffer (OBUFE4) with an active-high enable. When ENBL is at logic 1, the outputs from the D2_4E decoder can drive segments of the LED digit. When ENBL is pulled low, any changes on IN1 and IN0 will still affect outputs D3..D0 of the decoder, but these outputs will not appear on the I/O pins of the FPLD.

Note that the D2_4E module also has an active-high enable input (E) which we have tied high to VCC. When low, this enable input causes the outputs to all drive to logic 0. The outputs are not tristated as they are when we use the ENBL signal. A simulation of the circuit is shown in Figure 5.10. As expected, the outputs float when ENBL=0, but they are driven to valid 0 and 1 logic levels when ENBL=1.

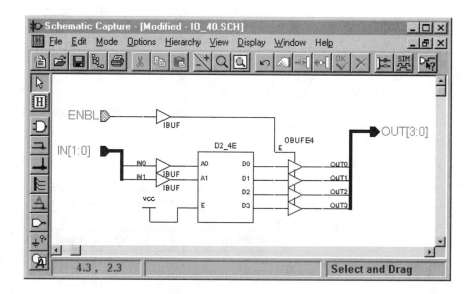

Figure 5.9   Using a 4-input tristate buffer.

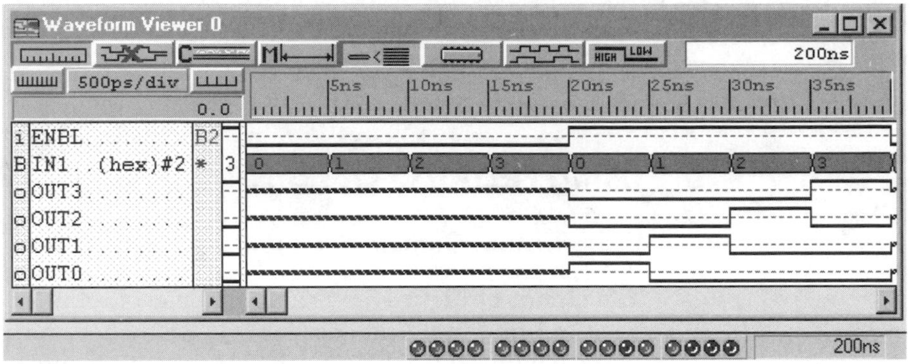

Figure 5.10 A simulation showing the effect of tristating the outputs.

To map this design to the XS40 Board, use the following IO_40.UCF constraint file:

```
NET IN<0>      LOC=P44;
NET IN<1>      LOC=P45;
NET ENBL       LOC=P46;
NET OUT<0>     LOC=P25;    # LED segment S0
NET OUT<1>     LOC=P26;    # LED segment S1
NET OUT<2>     LOC=P24;    # LED segment S2
NET OUT<3>     LOC=P20;    # LED segment S3
```

Or place these constraints for an XS95 Board into the IO_95.UCF file:

```
NET IN<0>      LOC=P46;
NET IN<1>      LOC=P47;
NET ENBL       LOC=P48;
NET OUT<0>     LOC=P21;    # LED segment S0
NET OUT<1>     LOC=P23;    # LED segment S1
NET OUT<2>     LOC=P19;    # LED segment S2
NET OUT<3>     LOC=P17;    # LED segment S3
```

Once the design is downloaded into the XS40 or XS95 Board, the following association between the XSPORT arguments and the circuit inputs is in effect:

XSPORT Bit	IO_40 Input
b_0	IN0
b_1	IN1
b_2	ENBL
b_3–b_7	*** Not Used ***

Applying test inputs to the XS40 or XS95 Board will produce the following results (note that tristated outputs leave the LED segments dark just as if they were driven to a logic 0):

Command	ENBL	IN1	IN0	D3	D2	D1	D0	Comment
XSPORT 000	0	0	0	—(dark)	—(dark)	—(dark)	—(dark)	Tristated
XSPORT 001	0	0	1	—(dark)	—(dark)	—(dark)	—(dark)	Tristated
XSPORT 010	0	1	0	—(dark)	—(dark)	—(dark)	—(dark)	Tristated
XSPORT 011	0	1	1	—(dark)	—(dark)	—(dark)	—(dark)	Tristated
XSPORT 100	1	0	0	0 (dark)	0 (dark)	0 (dark)	1 (bright)	Driven
XSPORT 101	1	0	1	0 (dark)	0 (dark)	1 (bright)	0 (dark)	Driven
XSPORT 110	1	1	0	0 (dark)	1 (bright)	0 (dark)	0 (dark)	Driven
XSPORT 111	1	1	1	1 (bright)	0 (dark)	0 (dark)	0 (dark)	Driven

Adding a Blanking Control to an LED Decoder for the XS95 Board

Tristate outputs can also be used in an ABEL-based design (BLANK_95). The LED decoder modified to add a blanking input is shown in Listing 5.1. The only changes to the design are as follows:

Line 6: This line declares the BLANK input which tristates the outputs that drive the LED segments.

Line 14: This line introduces something new: a signal extension. The signal extension .OE gives us access to the tristate control for the output buffers. A high level on a .OE input will enable the output while a low level makes the output float. Vector notation lets us use a single statement to control the output enable inputs of all 7 LED segment drivers.

Place the following constraints in the BLANK_95.UCF file and compile the LEDBLANK.ABL file for the XS95 Board:

```
NET  D0        LOC=P46;
NET  D1        LOC=P47;
NET  D2        LOC=P48;
NET  D3        LOC=P50;
NET  BLANK     LOC=P51;
NET  S0        LOC=P21;
NET  S1        LOC=P23;
NET  S2        LOC=P19;
NET  S3        LOC=P17;
NET  S4        LOC=P18;
NET  S5        LOC=P14;
NET  S6        LOC=P15;
```

Listing 5.1   LED decoder with blanking input and tristate outputs.

```
001- MODULE LEDBLANK
002- TITLE 'LED decoder WITH blanking input'
003-
004- DECLARATIONS
005-
006- BLANK PIN;   "blank LEDs when high
007- D3..D0 PIN;
008- D = [D3..D0];
009- S6..S0 PIN ISTYPE 'COM';
010- S = [S6..S0];
011-
012- EQUATIONS
013-
014- [S6..S0].OE = !BLANK; "disable LED drivers when BLANK=1
015-
016- TRUTH_TABLE
017-    (D -> [S6, S5, S4, S3, S2, S1, S0])
018-     0 -> [1,  1,  1,  0,  1,  1,  1 ];
019-     1 -> [0,  0,  1,  0,  0,  1,  0 ];
020-     2 -> [1,  0,  1,  1,  1,  0,  1 ];
021-     3 -> [1,  0,  1,  1,  0,  1,  1 ];
022-     4 -> [0,  1,  1,  1,  0,  1,  0 ];
023-     5 -> [1,  1,  0,  1,  0,  1,  1 ];
024-     6 -> [1,  1,  0,  1,  1,  1,  1 ];
025-     7 -> [1,  0,  1,  0,  0,  1,  0 ];
026-     8 -> [1,  1,  1,  1,  1,  1,  1 ];
027-     9 -> [1,  1,  1,  1,  0,  1,  1 ];
028-
029- END LEDBLANK
```

Once the design is compiled and converted to SVF, then download it to an XS95 Board like so:

```
C:\XCPROJ\BLANK_95> XSLOAD BLANK_95.SVF
```

After the download is complete, the following mapping between the XSPORT arguments and the circuit inputs is in effect:

XSPORT Bit	LED Decoder Input
b_0	D0
b_1	D1
b_2	D2
b_3	D3
b_4	BLANK
b_5–b_7	*** Not Used ***

The following commands will test the blanking circuitry:

```
C:\XCPROJ\BLANK_95> XSPORT 00111
C:\XCPROJ\BLANK_95> XSPORT 10111
```

The first command should display a numeral 7 on the 7-segment LED. The second command will extinguish all the LED segments because the drivers are disabled.

Note that you could get the same observable effect without using the buffer's output enable. Just rework the LED decoder's truth-table so the outputs all go to logic 0 if the BLANK input is ever high. But with this method, the outputs are actually driven to valid 0 levels instead of floating.

Adding a Blanking Control to an LED Decoder for the XS40 Board

Can you use the LEDBLANK.ABL file with the XS40 Board? Yes, if your Foundation Series software is version 1.5 or above. For earlier versions, Foundation does not compile signals with the .OE extensions into tristate outputs on an XC4000-series FPGA. In this case, you have to embed an LED decoder in a schematic and connect tristate buffers to its outputs like so:

1. Open the LEDDCD.ABL in an **HDL Editor** window. Select the **Synthesis → Options...** menu item and click on the **Macro** button in the **ABEL6** window that appears. Then select the **Project → Create Macro** menu item to initiate the synthesis process and add the netlist to the project library.

2. Open a **Schematic Editor** window and select the **LEDDCD** symbol from the **SC Symbols** parts-list window. Drop the LED decoder into the schematic drawing area.

3. Connect the outputs of the LEDDCD module through a 7 bit bus to the inputs of an OBUFT8 buffer (this is a group of 8 tristate buffers). Connect the buffer output to a 7-bit output bus.

4. Connect an input terminal named BLANK through an IBUF to the T control input of the OBUFT8 symbol. When BLANK is driven high, it will cause all the tristate buffers to enter a high-impedance state.

5. Connect terminals through IBUF input buffers to the data inputs of the LEDDCD symbol.

The schematic that results from these steps is shown in Figure 5.11.

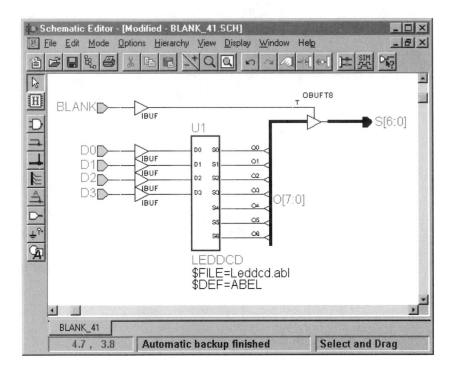

Figure 5.11   Connecting an LED decoder with tristate buffers.

Once the schematic is complete, save it and do the **Options → Create Netlist** and **Options → Export Netlist...** operations. Before compiling the netlist for the XC4005XL FPGA, place the following constraints in the BLANK_40.UCF file (do not forget the <...> around the indices of the output bus bits that drive the LED segments):

```
NET D0          LOC=P44;
NET D1          LOC=P45;
```

```
NET  D2        LOC=P46;
NET  D3        LOC=P47;
NET  BLANK     LOC=P48;
NET  S<0>      LOC=P25;
NET  S<1>      LOC=P26;
NET  S<2>      LOC=P24;
NET  S<3>      LOC=P20;
NET  S<4>      LOC=P23;
NET  S<5>      LOC=P18;
NET  S<6>      LOC=P19;
```

Once the design is compiled, you can download it to an XS40 Board:

```
C:\XCPROJ\BLANK_40> XSLOAD BLANK_40.BIT
```

After the download is complete, the following mapping between the XSPORT arguments and the circuit inputs is in effect:

XSPORT Bit	LED Decoder Input
b_0	D0
b_1	D1
b_2	D2
b_3	D3
b_4	BLANK
b_5–b_7	*** Not Used ***

The following commands will test the blanking circuitry:

```
C:\XCPROJ\BLANK_40> XSPORT 01001
C:\XCPROJ\BLANK_40> XSPORT 11001
```

The first command should display a numeral 9 on the 7-segment LED. The second command will extinguish all the LED segments because the drivers are disabled.

Adding a Blanking Control to an LED Decoder Using VHDL

Listing 5.2 shows the LED decoder with blanking control written in VHDL. (I have placed the VHDL code for the LED decoder into the XSE.VHD library source file to create the led package shown in Listing 5.3). The LED decoder module is instantiated on line 19 of the LEDBLANK.VHD file. The output of the LED decoder is connected to the tmp signal array. Line 20 passes the value on tmp to the s outputs when the blank input is not high (/= is the VHDL "not equals" operator). But when the blank input is at logic 1, then the 7-element s output array is set to ZZZZZZZ where Z is the VHDL symbol for a tristated logic value.

Listing 5.2   VHDL version of an LED decoder with blanking input and tristate outputs.

```
001- -- LED Blanking (LEDBLANK.VHD)
002-
003- LIBRARY IEEE,xse;
004- USE IEEE.std_logic_1164.ALL;
005- USE xse.led.ALL;
006-
007- ENTITY ledblank IS
008- PORT
009- (
010- d: IN STD_LOGIC_VECTOR (3 DOWNTO 0); -- 4-bit hex value
011- blank: IN STD_LOGIC;-- active-high blanking input
012- s: OUT STD_LOGIC_VECTOR (6 DOWNTO 0) -- drivers for LED
013- );
014- END ledblank;
015-
016- ARCHITECTURE ledblank_arch OF ledblank IS
017- SIGNAL tmp: STD_LOGIC_VECTOR (6 DOWNTO 0);
018- BEGIN
019-     u0: leddcd PORT MAP(d=>d,s=>tmp);
020-     s<=tmp WHEN blank/='1' ELSE "ZZZZZZZ";
021- END ledblank_arch;
```

The LEDBLANK.VHD and XSE.VHD files are placed in the BLANK_95 and BLANK_40 HDL mode projects in the XCPROJ-V directory. Add the LED-BLANK.VHD file to the project, and create a new library in the project and add the XSE.VHD to it. Then place the following constraints into the BLANK_40.UCF file if you are targeting the XS40 Board:

```
NET D<0>          LOC=P44;
NET D<1>          LOC=P45;
NET D<2>          LOC=P46;
NET D<3>          LOC=P47;
NET BLANK         LOC=P48;
NET S<0>          LOC=P25;
NET S<1>          LOC=P26;
NET S<2>          LOC=P24;
NET S<3>          LOC=P20;
NET S<4>          LOC=P23;
NET S<5>          LOC=P18;
NET S<6>          LOC=P19;
```

Listing 5.3   The led package and the LED decoder module description from the XSE.VHD library VHDL source code.

```
001- -- Xilinx Student Edition Library
002-
003- -- package declaration for 7-segment LED decoder component
004- LIBRARY IEEE;
005- USE IEEE.std_logic_1164.ALL;
006-
007- PACKAGE led IS
008- COMPONENT leddcd
009- PORT
010- (
011- d : IN STD_LOGIC_VECTOR (3 DOWNTO 0);
012- s : OUT STD_LOGIC_VECTOR (6 DOWNTO 0)
013- );
014- END COMPONENT;
015- END led;
016-
017- -- interface and architecture definitions for the LED decoder
018- LIBRARY IEEE;
019- USE IEEE.std_logic_1164.ALL;
020-
021- ENTITY leddcd IS
022- PORT(
023- d : IN STD_LOGIC_VECTOR (3 DOWNTO 0);
024- s : OUT STD_LOGIC_VECTOR (6 DOWNTO 0)
025-     );
026- END leddcd;
027-
028- ARCHITECTURE leddcd_arch OF leddcd IS
029- BEGIN
030- WITH d SELECT
031- s <= "1110111" WHEN "0000",
032-       "0010010" WHEN "0001",
033-       "1011101" WHEN "0010",
034-       "1011011" WHEN "0011",
035-       "0111010" WHEN "0100",
036-       "1101011" WHEN "0101",
037-       "1101111" WHEN "0110",
038-       "1010010" WHEN "0111",
039-       "1111111" WHEN "1000",
040-       "1111011" WHEN "1001",
041-       "1111110" WHEN "1010",
```

Listing 5.3 The led package and the LED decoder module description from the XSE.VHD library VHDL source code. (Cont'd.)

```
042-      "0101111" WHEN "1011",
043-      "1100101" WHEN "1100",
044-      "0011111" WHEN "1101",
045-      "1101101" WHEN "1110",
046-      "1101100" WHEN "1111",
047-      "0000000" WHEN OTHERS;
048- END leddcd_arch;
```

Or put these constraints into the BLANK_95.UCF file if you are going to use the XS95 Board:

```
NET D<0>        LOC=P46;
NET D<1>        LOC=P47;
NET D<2>        LOC=P48;
NET D<3>        LOC=P50;
NET BLANK       LOC=P51;
NET S<0>        LOC=P21;
NET S<1>        LOC=P23;
NET S<2>        LOC=P19;
NET S<3>        LOC=P17;
NET S<4>        LOC=P18;
NET S<5>        LOC=P14;
NET S<6>        LOC=P15;
```

Once the design is compiled, you can download it to the XS40 Board:

```
C:\XCPROJ-V\BLANK_40> XSLOAD BLANK_40.BIT
```

Or compile the design, create the SVF file, and download to the XS95 Board:

```
C:\XCPROJ-V\BLANK_95> XSLOAD BLANK_95.SVF
```

After the download is complete, the following mapping between the XSPORT arguments and the circuit inputs is in effect:

XSPORT Bit	LED Decoder Input
b_0	D<0>
b_1	D<1>
b_2	D<2>
b_3	D<3>
b_4	BLANK
b_5–b_7	*** Not Used ***

The following commands will test the blanking circuitry:

```
C:\XCPROJ-V\BLANK_40> XSPORT 00110
C:\XCPROJ-V\BLANK_40> XSPORT 10110
```

The first command should display a numeral 6 on the 7-segment LED. The second command will disable all the drivers and extinguish the LED segments.

Measuring Gate Delay

Figure 5.12 shows a simple circuit, DELAY, that demonstrates gate delays. When INP transitions to a logic 0 value, the outputs should all go to logic 0 after 1 gate delay. Then when INP rises back to logic 1, OUT1 will go high after 1 gate delay. This places a logic 1 on both inputs to the second AND gate so OUT2 will go high 1 gate delay after OUT1 did. And OUT3 will go to logic 1 after another gate delay, followed by OUT4 1 gate delay later.

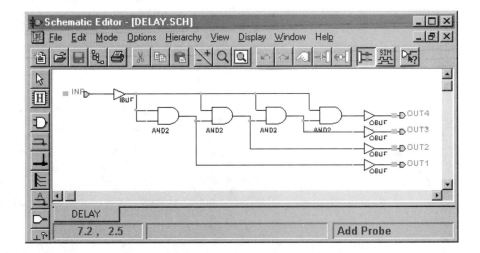

Figure 5.12   A simple circuit for demonstrating propagation delay.

Note the small boxes attached to each input and output terminal. These are probes which monitor the net or terminal they are attached to and send the logic values to the **Waveform Viewer** window in the simulator. They were added as follows:

1. Select the **Mode** → **Test Points** menu item in the **Schematic Capture** window.

2. Click on each net or terminal which you want to appear in the **Waveform Viewer** window of the simulator.

Once the probes are added to the schematic, the signals will appear in the **Waveform Viewer** window as soon as we start the simulator.

A functional simulation of the DELAY circuit is not very interesting - the outputs OUT4..OUT1 move in exact synchrony with the INP input! In order to see the effect of gate delays, click on the drop-down menu for the simulation mode in the toolbar of the **Logic Simulator** window and select **Unit** instead of **Functional**. This puts the simulator in the unit-delay mode where the output of a gate changes 1 time unit after the inputs change.

But how long is a gate delay? That is determined by the simulation precision used by the simulator algorithm. The simulation precision is the smallest increment of time between two non-simultaneous signal transitions. You can set the precision by selecting the **Options → Preferences...** menu item in the **Logic Simulator** window. In the **Preferences** window that appears (Figure 5.13), type the desired precision in the **Simulation Precision** box of the **Simulation** tab. I chose 1 nanosecond in this example. I also set the period of the B0 stimulator to 10 nanoseconds and connected the B2 bit of the stimulator to the INP input. That means the input to the DELAY circuit will change every twenty nanoseconds which gives the outputs plenty of time to settle before the input changes again.

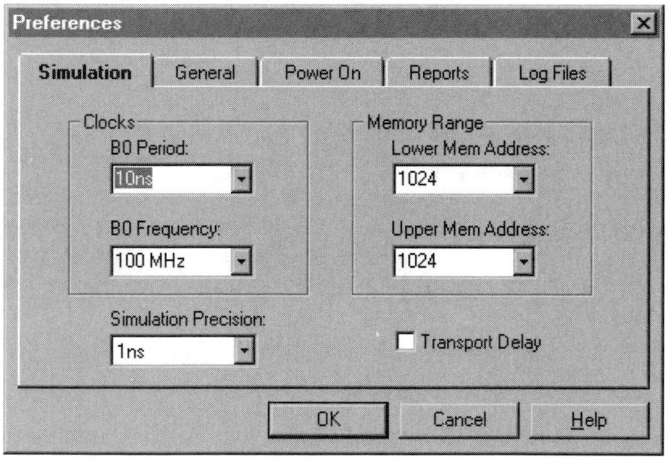

Figure 5.13   Setting the simulation precision.

Once the simulation precision and the input stimulator are set, we can simulate the design and view the effects of unit gate delays. The results are shown in Figure 5.14. As expected, the outputs are each offset by 1 gate delay on a rising transition, and they all fall at the same time on a negative-going transition. But note that OUT1 changes 3 ns (i.e., 3 gate delays) after a signal transition on the INP input. The 2 additional gate delays arise from the delays incurred by going through the input and output buffers.

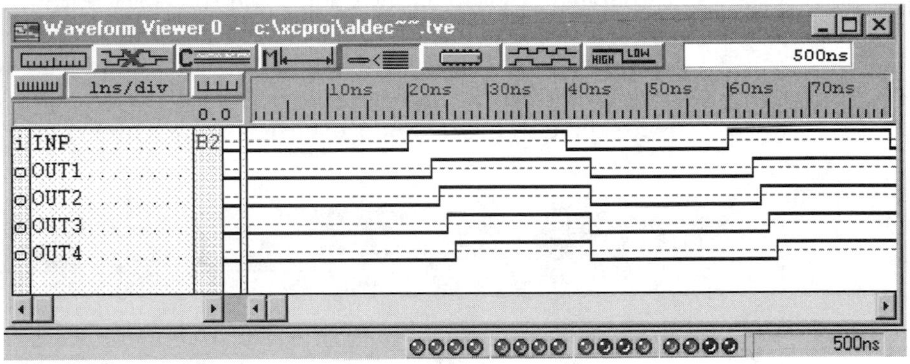

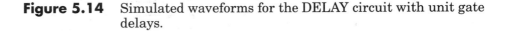

Figure 5.14   Simulated waveforms for the DELAY circuit with unit gate delays.

The actual circuit delays in the DELAY circuit cannot be tested from your PC keyboard or observed using the LED digit. Instead, you have to drive the input with the oscillator on the XS40 or XS95 Boards and observe the outputs with a dual-trace oscilloscope. Use the following pin assignments for the XS40 Board:

```
NET  INP        LOC=P13;      # the oscillator signal enters on pin 13 of the XS40
NET  OUT1       LOC=P25;
NET  OUT2       LOC=P26;
NET  OUT3       LOC=P24;
NET  OUT4       LOC=P20;
```

And use these pin assignments for the XS95 Board:

```
NET  INP        LOC=P9;       # the oscillator signal enters on pin 9 of the XS95
NET  OUT1       LOC=P21;
NET  OUT2       LOC=P23;
NET  OUT3       LOC=P19;
NET  OUT4       LOC=P17;
```

After compiling and downloading the DELAY design, attach one channel of the oscilloscope to the pin driven by the on-board oscillator of the XS40 or XS95 Board. Attach the other channel to one of the pins driven by the outputs of the DELAY circuit. After some adjustment of the scope, you should see waveforms similar to those of Figure 5.14.

You can use the simulator with unit-gate delays to observe the delay through the 8-bit adder we built in the last chapter. The maximum delay through the carry adder chain occurs when one of the operands is 11111111 (0xFF in hexadecimal), the other operand is 00000000 and the carry input flips from 0 to 1. Figure 5.15 shows the waveforms and delays for this case. The carry input goes to a logic 1 at 40 nanoseconds, and the carry output follows at 58 nanoseconds. That is 18 gate delays since the simulation precision was set to 1 nanosecond. From our previous discussions, we calculated this delay to be 16 gate delays, not 18. Once again, the 2 extra gate delays come from the IBUF input buffer for the CIN signal and the OBUF output buffer for the COUT signal.

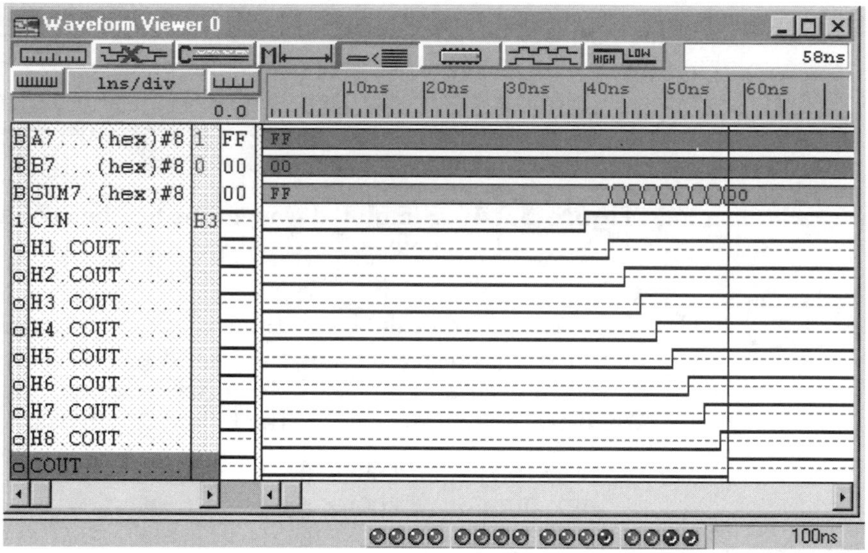

Figure 5.15   Waveforms for a simulation of the maximum ripple-carry chain delay of an 8-bit adder.

You can even examine the delays as the intermediate carry bits flip. This requires you to add signals from lower levels of the design hierarchy to the **Waveform Viewer**. Just select **Signal** → **Add Signals...** to bring up the **Component Selection for Waveform Viewer** window shown in Figure 5.16. Click on one of the **MY_ADD1** adder bits in the **Chip Selection** panel. Then click on the **COUT** pin in the right-most panel and **Add** it to your **Waveform Viewer** window. Once all the intermediate carry signals are displayed, you can see the cascade effect as they flip from 0 to 1.

This is nice and simple. But in the real-world not all gate delays are equal. And the routing of signals between macrocells or CLBs in an FPLD also contributes to the total delay. In order to simulate these delays, you need to use the **Timing** mode of the simulator. But the Timing mode cannot do an accurate simulation unless it has the gate and routing delay information that comes

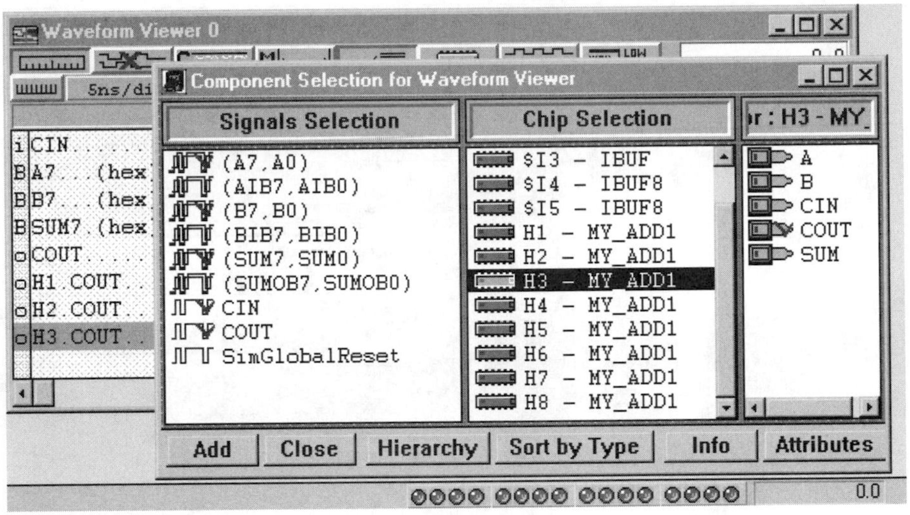

Figure 5.16 Displaying hierarchical signals in the 8-bit adder.

from how your circuit was mapped into the FPLD chip. You can tell the Foundation Series Implementation tools to generate this information for either schematic or HDL mode projects by selecting the **Implementation** → **Options...** menu item in the **Project Manager** window. Then check the **Produce Timing Simulation Data** box in the **Options** window (see Figure 5.17). Then run the Foundation Implementation tool and it will extract the routing and gate delays as it maps the netlist into the FPLD.

Once you have run the Foundation Implementation tools, you can click on the ![button] button in the **Flow** tab of the **Project Manager** window. This brings up the **Logic Simulator** window with the **Timing** mode selected. Other than that, you use the same steps to add signals and do simulations as you did for functional simulations.

The resulting waveforms for the 8-bit ripple-carry adder are shown in Figure 5.18. In this case, we compiled the adder for an XC4005XL FPGA with a −3 speed rating. In the simulation, the CIN input transitions at the 80 ns mark. The COUT output transitions in response at 121 ns. Thus, the delay through the 8 adder bits (including gate, I/O buffer, and wiring delays) is 41 ns. The worst-case combinatorial delays through an XC4005XL-3 CLB range from 1.6 ns to 3.2 ns. That would make the total gate delay through the 8 CLBs of the carry chain between 12.8 ns and 25.6 ns. The remainder of the delay is buffer propagation time and wiring delays.

All of our experiments have been done on ripple-carry adders. The ABEL and VHDL descriptions for a 4-bit carry-lookahead adder are shown in Listings 5.4 and 5.5, respectively. You should be able to see the correspondence between the

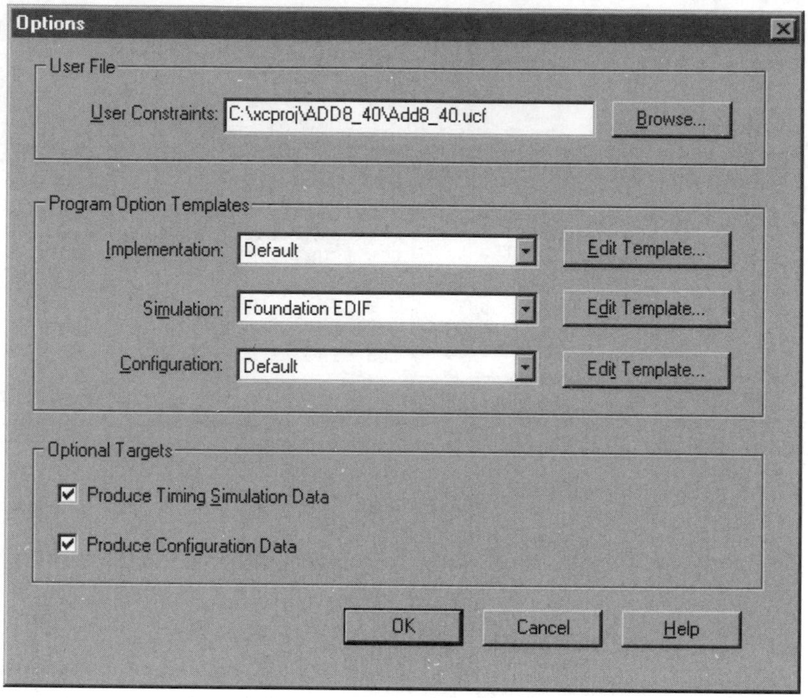

Figure 5.17   Configuring the Foundation Series Implementation tools to produce timing simulation data.

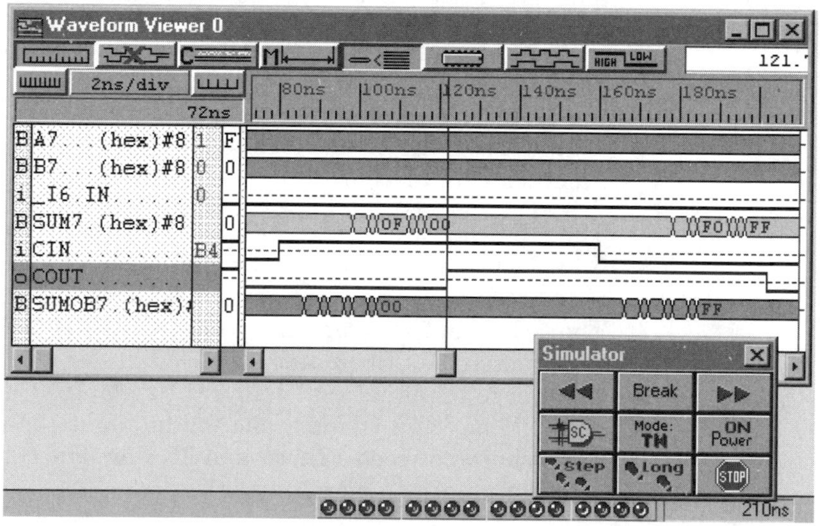

Figure 5.18   An accurate timing simulation of the 8-bit ripple-carry adder in an XC4005XL-3 FPGA.

code and the equations for c_i, G_i, and P_i presented earlier in this chapter. You will compare the propagation delays of the carry-lookahead and ripple-carry adders in the project at the end of this chapter.

Listing 5.4   ABEL code for a 4-bit carry-lookahead adder (LKADD.ABL).

```
001- MODULE lkadd
002- TITLE '4-bit carry-lookahead adder'
003-
004- DECLARATIONS
005-
006- cin PIN;
007- a3..a0 PIN;
008- b3..b0 PIN;
009- sum3..sum0 PIN ISTYPE 'COM';
010- cout PIN ISTYPE 'COM';
011- g3..g0 NODE ISTYPE 'COM'; "carry generate bits
012- p3..p0 NODE ISTYPE 'COM'; "carry-propagate bits
013- c2..c0 NODE ISTYPE 'COM'; "internal carry bits
014-
015- EQUATIONS
016-
017- g0 = a0 & b0;"first adder bit
018- p0 = a0 $ b0;
019- c0 = g0 # p0&cin;
020- sum0 = p0 $ cin;
021- g1 = a1 & b1;"second adder bit
022- p1 = a1 $ b1;
023- c1 = g1 # p1&g0 # p1&p0&cin;
024- sum1 = p1 $ c0;
025- g2 = a2 & b2;"third adder bit
026- p2 = a2 $ b2;
027- c2 = g2 # p2&g1 # p2&p1&g0 # p2&p1&p0&cin;
028- sum2 = p2 $ c1;
029- g3 = a3 & b3;" fourth adder bit
030- p3 = a3 $ b3;
031- cout = g3 # p3&g2 # p3&p2&g1 # p3&p2&p1&g0 #
032- p3&p2&p1&p0&cin;
033- sum3 = p3 $ c2;
034-
035- END lkadd
```

Experimental

Listing 5.5   VHDL code for a 4-bit carry-lookahead adder (LKADD.VHD).

```
001- -- Look-ahead Adder
002-
003- LIBRARY IEEE;
004- USE IEEE.std_logic_1164.ALL;
005-
006- -- Look-ahead Adder interface description
007- ENTITY lkadd IS
008-     PORT
009-     (
010-         a: IN STD_LOGIC_VECTOR (3 DOWNTO 0);-- 4-bit addend
011-         b: IN STD_LOGIC_VECTOR (3 DOWNTO 0);-- 4-bit addend
012-         cin: IN STD_LOGIC;-- input carry
013-         sum: OUT STD_LOGIC_VECTOR (3 DOWNTO 0);-- 4-bit sum
014-         cout: OUT STD_LOGIC-- carry output
015-     );
016- END lkadd;
017-
018- -- Look-ahead Adder architecture description
019- ARCHITECTURE lkadd_arch OF lkadd IS
020- -- signal vectors for transferring the carry-propagate,
021- -- carry-generate, and carry signals between the adder stages
022- SIGNAL g: STD_LOGIC_VECTOR (3 DOWNTO 0);-- carry-generate
023- SIGNAL p: STD_LOGIC_VECTOR (3 DOWNTO 0);-- carry-propagate
024- SIGNAL c: STD_LOGIC_VECTOR (3 DOWNTO 0);-- carry
025- BEGIN
026- PROCESS (a,b,cin,g,p,c)
027- BEGIN
028- -- use a FOR loop to iteratively connect the logic operations
029- -- for the carry-generate, carry-propagate, and sum signals
030- FOR i IN 0 TO 3
031- LOOP
032- g(i) <= a(i) AND b(i);
033- p(i) <= a(i) XOR b(i);
034- sum(i) <= p(i) XOR c(i);
035- END LOOP;
036- -- compute the carry bits for each stage of the adder
037- c(0) <= cin;
038- c(1) <= g(0) OR (p(0) AND c(0));
039- c(2) <= g(1) OR (p(1) AND g(0)) OR
040- (p(1) AND p(0) AND c(0));
041- c(3) <= g(2) OR (p(2) AND g(1)) OR
042- (p(2) AND p(1) AND g(0)) OR
```

Listing 5.5   VHDL code for a 4-bit carry-lookahead adder (LKADD.VHD). (Cont'd.)

```
043-  (p(2) AND p(1) AND p(0) AND c(0));
044-  cout <= g(3) OR (p(3) AND g(2)) OR
045-  (p(3) AND p(2) AND g(1)) OR
046-  (p(3) AND p(2) AND p(1) AND g(0)) OR
047-  (p(3) AND p(2) AND p(1) AND p(0) AND c(0));
048-  END PROCESS;
049-  END lkadd_arch;
```

Projects

1. Compile the 8-bit ripple-carry adder from the previous chapter. Create a simple input pattern that will make the adder exhibit its worst-case delay when a clock waveform is input to the adder as c_{in}. Simulate the adder and observe the delay. Download the adder to your XS40 or XS95 Board and measure the delay with an oscilloscope.

2. Design an 8-bit carry-lookahead adder using the 4-bit carry-lookahead module. Apply the same input pattern to this adder. Simulate the adder and observe the worst-case delay. Download the adder to your XS40 or XS95 Board and measure the delay with an oscilloscope. Compare the worst-case delay and size of this adder with the ripple-carry adder (use the report file). Does a faster design take more space in your FPGA?

Flip-Flops

Objectives

- Show how flip-flops work.
- Show the different types of flip-flops.
- Show how a counter is built from flip-flops.

Discussion

So far we have experimented with combinatorial circuits that generate their outputs in response only to the current input. When the input changes, the output immediately changes (after some gate delay, of course). Combinatorial circuits have no way to remember what has happened before; they are creatures of the present.

This chapter will introduce you to sequential circuits. Sequential circuits are able to remember what has happened to them and can use it to alter their future behavior. A counter is an example of a sequential circuit because it remembers its current value and can add one to it to get to its next value. The current value of the counter is referred to as its state. The current state determines what the next state of the counter will be. For example, if the counter currently stores the number five, then the next time the counter is commanded to increment it will store the value six. The change in state from five to six is called a state transition.

The concept of state is easy to understand, but we do not know enough yet to build a counter. We must start with something simpler. The simplest sequential circuit is an S-R flip-flop (SRFF). An SRFF can remember whether a single bit is one or zero. This single-bit memory can be used as the basis of all other sequential circuits.

The operation of an SRFF is shown in Figure 6.1. The SRFF is built from two 2-input NAND gates, each of whose outputs is fed back to one of the inputs of the other gate. The remaining inputs of the NAND gates are labeled /S and /R, while the outputs are labeled Q and /Q . In Figure 6.1a, both /S and /R are held at +5 V while Q = 1 and /Q = 0. Note that with /S = 1 and /Q = 0, the Q output should be 1 since one of the inputs to its NAND gate is low. Note also that with

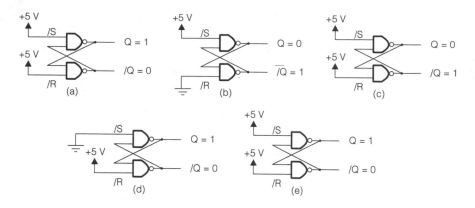

Figure 6.1   The operation of an S-R flip-flop.

/R = 1 and Q = 1, the /Q output should be 0 since both inputs to its NAND gate are high. So everything is consistent.

Suppose that /R is pulled to ground (Figure 6.1b). This will immediately cause /Q to go to 1 because both inputs to its NAND gate are no longer high. The high level on Q enters the other NAND gate which already has /S = 1 and causes Q to go low to logic 0. So the logical values of Q and /Q have been inverted. If the /R input is returned to +5 V (Figure 6.1c), the /Q output still remains high because the low level on the Q output keeps it that way. And the high level on the /Q output and the /S input keeps the Q output at a low level. So each of the outputs helps to keep the other output at its current level. In this case, the SRFF is said to be in the reset state because the Q output is 0. The flip-flop enters the reset state whenever the /R input is pulled to ground. It will stay in this state as long as both /S and /R inputs are held high.

Notice that both /S and /R inputs are high in both Figures 6.1a and 6.1c but the Q and /Q have switched values. For a given set of inputs a combinatorial circuit always generates the same output, but a sequential circuit does not have to because it has an internal state that lets it remember what has happened in the past and act differently in the future. Essentially, the state of a sequential circuit acts like another set of inputs that affects the operation of the circuit. Figures 6.1d and 6.1e demonstrate that the /S input can be pulled low to set the Q output to a logic 1. Thus, the /S and /R inputs can be used to set or reset the flip-flop, respectively.

What happens if both /S and /R are pulled low at the same time? This will cause both Q and /Q to go to logic 1 levels. Then if the /S input is pulled back to +5 V, the situation reverts to the one shown in Figure 6.1b and the SRFF is reset. On the other hand, if the /R input were to go high first, then the situation changes to the one shown in Figure 6.1d and the SRFF is set. What will happen if we try to pull both /S and /R back to +5 V simultaneously? It is very hard to do two things at *exactly* the same time, especially when you are dealing with circuits that respond within nanoseconds. Therefore, the final state of the SRFF could have either Q = 1 or Q = 0 depending on whether the /S or /R input was slower getting to +5 V, respectively. This is an example of a race condition that occurs when the state of a flip-flop is determined by unpredictable delays in the circuit. Race conditions cause intermittent failures in logic circuits, and they can be hard to find.

The SRFF will respond any time the /S or /R input changes values. Sometimes we would like a flip-flop to change state only at certain times. Figure 6.2 shows a modified SRFF that does this. When the C input is low, the outputs of the two initial NAND gates are held high. The high outputs make the attached SRFF hold onto whatever value it was loaded with. When C goes high, the /S and /R inputs can pass through the NAND gates and change the state of the SRFF. (The extra inverters are added to correct for the inversion that occurs when /S and /R pass through the NAND gates.) Thus, this clocked SRFF will only change state when the clock C is high.

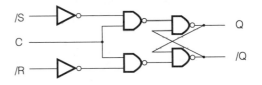

Figure 6.2   A clocked S-R flip-flop.

Many times we just want to store the value of a data bit in a flip-flop. Figure 6.3 shows the circuit for such a clocked data flip-flop (DFF) built from a clocked SRFF. When D = 0, then /S = 1 and /R = 0. This makes Q = 0 when the clock is high. When D = 1, then Q = 1. So when the clock is high, Q = D. When the clock is low, the DFF stores the last value that was on D when the clock was high.

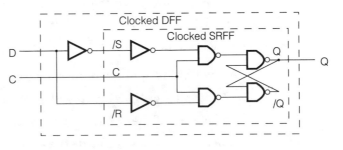

Figure 6.3   A clocked D flip-flop.

We can be even more stringent in our requirements by stating that the DFF should only change state at the instant when the clock falls from a high level to a low level. This is equivalent to saying that the flip-flop should be negative edge-triggered. The circuit for a negative edge-triggered DFF is shown in Figure 6.4. It consists of two clocked DFFs with an extra inverter to make the DFFs operate on opposite phases of the clock. When the clock is high (C = 1 in Figure 6.5a), the output of the master DFF reacts to any changes in the D input. The change on this output will not affect the slave DFF because its clock signal is low and the NAND gates on its inputs are turned off. When the clock goes low (C = 0 in Figure 6.5b), the output of the slave DFF will begin to react to any change on the output of the master DFF. But the output of the master DFF is no longer changing because its clock input is now low. So the output of the slave DFF will change at most once to reflect the values that were in force at the time C transitioned from 1 to 0.

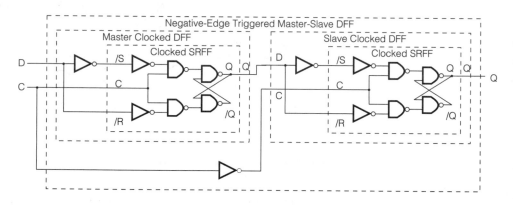

Figure 6.4   A negative edge-triggered master-slave D flip-flop.

The DFF is great for storing a state, but the problem arises as to what state it starts in when power is first applied to the circuit. Is it storing a 1 or a 0? We need a way to initialize the flip-flop to a known state. That is what the additional clear circuits of Figure 6.6 do for us.

The synchronous clear input in Figure 6.6 works as follows:

1. A low-level is applied to the synchronous clear input.
2. The outputs of NAND gates SR1 and SR2 go to a logic 1.
3. The output of the Q NAND gate in the master flip-flop goes to a logic 0.
4. A negative clock edge loads the output of the master flip-flop into the slave flip-flop.
5. The Q output of the slave flip-flop goes to logic 0. The flip-flop is cleared.

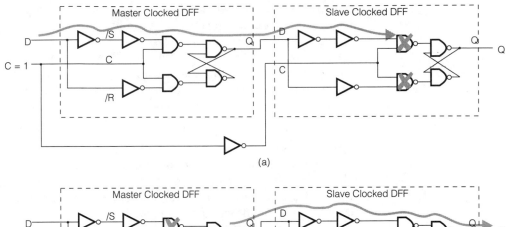

(a)

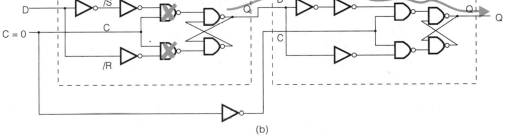

(b)

Figure 6.5 The operation of a negative edge-triggered master-slave D flip-flop.

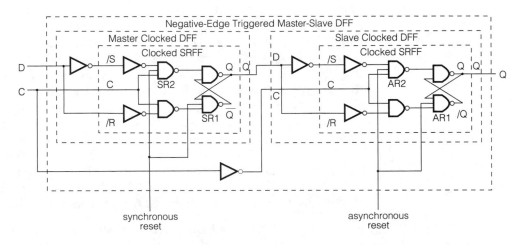

Figure 6.6 A DFF with asynchronous and synchronous clear inputs.

Now it is clear why we call it a synchronous clear operation: The circuit needs a clock signal before the output of the flip-flop will go to logic 0. What if we need to clear the flip-flop *right now* without a clock edge? That is when the circuitry for an asynchronous clear is needed. The asynchronous circuit looks just like the synchronous circuit, but since it is in the slave flip-flop its results appear immediately on the Q output. No clock is needed before we get the desired result.

But there is one difficulty: When we bring the asynchronous clear input high again, what prevents the master flip-flop from reloading the slave and undoing our clear operation? This is not a problem if the clock is at logic 1 because the NAND gates at the entrance to the slave will prevent the master flip-flop output from getting in. But the master's output can enter if the clock is at logic 0. For that reason, we need to clear both the master and slave flip-flops if we want to do an asynchronous clear. This does not prevent the master flip-flop from reloading the slave, but it does guarantee that the slave will be reloaded with a logic 0. Therefore, an asynchronous clear input has to trigger both the asynchronous and synchronous clear circuitry shown in Figure 6.6.

Once all the major circuitry for the flip-flop is built, you can begin to ask questions about finer details of its operation, such as, "What happens if the D input changes at exactly the same time as the clock goes from 1 to 0?" Our previous discussion of gate propagation delays leads us to believe that such an occurrence would have no effect because the input value would not have enough time to propagate into the rest of the circuit before the first two clocked NAND gates were shut off by the low clock.

Then you might ask, "How close can I get to the clock edge before the signal changes are ignored?" This brings up the concept of setup time, often denoted as t_{su}. You must have the inputs to a flip-flop stabilized at least t_{su} seconds before the clock edge occurs. The exact value of t_{su} will vary depending upon the propagation delays in the input stages of your logic. If your signals have to go through a lot of combinatorial logic before reaching the clocked gates, then your t_{su} will have to be longer in order to provide enough time for propagation. For the XC95108 CPLD, the setup time is computed for changes on one of the package pins relative to a clock transition on one of the global clock inputs, and t_{su} ranges from 6.5 to 10 ns depending on the speed grade of the device. This value reflects both the gate and internal-wiring propagation delays. For the XC4005XL-3 FPGA, t_{su} ranges from 1.1 to 3.5 ns within a CLB and has a worst-case value of 6.8 ns when measured from the package pins.

Finally, if you are really inquisitive, you might ask, "How soon can I change my inputs after the clock edge has passed?" Now you are addressing the issue of hold-time, or t_h. It is possible that the SRFF could react to a change in an input to one of the clocked NAND gates if it occurred very soon after the clock has gone low. This is an effect caused by propagation delays within the clock wiring and clocked NAND gates. So you might have to hold your inputs to a flip-flop steady for a few nanoseconds after the clock edge occurs. Both the XC4005XL FPGA and XC95108 CPLD have hold times of 0 ns.

It is possible to trade off between t_{su} and t_h. You can reduce the hold time by placing extra gates on the inputs to the flip-flop. The extra gates will delay the propagation of any changes to the flip-flop's inputs. Therefore, you can change your inputs sooner after a clock edge because the changes will not propagate into the rest of the circuit fast enough to cause a problem. The trade-off occurs when you realize that now you must set up your inputs farther in advance of the clock edge so they can propgate through the extra gates. In most cases, the sum of t_{su} and t_h is approximately constant.

How do you know if you have violated the setup and hold times for a flip-flop? Usually you do not know until it's too late. Often the circuit appears to work just fine, but every now and then the flip-flop moves into the wrong state. In some cases the flip-flop enters a metastable state. In the metastable state, the outputs of the flip-flop are converging slowly to valid 1 or 0 logic levels, but in the meantime they are at invalid levels. Timing problems are difficult to debug in practice because it is hard to repeat the exact conditions that cause them. Timing simulators can help since they usually will warn you if they notice a change to the inputs of a flip-flop that violates t_{su} or t_h. A bit of luck helps as well.

You can build a positive edge-triggered DFF by simply inverting the clock signal that drives both DFFs. Given that you know how a positive edge-triggered DFF works, you can use it to make a toggle flip-flop (Figure 6.7a). A toggle flip-flop (TFF) is built from a DFF by feeding the inverted Q output of the DFF back to its D input. Then whenever there is a rising clock edge, the DFF loads itself with a value that is the inverse of what it was storing. The T output of a TFF as it responds to a clock waveform is shown in Figure 6.7b. Note that the frequency of the T output is half that of the clock waveform (i.e., the period is twice as long). The TFF acts as a frequency divider. If several TFFs are joined together with the T output of each TFF serving as the clock input to the next TFF (Figure 6.7c), then the waveforms shown in Figure 6.7d are generated. The output of TFF i is half the frequency of the output from TFF $i - 1$. By looking at these output waveforms as binary numbers with T2 as the most-significant bit and T0 as the least-significant bit, you can see the three TFFs form a 3-bit down counter.

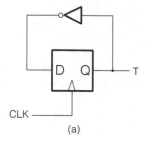

(a)

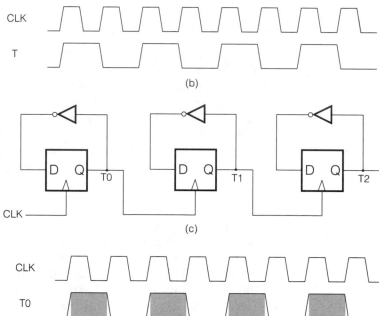

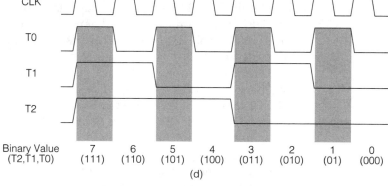

Figure 6.7   Building a counter from toggle flip-flops.

Experimental

An S-R Flip-Flop

We will start by building a simple S-R flip-flop. Listing 6.1 shows the ABEL code for this. Lines 11 and 12 describe the connections for a pair of cross-coupled NAND gates. Listing 6.2 shows the same S-R flip-flop written in VHDL with the cross-coupled NAND gates on lines 18 and 19.

(Note that I will use names ending with a B to denote signals which are active when they are at logic 0. So /S is named SB when it appears in an ABEL or VHDL file. This is just my personal convention, so you do not have to do it this way.)

Listing 6.1   ABEL code for a simple SR flip-flop.

```
001- MODULE SRFF1
002- TITLE 'simple SR flip-flop'
003-
004- DECLARATIONS
005- SBPIN;               "/S input
006- RBPIN;               "/R input
007- Q PIN ISTYPE 'COM'; "Q output
008- QBPIN ISTYPE 'COM'; "/Q output
009-
010- EQUATIONS
011- Q  = !(SB & QB);
012- QB = !(RB & Q);
013- END SRFF1
```

After synthesizing the netlist for this design, we can perform a simulation to check its operation. For this example, we will use the **Stimulator Selection** window to attach the SB and RB inputs to the keys of our keyboard. Just highlight the name of the SB input in the **Waveform Viewer** window and then click the mouse on the **Q** keyboard symbol in the **Stimulator Selection** window. This will bind the input to the that keyboard key. Repeat this operation to bind the RB input to the W key. Then close the **Stimulator Selection** window.

Now we can toggle the values on the SB or RB inputs by pressing the Q and W keys. Clicking the ▣ button on the toolbar of the **Logic Simulator** window performs a simulation step with the current inputs. Then we press the Q and W keys on the keyboard to change the inputs and do another simulation step. The simulated operation of the SRFF is shown in Figure 6.8.

Listing 6.2   VHDL code for a simple SR flip-flop.

```
001- -- S-R Flip-Flop
002-
003- LIBRARY IEEE;
004- USE IEEE.std_logic_1164.ALL;
005-
006- ENTITY srff1 IS
007- PORT
008- (
009-   sb: IN STD_LOGIC;      -- active-low set input
010-   rb: IN STD_LOGIC;      -- active-low reset input
011-   q: BUFFER STD_LOGIC;   -- data output
012-   qb: BUFFER STD_LOGIC   -- inverted data output
013- );
014- END srff1;
015-
016- ARCHITECTURE srff1_arch OF srff1 IS
017- BEGIN
018- q  <= NOT(sb AND qb);    -- cross-coupled NAND gates
019- qb <= NOT(rb AND q);
020- END srff1_arch;
```

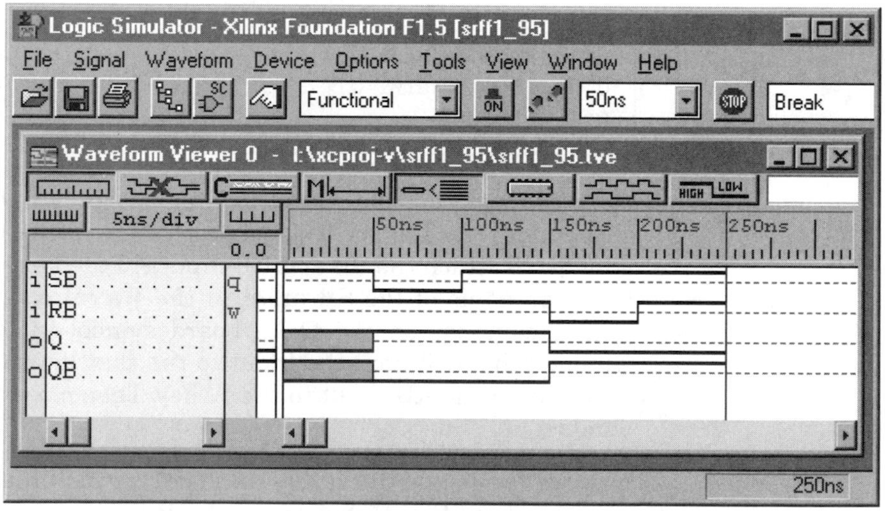

Figure 6.8   Simulated waveforms for a simple SR flip-flop.

Note that the simulation starts off with the Q and QB outputs in unknown states. The feedback between the NAND gates thwarts the simulator's attempts to figure out what the initial state of the flip-flop should be. But once the SB input is lowered to 0, the output of the Q NAND gate must stabilize at logic 1. With RB and Q now both at logic 1, the QB output goes to logic 0. So now the state of the flip-flop is no longer unknown.

When the SB input returns to 1, the Q output remains at 1. Thus, the SRFF is storing the logic 1 that was loaded in by the SB input. Then when the RB input goes low, the QB output is set to logic 1 and the Q output goes low. This situation persists when the RB input returns to logic 1, so now the SRFF is storing a logic 0. It appears the SR flip-flop is working, at least in the simulation.

Now we can download the SRFF to the XS40 or XS95 Board for an actual test. Create a user-constraint file which assigns the inputs and outputs as follows for the XS40 Board:

```
NET sb      LOC=P44;    # B0 argument of XSPORT
NET rb      LOC=P45;    # B1 argument of XSPORT
NET q       LOC=P25;    # S0 LED segment
NET qb      LOC=P26;    # S1 LED segment
```

Or use the following constraints for the XS95 Board:

```
NET sb      LOC=P46;    # B0 argument of XSPORT
NET rb      LOC=P47;    # B1 argument of XSPORT
NET q       LOC=P21;    # S0 LED segment
NET qb      LOC=P23;    # S1 LED segment
```

After compiling and downloading the SRFF1 design, you should see the following results when you exercise the design on the XS40 or XS95 Board with the given sequence of commands:

Command	RB	SB	LED S0 (Q)	LED S1 (Q_)
XSPORT 10	1	0	Bright (1)	Dark (0)
XSPORT 11	1	1	Bright (1)	Dark (0)
XSPORT 01	0	1	Dark (0)	Bright (1)
XSPORT 11	1	1	Dark (0)	Bright (1)

These results match those found in the simulation.

The ABEL code in Listing 6.1 uses 2 macrocells or LUTs in an FPLD because we generate both the Q output and its complement. If you only need the Q output, the ABEL code in Listing 6.3 will suffice and it uses only 1 macrocell or LUT. Listing 6.3 replaces the QB in line 11 of Listing 6.1 with the right-half of the equation of line 12. This SRFF works identically to the one described by Listing 6.1 except there is no QB output.

Listing 6.3   ABEL code for a single macrocell or LUT SR flip-flop.

```
001-  MODULE SRFF2
002-  TITLE 'simple SR flip-flop'
003-
004-  DECLARATIONS
005-  SB PIN;                  "/S input
006-  RB PIN;                  "/R input
007-  Q PIN ISTYPE 'COM';  "Q output
008-
009-  EQUATIONS
010-  Q  = !(SB & !(RB & Q));
011-
012-  END SRFF2
```

The VHDL code for an SRFF with no QB output is shown in Listing 6.4. This SRFF consumes just 1 LUT or CLB, but it is expressed in a behavioral style rather than as Boolean equations. A behavioral logic description typically has one or more processes surrounded by PROCESS...END PROCESS statements. The PROCESS keyword is followed by a parenthesized sensitivity list. If any of the signals in the list changes its value, then the statements in the process block are evaluated.

The process starting on line 17 of Listing 6.4 is initiated if either the sb or qb inputs or the q output changes state. In general, the sensitivity list contains all the signals which appear as operands in conditional expressions or on the right-hand side of the <= assignment operator of statements in the process block.

The IF...ELSIF...ELSE conditional statement on lines 19 through 25 describes the actual operations of the S-R flip-flop. Line 20 sets the q output to 1 if the sb input is ever low. But if sb is not 0 and rb is, then line 22 will reset q to 0. Finally, line 24 states that q will not change its value if neither sb or rb is low. This is exactly the operation we want in an S-R flip-flop.

Note that we do not even need the statements on lines 23 and 24 and they could be removed without affecting the operation of the flip-flop. Consider the case when sb=0 and rb=1. The q output will be 1 after the process executes line 20 and the process terminates. Then if sb goes high, the change will initiate the process once more. But this time neither sb or rb is low, so neither of the assignment statements on lines 20 or 22 will be executed. With no explicit statement to set its value, the q output will retain its previous value which was 1. When an output retains its last value because none of the assignment statements are executed in a process, then the output has an implied latch. The S-R flip-flop with an implied latch operates identically to the flip-flop in Listing 6.4 with an explicit latch on lines 23 and 24.

One other new feature in Listing 6.4 is the declaration of the q output as a BUFFER rather than an OUT on line 11 of the interface description. VHDL requires that values on an OUT signal can only exit from a module and cannot be read from within the module ARCHITECTURE section. Thus if q were declared as an OUT, it would be illegal to use it on the right-hand side of an assignment as on line 24. VHDL provides the BUFFER and INOUT keywords to declare signals which can serve as outputs but can also feed their values back into the module.

Listing 6.4   Behavioral VHDL code for a single macrocell or LUT SR flip-flop.

```
001- -- S-R Flip-Flop
002-
003- LIBRARY IEEE;
004- USE IEEE.std_logic_1164.ALL;
005-
006- ENTITY srff2 IS
007- PORT
008- (
009-   sb: IN STD_LOGIC;      -- active-low set input
010-   rb: IN STD_LOGIC;      -- active-low reset input
011-   q: BUFFER STD_LOGIC    -- data output
012- );
013- END srff2;
014-
015- ARCHITECTURE srff2_arch OF srff2 IS
016- BEGIN
017- PROCESS (sb,rb,q)
018- BEGIN
019- IF (sb='0') THEN
020-   q  <= '1';
021- ELSIF (rb='0') THEN
022-   q  <= '0';
023- ELSE
024-   q  <= q;
025- END IF;
026- END PROCESS;
027- END srff2_arch;
```

A Clocked S-R Flip-Flop

Next we move onto the clocked SRFF (Listing 6.5). All we have done is add a set of NAND gates between the original set of cross-coupled NAND gates and the SB and RB inputs. We also added inverters to the SB and RB inputs to counteract the inversion of the added NAND gates.

Listing 6.5   ABEL code for a clocked SRFF.

```
001- MODULE CSRFF
002- TITLE 'clocked SR flip-flop'
003-
004- DECLARATIONS
005- C PIN;                "clock input
006- SBPIN;               "/S input
007- RBPIN;               "/R input
008- Q PIN ISTYPE 'COM'; "Q output
009- QBPIN ISTYPE 'COM'; "/Q output
010-
011- EQUATIONS
012- Q  = !( !(!SB & C) & QB); "gated SB and RB inputs
013- QB = !( !(!RB & C) & Q);  "with the clock input
014-
015- END CSRFF
```

Listing 6.6 is a behavioral VHDL description of a clocked SRFF. We have added a new clock input to the interface description (line 8) and added the clock to the process sensitivity list (line 17). The IF statement on line 19 prevents the execution of the statements on lines 20 through 24 unless the clock input is high. Thus the q output can only change when c=1. (Note that the q output is treated as an implied latch.)

The simulated waveforms for the clocked SRFF are shown in Figure 6.9. When the clock input, C, is high the waveforms are identical to those of Figure 6.8. When C=0, the SB and RB inputs no longer have any affect on the Q output. The simulated operation of the clocked SRFF is correct.

To download the clocked SRFF to the XS40 or XS95 Board for an actual test, create a user-constraint file which assigns the inputs and outputs as follows for the XS40 Board:

```
NET c       LOC=P44;    # B0 argument of XSPORT
NET sb      LOC=P45;    # B1 argument of XSPORT
NET rb      LOC=P46;    # B2 argument of XSPORT
NET q       LOC=P25;    # S0 LED segment
NET qb      LOC=P26;    # S1 LED segment
```

Listing 6.6 Behavioral VHDL code for a clocked SRFF.

```
001- -- Clocked S-R Flip-Flop
002-
003- LIBRARY IEEE;
004- USE IEEE.std_logic_1164.ALL;
005-
006- ENTITY csrff IS
007- PORT (
008-   c: IN STD_LOGIC;        -- clock
009-   sb: IN STD_LOGIC;       -- active-low set input
010-   rb: IN STD_LOGIC;       -- active-low reset input
011-   q: BUFFER STD_LOGIC;    -- stored data bit value
012-   qb: OUT STD_LOGIC       -- inverted output
013- );
014- END csrff;
015-
016- ARCHITECTURE csrff_arch OF csrff IS
017- BEGIN
018- PROCESS (c,sb,rb,q)
019- BEGIN
020- IF c='1' THEN -- change state only when clock is high
021-   IF (sb='0') THEN
022-            q  <= '1';
023-   ELSIF (rb='0') THEN
024-            q  <= '0';
025-   END IF;
026- END IF;
027- END PROCESS;
028- qb <= q;    -- generate the inverted output
029- END csrff_arch;
030-
```

or use the following constraints for the XS95 Board:

```
NET c       LOC=P46;    # B0 argument of XSPORT
NET sb      LOC=P47;    # B1 argument of XSPORT
NET rb      LOC=P48;    # B2 argument of XSPORT
NET q       LOC=P21;    # S0 LED segment
NET qb      LOC=P23;    # S1 LED segment
```

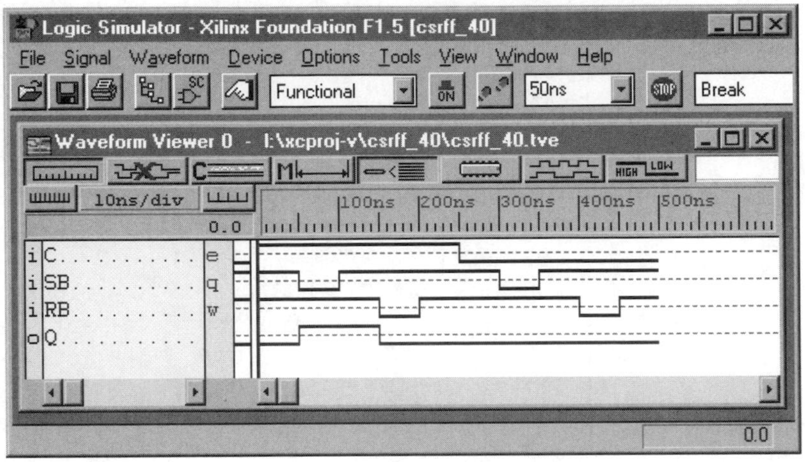

Figure 6.9　Simulated waveforms for a clocked SR flip-flop.

After compiling and downloading the CSRFF design, you should see the following results when you exercise the design on the XS40 or XS95 Board with the given sequence of commands:

Command	RB	SB	C	LED S0 (Q)	LED S1 (QB)
XSPORT 101	1	0	1	Bright (1)	Dark (0)
XSPORT 111	1	1	1	Bright (1)	Dark (0)
XSPORT 011	0	1	1	Dark (0)	Bright (1)
XSPORT 111	1	1	1	Dark (0)	Bright (1)
XSPORT 100	1	0	0	Dark (0)	Bright (1)
XSPORT 110	1	1	0	Dark (0)	Bright (1)
XSPORT 010	0	1	0	Dark (0)	Bright (1)
XSPORT 110	1	1	0	Dark (0)	Bright (1)

These results show the clock input is passing and blocking the SB and RB inputs as desired.

An Edge-Triggered D Flip-Flop

Now we can use the clocked SRFF to build an edge-triggered D flip-flop. The ABEL code is shown in Listing 6.7. Lines 13 and 14 describe the master clocked DFF and lines 15 and 16 describe the slave DFF that follows. The slave clock is an inverted copy of the master clock.

Listing 6.7   ABEL code for an edge-triggered, master-slave D flip-flop.

```
001- MODULE MSDFF
002- TITLE 'edge-triggered D flip-flop'
003-
004- DECLARATIONS
005- C          PIN;               "clock input
006- D          PIN;               "D input
007- MQ         PIN ISTYPE 'COM';  "master Q output
008- MQB        PIN ISTYPE 'COM';  "master /Q output
009- SQ         PIN ISTYPE 'COM';  "slave Q output
010- SQB        PIN ISTYPE 'COM';  "slave /Q output
011-
012- EQUATIONS
013- MQ  = !( !(!!D  & C) & MQB);  "master clocked DFF
014- MQB = !( !(!D   & C) & MQ );
015- SQ  = !( !(!!MQ & !C) & SQB); "slave clocked DFF
016- SQB = !( !(!MQ  & !C) & SQ ); "with inverted clock
017-
018- END MSDFF
```

Listing 6.8 is a behavioral VHDL description of an edge-triggered D flip-flop. The process is now sensitive to changes on the data (d) and clock (c) inputs (line 18). We need a way to describe a falling edge on the clock input so we can trigger a change in the sq output. This is done using the VHDL attribute 'event which can be attached to a signal and will evaluate to a TRUE condition when the value of the signal is different than the last time the process was executed. On line 20, the conditional in the IF statement is true if the c signal was high during the previous execution of the process but is now low. Then line 21 will execute and transfer the value on the d input to the sq output. Now if the d input changes value but c does not (i.e., c is still low), the process will be executed once more. But line 21 will not be executed because c'event is FALSE (i.e., there was no change in the value of c). So the flip-flop described by Listing 6.8 is edge-triggered rather than level-sensitive.

The simulated waveforms for the ABEL version of the edge-triggered DFF are shown in Figure 6.10. When the simulation starts, both the master and slave flip-flop are in unknown states. As soon as the clock rises to a high level, the master flip-flop accepts the logic 0 present on the D input. The slave flip-flop is still in an unknown state, however, because its clock input is low (the inverse of the C input) which blocks the output of the master flip-flop from entering. Once the clock falls to logic 0, the slave flip-flop output is set to logic zero. Note how the master and slave flip-flops pass the D input along in a type of "bucket-brigade."

Listing 6.8 Behavioral VHDL code for an edge-triggered, master-slave D flip-flop.

```
001- -- Edge-triggered D Flip-flop
002-
003- LIBRARY IEEE;
004- USE IEEE.std_logic_1164.ALL;
005-
006- ENTITY msdff IS
007- PORT
008- (
009-    c: IN STD_LOGIC;        -- clock
010-    d: IN STD_LOGIC;        -- data input
011-    sq: OUT STD_LOGIC;      -- data output
012-    sqb: OUT STD_LOGIC      -- inverted data output
013- );
014- END msdff;
015-
016- ARCHITECTURE msdff_arch OF msdff IS
017- BEGIN
018- PROCESS (c,d)
019- BEGIN
020- IF (c'event AND c='0') THEN-- true on a falling edge of c
021-    sq  <= d;
022- END IF;
023- END PROCESS;
024- sqb <= sq; -- generate the inverted output
025- END msdff_arch;
026-
```

When the D input rises to a logic 1, this is not reflected on the slave flip-flop's output until the next negative clock edge at the 250 ns point. In all cases, the value output by the slave always matches the value on the D input at the time the negative-going edge on the clock input occurs. So the basic functions of the edge-triggered DFF seem to be working.

A constraint file for the MSDFF design should assign the inputs and outputs like so for the XS40 Board (remove the MQ and MQB assignments for the VHDL version of the flip-flop):

```
NET c      LOC=P44;    # B0 argument of XSPORT
NET d      LOC=P45;    # B1 argument of XSPORT
NET sq     LOC=P25;    # S0 LED segment
NET sqb    LOC=P26;    # S1 LED segment
NET mq     LOC=P24;    # S2 LED segment (not used in VHDL version)
NET mqb    LOC=P20;    # S3 LED segment (not used in VHDL version)
```

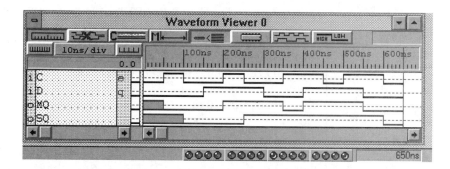

Figure 6.10 Simulated waveforms for an edge-triggered, master-slave D flip-flop.

Or use the following constraints for the XS95 Board:

```
NET c      LOC=P46;   # B0 argument of XSPORT
NET d      LOC=P47;   # B1 argument of XSPORT
NET sq     LOC=P21;   # S0 LED segment
NET sqb    LOC=P23;   # S1 LED segment
NET mq     LOC=P19;   # S2 LED segment (not used in VHDL version)
NET mqb    LOC=P17;   # S3 LED segment (not used in VHDL version)
```

After compiling and downloading the MSDFF design, you should see the following results when you exercise the design on the XS40 or XS95 Board with the given sequence of commands:

Command	D	C	LED S0 (SQ)
XSPORT 01	0	1	???
XSPORT 00	0	0	Dark (0)
XSPORT 11	1	1	Dark (0)
XSPORT 10	1	0	Bright (1)
XSPORT 01	0	1	Bright (1)

These results show the flip-flop output is only reacting to the input when a negative-going clock edge occurs.

An Edge-Triggered D Flip-Flop with Asynchronous Clear

Listing 6.9 shows the modifications made to add an asynchronous clear to the ABEL code for the edge-triggered DFF. As we discussed earlier, the asynchronous clear circuitry has to be added to both the master and slave flip-flops in order to work properly under all conditions.

Listing 6.9   **Listing 6.9:** ABEL code for a D flip-flop with asynchronous clear.

```
001- MODULE MSDFFAC
002- TITLE 'edge-triggered D flip-flop with async. clear'
003-
004- DECLARATIONS
005- C          PIN;                 "clock input
006- D          PIN;                 "D input
007- ACLRB      PIN;                 "async. clear
008- MQ         PIN ISTYPE 'COM';    "master Q output
009- MQB        PIN ISTYPE 'COM';    "master /Q output
010- SQ         PIN ISTYPE 'COM';    "slave Q output
011- SQB        PIN ISTYPE 'COM';    "slave /Q output
012-
013- EQUATIONS
014- MQ  = !( !(!!D  &  C  & ACLRB) & MQB); "master FF with
015- MQB = !( !(!D   &  C) & ACLRB  & MQ ); "sync. clear
016- SQ  = !( !(!!MQ & !C  & ACLRB) & SQB); "slave FF with
017- SQB = !( !(!MQ  & !C) & ACLRB  & SQ ); "async. clear
018-
019- END MSDFFAC
```

Listing 6.10 shows the equivalent modifications made to the VHDL of Listing 6.8 to add an asynchronous clear. The clear input is added to the interface definition on line 11 and the process sensitivity list on line 19. The edge-triggered loading of the sq output on lines 23 and 24 is subordinated to the clearing of the flip-flop on lines 21 and 22. Thus, the flip-flop is cleared if aclrb=0 regardless of whether or not there is a falling edge on the clock input.

The simulated waveforms for the DFF are shown in Figure 6.11. With the asynchronous clear input, ACLRB, held high, the DFF operates like it did in the last section. But when ACLRB goes low at 250 ns, the slave output, SQ, drops to a logic 0 immediately even though no clock edge has occurred.

Also note that the clear input has priority over the D input. When a negative clock edge occurs at 350 ns with D=1, the SQ output stays at logic 0 because the ACLRB input is still asserted.

A user-constraint file for the MSDFFAC design should assign the inputs and outputs for the XS40 Board as follows (remove the MQ and MQB assignments for the VHDL version of the flip-flop):

```
NET c     LOC=P44;    # B0 argument of XSPORT
NET d     LOC=P45;    # B1 argument of XSPORT
NET aclrb LOC=P46;    # B2 argument of XSPORT
```

Listing 6.10   Behavioral VHDL code for a D flip-flop with asynchronous clear.

```
001- -- edge-triggered D flip-flop with asynchronous clear
002-
003- LIBRARY IEEE;
004- USE IEEE.std_logic_1164.ALL;
005-
006- ENTITY msdac IS
007- PORT
008- (
009-   c: IN STD_LOGIC;        -- clock
010-   d: IN STD_LOGIC;        -- data input
011-   aclrb: IN STD_LOGIC;    -- active-low asynchronous clear
012-   sq: BUFFER STD_LOGIC;   -- data output
013-   sqb: OUT STD_LOGIC      -- inverted data output
014- );
015- END msdac;
016-
017- ARCHITECTURE msdac_arch OF msdac IS
018- BEGIN
019- PROCESS (c,d,aclrb)
020- BEGIN
021-     IF aclrb='0' THEN -- asynch clear takes precedence
022-            sq <= '0';
023-        ELSIF (c'event AND c='0') THEN -- over falling clock edge
024-            sq  <= d;
025- END IF;
026- END PROCESS;
027- sqb <= sq; -- generate the inverted output
028- END msdac_arch;
029-
```

```
NET sq     LOC=P25;   # S0 LED segment
NET sqb    LOC=P26;   # S1 LED segment
NET mq     LOC=P24;   # S2 LED segment (not used in VHDL version)
NET mqb    LOC=P20;   # S3 LED segment (not used in VHDL version)
```

Or use the following constraints for the XS95 Board:

```
NET c      LOC=P46;   # B0 argument of XSPORT
NET d      LOC=P47;   # B1 argument of XSPORT
NET aclrb LOC=P48;    # B2 argument of XSPORT
NET sq     LOC=P21;   # S0 LED segment
NET sqb    LOC=P23;   # S1 LED segment
NET mq     LOC=P19;   # S2 LED segment (not used in VHDL version)
NET mqb    LOC=P17;   # S3 LED segment (not used in VHDL version)
```

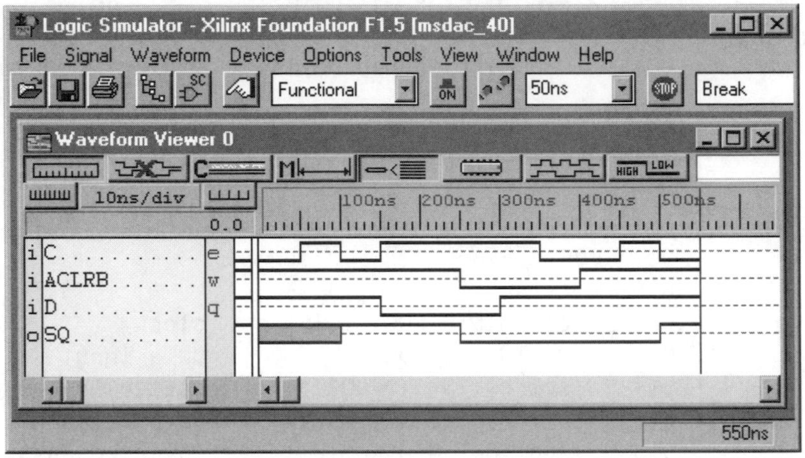

Figure 6.11   Simulated waveforms for a D flip-flop with asynchronous clear.

After compiling and downloading the MSDFFAC design to an XS40 or XS95 Board, the following sequence of commands should give you the indicated results:

Command	ACLRB	D	C	LED S0 (SQ)
XSPORT 101	1	0	1	???
XSPORT 110	1	1	0	Bright (1)
XSPORT 010	0	1	0	Dark (0)
XSPORT 011	0	1	1	Dark (0)
XSPORT 010	0	1	0	Dark (0)

These results show the flip-flop reacts to the asynchronous clear without the need for a clock edge and that the clear input has priority over the data input.

Built-In D Flip-Flops

We have built some flip-flops with various capabilities using basic logic gates. But an edge-triggered DFF takes at least 2 macrocells or LUTs even if we use the condensed form of Listing 6.2. To improve the utilization of resources in its FPLDs, XILINX has included built-in flip-flops in both its XC95108 CPLD and XC4005XL FPGA. Now all we have to do is learn how to use them with ABEL.

Listing 6.11 replicates the functions of our edge-triggered DFF with asynchronous clear. Line 8 declares the actual D flip-flop and gives it the name Q. The use of the type REG instead of COM distinguishes Q as a registered output rather than a combinatorial output like we have always used.

On line 12, the flip-flop is loaded with the value on the D input. Note the use of the := instead of =. The := implies that the Q output will only take on the value of the D input when a transition occurs on the clock input.

Line 13 specifies which signal serves as the clock input that controls the loading of the Q output. We have chosen to connect the inversion of C to the clock input of the Q DFF. This connection is specified by using the .CLK signal extension. Using the inverted C input makes the output of the Q flip-flop change on negative-going edges of the C input. (The normal operation of the built-in flip-flops is to change states on the rising edge.)

Line 14 specifies that the inversion of the ACLRB input is attached to the asynchronous clear input of the built-in flip-flop. The inversion is necessary because the built-in DFF is cleared when its asynchronous clear input is high, but we want the clearing to occur when ACLRB is low.

Listing 6.11 ABEL code for using a built-in D flip-flop with asynchronous clear.

```
001- MODULE BIDFF
002- TITLE 'using a built-in DFF'
003-
004- DECLARATIONS
005- C         PIN;               "clock input
006- D         PIN;               "D input
007- ACLRB     PIN;               "clear input
008- Q         PIN ISTYPE 'REG'; "DFF
009-
010-
011- EQUATIONS
012- Q    := D;      "connect to the D input of the DFF
013- Q.CLK = !C;     "connect the clock with inversion
014- Q.ACLR = !ACLRB; "connect the asynchronous clear
015-
016- END BIDFF
```

In VHDL we really do not have to do anything special to make use of the built-in flip-flops of the XC9500 or XC4000 chips. The Foundation Series synthesis tools can automatically extract the latches implied by the behavioral VHDL code and create a netlist which uses the built-in flip-flops. But if you did want to explicitly use a built-in DFF, then Listing 6.12 shows how to do it for the XC9500. The interface to the FDPC flip-flop contained in every XC9500 macrocell is declared as a component on lines 17 through 24 of the bidff ARCHITECTURE block. The FDPC module is a DFF which changes state on the rising edge of the clock (C). A logic 1 on the CLR or PRE input will clear or set the flip-flop, respectively.

The FDPC is instantiated on line 33. The clock and clear inputs of the FDPC are connected to the inversions of the clock and clear inputs to the bidff module. The preset input of FDPC is disabled by holding it low. This makes the bidff flip-flop change state on the falling edge of the clock and gives it an active-low clear input.

Where did I find the interface to the FDPC? It was stored in the FDPC.XNF file in the Xilinx\Synth\xilinx\macros\xc9500\m1_xnf directory of the foundation Series software. Listing 6.13 shows the contents of FDPC.XNF.

The built-in DFF works just like the ones we built from basic gates as can be seen from the simulation shown in Figure 6.12. It outputs the value on the D input when a negative clock edge occurs at 150 ns. But when ACLRB goes low at 200 ns, the Q output drops to a logic 0 immediately even though no clock edge has occurred. And the asynchronous clear has priority over the D input as can be seen when a clock edge occurs at 300 ns. Even though D=1, the asynchronous clear input keeps Q at logic 0.

A user-constraint file for the BIDFF design should assign the inputs and outputs for the XS40 Board as follows:

```
NET c        LOC=P44;   # B0 argument of XSPORT
NET d        LOC=P45;   # B1 argument of XSPORT
NET aclrb    LOC=P46;   # B2 argument of XSPORT
NET q        LOC=P25;   # S0 LED segment
```

Or use the following constraints for the XS95 Board:

```
NET c        LOC=P46;   # B0 argument of XSPORT
NET d        LOC=P47;   # B1 argument of XSPORT
NET aclrb    LOC=P48;   # B2 argument of XSPORT
NET q        LOC=P21;   # S0 LED segment
```

Listing 6.12 VHDL code for using an XC9500 built-in D flip-flop with asynchronous clear.

```
001- -- Built-in D Flip-Flops
002-
003- LIBRARY IEEE;
004- USE IEEE.std_logic_1164.ALL;
005-
006- ENTITY bidff IS
007- PORT
008- (
009-   c: IN STD_LOGIC;        -- clock
010-   d: IN STD_LOGIC;        -- data
011-   aclrb: IN STD_LOGIC;    -- active-low asynchronous clear
012-   q: OUT STD_LOGIC        -- stored data bit value
013- );
014- END bidff;
015-
016- ARCHITECTURE bidff_arch OF bidff IS
017- COMPONENT FDPC PORT     -- interface to the built-in DFF
018- (
019-   C: IN STD_LOGIC;
020-   CLR: IN STD_LOGIC;
021-   PRE: IN STD_LOGIC;
022-   D: IN STD_LOGIC;
023-   Q: OUT STD_LOGIC
024- );
025- END COMPONENT;
026- SIGNAL aclr: STD_LOGIC;
027- SIGNAL cb: STD_LOGIC;
028- SIGNAL pre: STD_LOGIC;
029- BEGIN
030- aclr <= NOT(aclrb);      -- invert sense of clear
031- cb <= NOT(c);            -- invert clock to falling edge
032- pre <= '0';              -- disable preset
033- u0: FDPC PORT MAP (C=>cb, D=>d, CLR=>aclr, PRE=>pre,
Q=>q);
034- END bidff_arch;
035-
```

Listing 6.13 XNF for an XC9500 CPLD built-in D flip-flop with asynchronous clear.

```
001- LCANET, 5
002- SYM, Q, DFF, SCHNM=FDPC, LIBVER=2.0.0
003- PIN, D, I, D
004- PIN, C, I, C
005- PIN, PRE, I, PRE
006- PIN, CLR, I, CLR
007- PIN, Q, O, Q
008- END
009- EOF
```

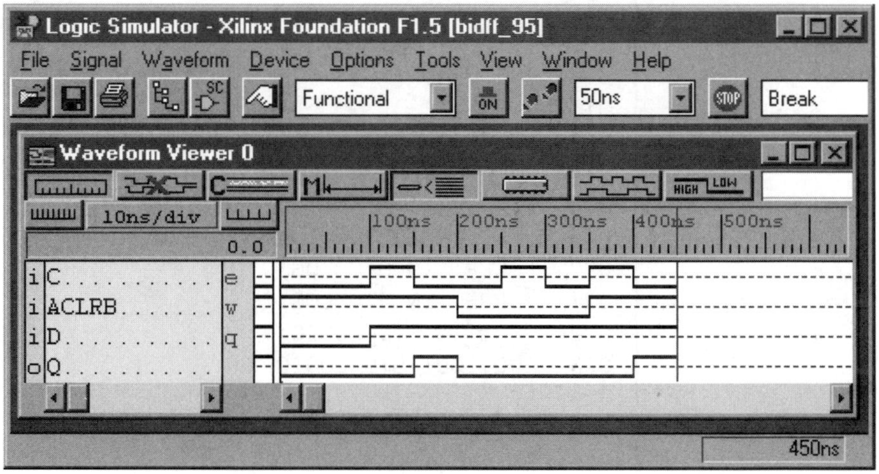

Figure 6.12 Simulated waveforms for a built-in D flip-flop.

After compiling and downloading the BIDFF design to an XS40 or XS95 Board, the following sequence of commands should give you the indicated results:

Command	ACLRB	D	C	LED S0 (SQ)
XSPORT 101	1	0	1	???
XSPORT 110	1	1	0	Bright (1)
XSPORT 010	0	1	0	Dark (0)
XSPORT 011	0	1	1	Dark (0)
XSPORT 010	0	1	0	Dark (0)

A Toggle Flip-Flop Counter

As our final example, Listing 6.14 shows a 3-bit counter built from toggle flip flops. We begin with line 7 which declares 3 DFFs T0, T1, and T2. These D flip-flops are transformed into toggle flip-flops by loading them with the inverse of their contents (line 11). Then the least-significant bit of the counter is clocked by positive-going edges of the input clock, C, on line 12. On line 13, the next bit of the counter is clocked by T0, the output of the LSB. In the same manner, the most-significant bit of the counter, T2, is clocked by T1 on line 14. Finally, on line 15 we provide an active-high asynchronous clear to each TFF so we can clear the counter at any time if needed.

Listing 6.14   ABEL code for a 3-bit counter built from toggle flip-flops.

```
001- MODULE TGLCNT
002- TITLE '3-bit toggle flip-flop counter'
003-
004- DECLARATIONS
005- C           PIN;                "clock input
006- ACLR        PIN;                "clear input
007- T2..T0      PIN ISTYPE 'REG'; "DFFs for counter
008-
009-
010- EQUATIONS
011- [T2..T0] := ![T2..T0]; "make DFFs into TFFs
012- T0.CLK = C;                "LS bit clocked by input clock
013- T1.CLK = T0;               "remaining bits clocked by output
014- T2.CLK = T1;               "of the preceding bit
015- [T2..T0].ACLR = ACLR;  "connect the asynchronous clear
016-
017- END TGLCNT
```

The equivalent VHDL code for the 3-bit counter is shown in Listing 6.15. The toggle flip-flops are implemented as 3 different processes. On the rising edge of the main clock input, the bit0 process (lines 27–35) toggles the t0 TFF output by loading it with its inverse (lines 32 and 33). The t0 output is the clock for the next bit of the counter described by the bit1 process (lines 37–45). And the t1 output from that bit is the clock for the most significant bit of the counter described by the bit2 process (lines 47–55). Each process will override the toggling and clear its bit if the asynchronous clear input is high.

The VHDL code in Listing 6.15 also has special modifications so it will run on the XS40 Board. When targeting an XC4000 FPGA, the Foundation Series synthesis software tries to use special clock buffers to drive clock signals to the flip-flops. These buffers minimize the skew between the times at which the clock edges reach the individual flip-flops. The Foundation Series Implementation tools also like to have the clock input come into the XC4000 FPGA through special pins which connect directly to these clock buffers. But for our example designs we use XSPORT to drive clock signals through the parallel port interface, and none of these pins directly connects to these special clock input pins. So we have to explicitly describe the connection of the clock input to one of the clock buffers. This is done on lines 24 and 25. On line 24, the main clock input is connected to an input buffer (IBUF) module. Then the output of the IBUF module is connected to the input of a general-purpose clock buffer (BUFG). The output of the BUFG, c_buf, becomes the clock input to the least-significant counter bit. (The IBUF and BUFG interfaces are described by the COMPONENT statements on lines 18–21.) You must use this clock input buffer circuitry when you are targeting the XC4000 FPGAs. For the XS95 Board and the XC9500 CPLDs, you can remove these buffers and use the clock input directly.

The simulated waveforms for this counter are shown in Figure 6.13. A high level on the ACLR input resets all the counter bits to zero. A series of rising edges on C moves the counter bits through the sequence 111 (7), 110 (6), 101 (5), 100 (4), 011 (3), 010 (2), 001 (1), 000 (0), 111 (7), and 110 (6). Then a logic 1 on the asynchronous clear input at 1.1 µs brings the counter back to 000 (0). All the features of the counter appear to be working.

A user-constraint file for the TGLCNT design should assign the following inputs and outputs for the XS40 Board:

```
NET c        LOC=P44;    # B0 argument of XSPORT
NET aclr     LOC=P45;    # B1 argument of XSPORT
NET t0       LOC=P25;    # S0 LED segment
NET t1       LOC=P26;    # S1 LED segment
NET t2       LOC=P24;    # S2 LED segment
```

Or use the following constraints for the XS95 Board:

```
NET c        LOC=P46;    # B0 argument of XSPORT
NET aclr     LOC=P47;    # B1 argument of XSPORT
NET t0       LOC=P21;    # S0 LED segment
NET t1       LOC=P23;    # S1 LED segment
NET t2       LOC=P19;    # S2 LED segment
```

For this experiment, the clock input, C, must be assigned to either B0 or B1. These inputs pass through a Schmitt-trigger buffer on the XS40 and XS95 Boards that removes any spurious glitches caused by the slow transitions of the PC parallel port signals. Using any of the B2-B7 inputs may cause the counter to decrement erratically.

Listing 6.15 VHDL code for a 3-bit counter built from toggle flip-flops.

```
001- -- Toggle Flip-Flop Counter
002-
003- LIBRARY IEEE;
004- USE IEEE.std_logic_1164.ALL;
005-
006- ENTITY tglcnt IS
007- PORT
008- (
009-    c: IN STD_LOGIC;        -- clock
010-    aclr: IN STD_LOGIC;     -- asynchronous active-high clear
011-    t0: BUFFER STD_LOGIC;  -- counter bit 0
012-    t1: BUFFER STD_LOGIC;  -- counter bit 1
013-    t2: BUFFER STD_LOGIC   -- counter bit 2
014- );
015- END tglcnt;
016-
017- ARCHITECTURE tglcnt_arch OF tglcnt IS
018- COMPONENT IBUF PORT(I:IN STD_LOGIC;
019-                     O:OUT STD_LOGIC); END COMPONENT;
020- COMPONENT BUFG PORT(I:IN STD_LOGIC;
021-                     O:OUT STD_LOGIC); END COMPONENT;
022- SIGNAL c_in, c_buf: STD_LOGIC;-- buffered clock signals
023- BEGIN
024- u0: IBUF PORT MAP(I=>c,O=>c_in); -- bring in clock through IBUF
025- u1: BUFG PORT MAP(I=>c_in,O=>c_buf); -- route it to clock buffer
026-
027- bit0: PROCESS (c_buf,t0,aclr)   -- process for counter LSB
028- BEGIN
029- IF aclr='1' THEN   -- async clear bit if clear is high
030-    t0 <= '0';
031- -- toggle the bit on the rising edge of the clock
032- ELSIF (c_buf'event AND c_buf='1') THEN
033-    t0  <= NOT t0;
034- END IF;
035- END PROCESS bit0;
036-
037- bit1: PROCESS (t0,t1,aclr)   -- process for middle counter bit
038- BEGIN
039- IF aclr='1' THEN
040- t1 <= '0';
```

Listing 6.15 VHDL code for a 3-bit counter built from toggle flip-flops.

```
041-  -- toggle this bit when the LSB toggles from 0 to 1
042-  ELSIF (t0'event AND t0='1') THEN
043-     t1   <= NOT t1;
044-  END IF;
045-  END PROCESS bit1;
046-
047-  bit2: PROCESS (t1,t2,aclr)   -- process for counter MSB
048-  BEGIN
049-  IF aclr='1' THEN
050-     t2 <= '0';
051-  -- toggle the MSB when the middle bit toggles from 0 to 1
052-  ELSIF (t1'event AND t1='1') THEN
053-     t2   <= NOT t2;
054-  END IF;
055-  END PROCESS bit2;
056-
057-  END tglcnt_arch;
058-
```

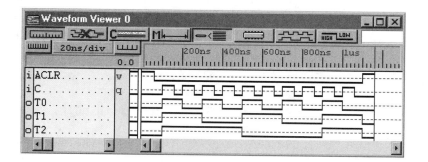

Figure 6.13 Simulated waveforms 3-bit counter built from toggle flip-flops.

The following sequence of commands will step the downloaded design through the same set of states as the simulation:

Command	ACLR	C	LED S2 (T2)	LED S1 (T1)	LED S0 (T0)
XSPORT 10	1	0	Dark (0)	Dark (0)	Dark (0)
XSPORT 00	0	0	Dark (0)	Dark (0)	Dark (0)
XSPORT 01	0	1	Bright (1)	Bright (1)	Bright (1)
XSPORT 00	0	0	Bright (1)	Bright (1)	Bright (1)
XSPORT 01	0	1	Bright (1)	Bright (1)	Dark (0)
XSPORT 00	0	0	Bright (1)	Bright (1)	Dark (0)
XSPORT 01	0	1	Bright (1)	Dark (0)	Bright (1)
XSPORT 00	0	0	Bright (1)	Dark (0)	Bright (1)
XSPORT 01	0	1	Bright (1)	Dark (0)	Dark (0)
XSPORT 00	0	0	Bright (1)	Dark (0)	Dark (0)
XSPORT 01	0	1	Dark (0)	Bright (1)	Bright (1)
XSPORT 00	0	0	Dark (0)	Bright (1)	Bright (1)
XSPORT 01	0	1	Dark (0)	Bright (1)	Dark (0)
XSPORT 00	0	0	Dark (0)	Bright (1)	Dark (0)
XSPORT 01	0	1	Dark (0)	Dark (0)	Bright (1)
XSPORT 00	0	0	Dark (0)	Dark (0)	Bright (1)
XSPORT 01	0	1	Dark (0)	Dark (0)	Dark (0)
XSPORT 00	0	0	Dark (0)	Dark (0)	Dark (0)
XSPORT 01	0	1	Bright (1)	Bright (1)	Bright (1)
XSPORT 00	0	0	Bright (1)	Bright (1)	Bright (1)
XSPORT 01	0	1	Bright (1)	Bright (1)	Dark (0)
XSPORT 00	0	0	Bright (1)	Bright (1)	Dark (0)
XSPORT 10	1	0	Dark (0)	Dark (0)	Dark (0)

Projects

1. Change the edge-triggered DFF so it has an asynchronous preset input that sets it to a logic 1.

2. Do a unit-delay simulation of the edge-triggered DFF. Do you see where the setup and hold times arise?

3. Use the schematic editor to re-create the toggle counter. (The D flip-flop is symbol **FD** in the **SC Symbols** window.)

4. Build a shift-register consisting of a string of seven D flip-flops, each of whose output attaches to the input of the next flip-flop. All the flip-flops receive the same clock. Create a user-constraints file for the XS40 or XS95 Board that assigns the clock input and the input of the initial D flip-flop in the string to pins that are driven by B0 and B1 of the XSPORT command. Also assign the outputs of the flip-flops to pins that drive the LED segments. Download the circuit to your board and use XSPORT to display different patterns on the LED segments. Could you use the outputs of the shift-register flip-flops to drive logic inputs to other circuits? If so, does the shift register give you a way to test other circuits with more than eight inputs that exceed the I/O capabilities of the PC parallel port?

5. Build a counter that "walks" around the LED digit by lighting each segment one at a time.

State Machine Design

Objectives

- Introduce the concepts of state machines.
- Discuss the different types of state machines.
- Build counters using D flip-flops and combinational logic.
- Show how a state machine is designed to solve a real problem.

Discussion

Simple State Machines

In the last chapter, you saw that you could build logic circuits that store previous inputs by altering their internal state. You also used these flip-flops to build a simple counter.

A simple state machine uses one or more flip-flops to store its internal state (Figure 7.1). The pattern of ones and zeros on the Q outputs of the flip-flops are the current state of the state machine. When a rising clock edge occurs, the current state is replaced by the next state. How is the next state determined? An easy way is to build a combinational logic circuit that accepts the current state as its input and computes the next state as its output. The output of the combinational logic is loaded into the flip-flops on the next rising clock edge. What is a good example of such a state machine? A counter! The current value of the count is held in the DFFs, and this value is passed to a combinational circuit that is just a simple incrementer. The output of the incrementer is clocked into the DFFs on the next rising clock edge.

So we already know everything we need to build state machines:

- We know how to build combinational logic circuits.
- We know how to store state information in D flip-flops.
- We know how to connect the combinational logic and the DFFs.

A state machine that only uses its current state to compute its next state is not very interesting (kind of like a person who only talks to himself or herself). A more useful variation allows inputs from the environment to affect the state transitions (see Figure 7.2). The external inputs interact inside the combina-

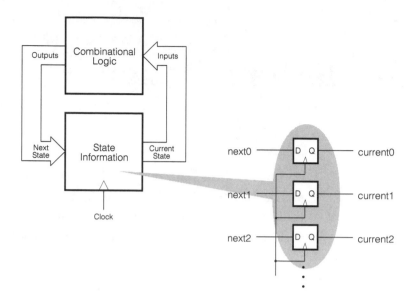

Figure 7.1   A simple state machine.

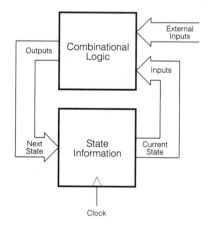

Figure 7.2   A state machine with external inputs.

tional logic with the current state to determine the next state. An example of this type of state machine is a loadable counter. Such a counter allows you to force a value into the counter so that the counting will proceed from this value. The external inputs let you pass in a new counter value and a load signal that indicates when the new value should be loaded.

The current state of the state machine can be used directly as an output (the counter value, for example). But often you would like to process the current

State Machine Design   Chapter 7

state to create a different encoding for the output (see Figure 7.3). Another combinational logic circuit generates the new external outputs based on the current state. A state machine with this type of configuration is called a Moore machine. A counter whose output is passed to an LED decoder is an example of a Moore machine. The LED decoder will light an LED digit to display the counter's value in a more understandable form.

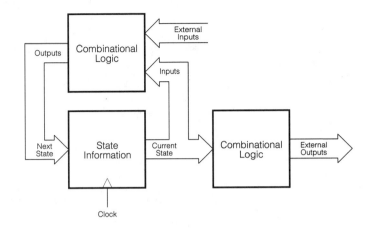

Figure 7.3   A Moore state machine.

As a final change to make the state machine even more flexible, you can pass external inputs directly into the combinational logic that drives the external outputs (see Figure 7.4). Now the external outputs are affected by both the current state and the external inputs. This type of configuration is called a Mealy machine. A Mealy machine is more general than a Moore machine. An example of a Mealy Machine is a counter whose output is passed to an LED decoder that can be blanked if the user wants to sleep. The blanking input goes directly to the LED decoder and disables the LED drivers, thus blanking the LEDs. Note that the blanking input affects the external outputs, but it has no affect on the current or next state (e.g., a clock does not forget what time it is just because the digits are not showing).

These are the basic state machine organizations, but how do you actually design a state machine to solve a specific problem? There are several methods (at least!), but Figure 7.5 shows one that is based on our standard logic design flow with the following modifications:

- In the second step of the design flow, all the possible states of the state machine are listed in addition to the circuit's inputs and outputs.
- A new step is added that assigns each state with a binary pattern that can be stored in a set of flip-flops.

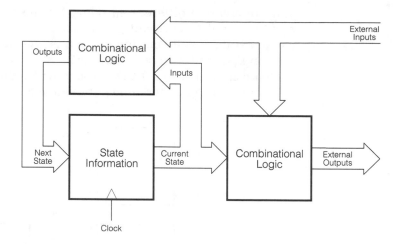

Figure 7.4 A Mealy state machine.

- Another new step is inserted in which all the possible movements from each current state to the next state are listed.

- The third and fourth steps are broken into two separate paths. One path develops the output truth table that describes what the outputs will be for each possible combination of inputs and current states of the circuit. The parallel path generates the state transition truth table that lists the next state of the state machine for each possible combination of inputs and current states.

Defining the states for a design requires you to examine the specifications and list all the possible states the machine could get into. As a simple example, suppose you were asked to design a counter that counts in the sequence 0, 1, 2, 0, 1, 2, ... You would see immediately that there were only three distinct states for the counter.

In more complicated designs, state definition typically proceeds in a hierarchical fashion: you determine what the major states of the machine are and then break each major state into a set of related states. For example, you might list the major states for a laser printer controller as "printing" and "not-printing." Upon closer examination, you might divide the "not-printing" state into states relating to initialization of the printer electronics, periodic warming of the toner-transfer roller, diagnostic checks, and just idling. Likewise, the "printing" state could be partitioned into states concerned with transferring data into a buffer, activating the paper feed mechanism, writing a pattern to the electrostatic roller, etc. Then you could take each of these more-detailed states and break them into a set of even more-detailed states. The process ends when you cannot break any state into a set of substates. For example, the state

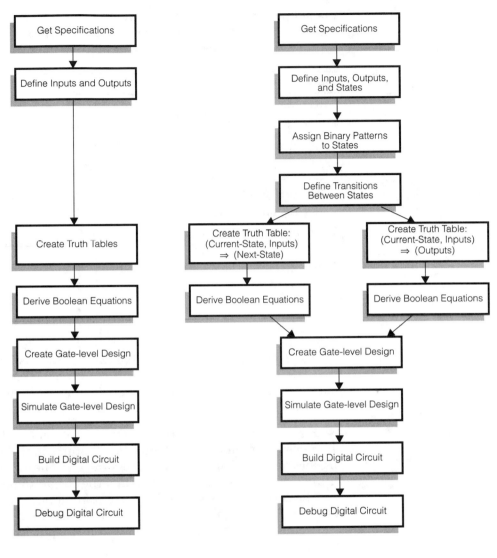

Figure 7.5 The logic design flow modified for state machine designs.

in which the toner-transfer roller is heated may not need to be expanded further because it just applies current to a heater coil until a time-out occurs.

After you determine what the states are, you have to assign a distinct binary pattern to each state. A set of D flip-flops in the state machine stores these patterns. By examining the contents of the DFFs, we can determine what state the circuit is in. For a counter, the assignment of binary patterns to states is usually simple: Just use the binary value of the counter as the state. For other designs, the binary patterns can be more arbitrary. For the laser printer example, you might use the following state assignment:

State	Binary Pattern
Initialization	000
Diagnostics	001
Toner-transfer warming	010
Idle	011
Data transfer to buffer	100
Paper feed activation	101
Write electrostatic pattern	110

If you have N total states, you will need at least n flip-flops such that $2^n \geq N$. For the printer example, $N = 7$ so $n \geq 3$. If we choose $n = 3$, then we are using a minimal encoding for the states because it uses the smallest possible number of flip-flops.

Once you have the states defined, you have to determine how your machine will move from state to state as the inputs change. You also need to specify what the values on the outputs will be at each state. You can do this with "bubble diagrams," as shown in Figure 7.6, but that soon becomes unmanageable when the number of states and possible transitions gets large.

As an alternative, you can use state transition tables like the one in Table 7.1. State transition tables list all the possible states in the leftmost column and all possible combinations of inputs across the top. The entry at row i and column j within the table lists the next state and the output that will be generated given that the current state is i and the current input pattern is j. For example, when the counter is in state 1 and the reset is not active (reset = 0) and the counter is incrementing (up = 1), then the current output is 01 and the next state is state 2.

The advantage of the state transition table is that it forces you to account for the actions of your state machine for each combination of current state and input pattern. Empty locations in the table immediately make it apparent that there is something you have forgotten to specify. The problem is that the table gets large, especially when the number of inputs increases. Still, it's a reasonable method for the problems we will handle here.

With a completed state transition table in hand, we can proceed to make a truth table that describes how the inputs and current state affect the outputs and next state. The truth table will have a row for each entry in the state transition table. The inputs for each row of the truth table are the current state and

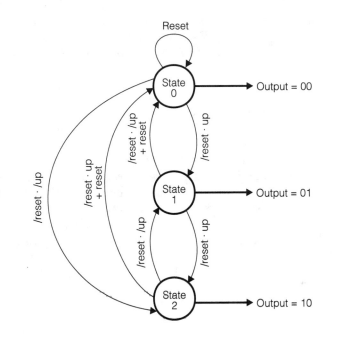

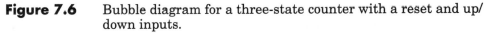

Figure 7.6   Bubble diagram for a three-state counter with a reset and up/down inputs.

input combination from the row and column containing the state transition table entry. The outputs for each row of the truth table are just the next state and output from the state transition table entry.

The complete truth table in Table 7.2 was generated from the state transition table in Table 7.1. The state can be stored using two flip-flops, q1 and q2, and we have used an obvious state assignment (state 0 is 00, state 1 is 01, and state 2 is 10). As an example, on row 6 the machine is in state 1 (q1 = 0, q0 = 1), the

Table 7.1   State transition table for the 3-state counter with reset and up/down inputs.

	Inputs			
Current State	**reset = 0 up = 0**	**reset = 0 up = 1**	**reset = 1 up = 0**	**reset = 1 up = 1**
state 0	state2, output = 00	state1, output = 00	state0, output = 00	state0, output = 00
state 1	state0, output = 01	state2, output = 01	state0, output = 01	state0, output = 01
state 2	state1, output = 10	state0, output = 10	state0, output = 10	state0, output = 10

reset is not asserted (reset = 0), and the counter is in the incrementing mode (up = 1). For this combination of the current state and inputs, the next state will be state 2 (q1=1, q0=0) and the outputs are identical to the current state (out1=0, out0=1). You can interpret the other rows of Table 7.2 in the same way.

Also notice that the original state transition table only had 12 entries, but the truth table has 16 rows. That is because the two flip-flops can store a total of four distinct states even though we only need three of them. But what happens if our state machine gets into the unused state with q1 = 1 and q0 = 1? This is obviously an error since our original state transition table makes no mention of a fourth state. But errors happen, especially in the real world, where voltage and current transients can upset the operation of a circuit. Our job is to make sure that we recover from these errors. The last four rows of the truth table perform error recovery by returning the state machine to state 0 (q1 = q0 = 0) no matter what combination of the inputs occurs. You could also output some distinct pattern on out1 and out0 to indicate an error occurred for diagnostic purposes. (We have not done that in this example.)

Table 7.2 Truth table for state transitions and outputs for the three-state counter.

Current State		Inputs		Next State		Outputs	
q1	q0	reset	up	q1	q0	out1	out0
0	0	0	0	1	0	0	0
0	0	0	1	0	1	0	0
0	0	1	0	0	0	0	0
0	0	1	1	0	0	0	0
0	1	0	0	0	0	0	1
0	1	0	1	1	0	0	1
0	1	1	0	0	0	0	1
0	1	1	1	0	0	0	1
1	0	0	0	0	1	1	0
1	0	0	1	0	0	1	0
1	0	1	0	0	0	1	0
1	0	1	1	0	0	1	0
1	1	0	0	0	0	0	0
1	1	0	1	0	0	0	0
1	1	1	0	0	0	0	0
1	1	1	1	0	0	0	0

Now that we have a truth table, we can derive the Boolean equations and gate-level diagrams for the combinational logic that generates the outputs and the next state. Or we can let the software do it for us. Either way, it is something we have done quite a few times before, so there's nothing new from this point on.

You might be concerned about the complexity of the logic circuitry needed to generate the next state. For complicated state machines, the state transition circuitry can consume a lot of gates with large fan-ins, and there can be an appreciable delay before the next state is available. This is especially true if we use minimal encodings for our states.

An alternative to minimal state encoding is called one-hot encoding. With one-hot encoding, a machine with N states will have N flip-flops. Only one flip-flop will store a logic 1 in each state; the rest will be reset to logic 0. For our three-state counter example, the state assignments with one-hot encoding would be

State	One-hot Pattern
state 0	001
state 1	010
state 2	100

It is always easy to tell what state a one-hot state machine is in: Just look for the flip-flop that has a 1 stored in it. This simple state encoding also makes the next-state generation circuitry smaller and faster and reduces its fan-in. The trade-off is that you need a lot of flip-flops with one-hot encodings. A 128-state machine needs 128 flip-flops with one-hot encoding, but only seven flip-flops if you use a minimal encoding.

FPGAs, like the XC4000 series, have hundreds to thousands of flip-flops but their LUTs have a limited fan-in of four or five inputs. One-hot encodings are a good match for FPGA architectures, especially if high circuit speeds are needed. CPLDs, like the XC95108, do not have a large number of flip-flops but they do have large amounts of fast sum-of-product circuits with wide fan-ins. That makes state machines with minimal encodings the best fit for CPLDs.

So far, we have discussed state machines that use a synchronous design philosophy: All state transitions occur on a single edge of a global master clock. Generally, this is a safe design approach. But can you get in trouble even when building a strictly synchronous design? Of course! The majority of your problems will come from violating setup and hold times for flip-flops. The basic constraints are shown in Figure 7.7. We need to get the current state from one or more flip-flops to some combinational logic that generates the next state. Then the next state has to arrive at the inputs of the flip-flops so it can be loaded in

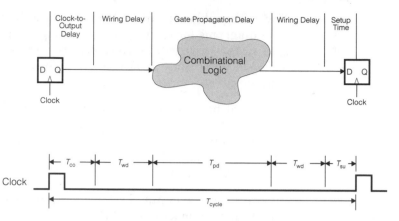

Figure 7.7   Timing concerns for synchronous state machines.

by the next clock edge. There are various time delays we need to account for in this trip:

T_{co}: This is the delay between the arrival of the clock edge and the change in the output of a flip-flop. This is somewhat analogous to propagation delay for combinational logic. (Clock-to-out delay)

T_{wd1}: This is the wiring delay that is introduced as the output of the flip-flop makes its way though the wiring and switch matrices of an FPLD. (Wiring delay 1)

T_{pd}: This is combinational gate propagation delay for the logic that computes the next state. We are already familiar with this. (Gate propagation delay)

T_{wd2}: More wiring delay as the next-state output from the combinational logic makes it to the inputs of the flip-flops. This is not necessarily the same duration as the previous wiring delay. (Wiring delay 2)

T_{su}: The next state has to be present on the inputs of the flip-flop for at least the setup time before the next clock edge arrives. Otherwise we might load garbage into the flip-flops or get into a metastable state. (Setup time)

The sum of all these times must be less than or equal to the cycle time of the master clock:

$$T_{co} + T_{wd1} + T_{pd} + T_{wd2} + T_{su} \leq T_{cycle}$$

If we satisfy this equation, we will not have setup violations. This inequality determines how fast our state machine can operate. Assuming we have one layer of combinational logic feeding flip-flops in another CLB and ignoring routing-dependent wiring delays, the inequality for the XC4005XL-3 FPGA is

$$1.9 \text{ ns} + 3.2 \text{ ns} + 1.2 \text{ ns} \leq T_{\text{cycle}}$$

So our clock frequency cannot exceed $1/T_{\text{cycle}} = 159$ MHz. That seems pretty fast, but wiring delay can easily add another 5 to 10 ns, which would bring the speed below 100 MHz.

We also have to make sure the inputs to the flip-flops hang around after the clock edge for a duration longer than the minimum hold time. This requirement is expressed by this inequality:

$$T_{\text{co}} + T_{\text{wd1}} + T_{\text{pd}} + T_{\text{wd2}} \geq T_h$$

The hold-time requirement is easy to satisfy since T_h is 0 ns for both the XC95108 and the XC4005XL devices.

Our timing analysis is valid but it is based on the assumption that all the flip-flops get the clock edge at exactly the same time. But the wiring to the clock inputs of different flip-flops can introduce different delays, so all the clock edges will not stay lined up (Figure 7.8). The difference in time between the clock edges is termed clock skew, and we will denote it by T_{skew}. Now we can modify our inequalities to account for the clock skew:

$$T_{\text{co}} + T_{\text{wd1}} + T_{\text{pd}} + T_{\text{wd2}} + T_{\text{su}} + T_{\text{skew}} \leq T_{\text{cycle}}$$

$$T_{\text{co}} + T_{\text{wd1}} + T_{\text{pd}} + T_{\text{wd2}} - T_{\text{skew}} \geq T_h$$

In general, we cannot be sure if the clock skew will be positive or negative for a given pair of flip-flops. But we assume clock skew is always going to make

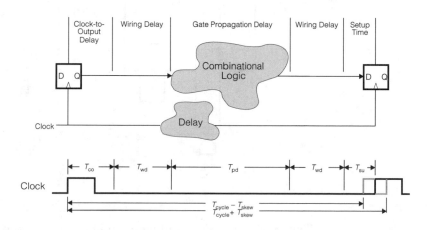

Figure 7.8   Clock skew.

things worse for us. So we assume the clock skew reduces the time we have to set up our inputs (negative clock skew) and that the skew also increases the time we need to hold the inputs steady (positive clock skew). The major effect of clock skew is that it forces us to increase T_{cycle} and reduces the maximum clock frequency for our design.

Clock skew in FPLDs arises mainly from routing delays as the clock signal passes through the various switching circuits. For this reason, FPLDs include global clock lines that reach all the flip-flops without going through the general-purpose routing switches. Clock edges on the global lines have very low skew. The XC95108 CPLD has three global clock inputs, and the XC4005XL has eight (four global clock inputs with minimum skew and four secondary global clock inputs with slightly more skew).

You can also introduce clock skew by choices you make in your design. For example, you might need a counter that only counts at certain times. You could build it using a standard counter whose clock is turned off when the counter is disabled (Figure 7.9a), or you could continually clock the counter but only allow it to load the incremented value when it is enabled (Figure 7.9b). The first option adds more skew to your design because the clock to this counter is delayed by having to pass through the AND gate. The second option does not add clock skew but does require more logic to build the multiplexer into the counter. The second option is the safest from a reliability standpoint. The first option is useful if you are running out of gates in your FPLD, the design does not have to run too fast, and you are willing to exhaustively check all the potential timing violations.

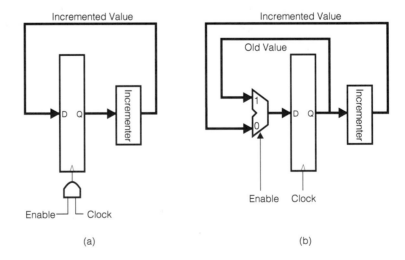

(a)   (b)

Figure 7.9   Skew-inducing and no-skew methods for building a counter with enable.

Experimental

An Up/Down Counter

Since we have already done all the design work, we might as well build the 3-state up/down counter. The ABEL code for it is shown in Listing 7.1.

The 2 DFFs for storing the state information, Q1 and Q0, are declared on line 9. The clock input is attached to them on line13.

Lines 14–31 hold the truth-table for the state machine. Line 15 lists the inputs and outputs for the truth-table logic. Note that Q1 and Q0 act both as inputs and outputs because the current state information is needed to determine the next state. Note also the use of two different types of output-list delimiters. We have already seen the -> delimiter used to denote the start of a list of combinatorial outputs. But now the :> delimiter is added to signal the start of a list of registered outputs. In this case, the registered outputs are the Q1 and Q0 flip-flop outputs. The :> delimiter indicates that the Q1 and Q0 outputs will only change to the values listed in the table when the flip-flops receive a rising clock edge (in this case). That is just what we want for a state machine.

Each of the remaining lines in the truth-table just replicates what we have already seen in Table 7.2.

The VHDL version of the 3-state up/down counter is given in Listing 7.2. On line 23, I define two 2-element signal arrays: one for holding the current state (curr_state) and another for holding the next state (next_state). Within the main body of the ARCHITECTURE section there are 2 processes that make use of these arrays. The first process takes the current state along with the reset and up/down control inputs and computes the value for the next state. For example, if the reset is active then the next state is set to all zeroes on line 37. In another case, if the reset is not active, the current state is 1 (curr_state=01) and the counter is counting up (up=1), then line 43 sets the next state to be 2 (next_state<=10). Note that this process is entirely combinational: the next state is not latched by this process. If there were other combinational outputs that were a function of the current state, we could place their equations into the IF...THEN...ELSE statements of this process. (In this example the outputs are actually placed outside the process on lines 29 and 30 since they are simple copies of the current state array elements.)

The second process on lines 55 through 60 just copies the state value held in the next_state array into the curr_state array on a rising edge of the clock. Thus, this process moves the state machine from the current state to the next state. It is in this process where the actual state latches are implied. Also, the next state is affected by the reset input in the first process, but does not take effect until the second process updates the current state. This makes the reset synchronous.

Listing 7.1   ABEL code for a 3-state up/down counter.

```
001- MODULE CNT3
002- TITLE '3-state counter with reset'
003-
004- DECLARATIONS
005- C          PIN;              "clock input
006- RESET      PIN;              "reset input
007- UP         PIN;              "count direction input
008- OUTP1..OUTP0 PIN ISTYPE 'COM'; "outputs
009- Q1..Q0     PIN ISTYPE 'REG'; "two DFFs for counter
010-
011- EQUATIONS
012-
013- [Q1..Q0].CLK = C; "connect input clock to counter DFFs
014- TRUTH_TABLE
015- ([Q1,Q0,RESET,UP] :> [Q1,Q0] -> [OUTP1,OUTP0])
016-  [ 0, 0,  0,   0 ] :> [1,  0] -> [ 0,    0 ]; "down : 0->2
017-  [ 0, 0,  0,   1 ] :> [0,  1] -> [ 0,    0 ]; "up   : 0->1
018-  [ 0, 0,  1,   0 ] :> [0,  0] -> [ 0,    0 ]; "reset: 0->0
019-  [ 0, 0,  1,   1 ] :> [0,  0] -> [ 0,    0 ]; "reset: 0->0
020-  [ 0, 1,  0,   0 ] :> [0,  0] -> [ 0,    1 ]; "down : 1->0
021-  [ 0, 1,  0,   1 ] :> [1,  0] -> [ 0,    1 ]; "up   : 1->2
022-  [ 0, 1,  1,   0 ] :> [0,  0] -> [ 0,    1 ]; "reset: 1->0
023-  [ 0, 1,  1,   1 ] :> [0,  0] -> [ 0,    1 ]; "reset: 1->0
024-  [ 1, 0,  0,   0 ] :> [0,  1] -> [ 1,    0 ]; "down : 2->1
025-  [ 1, 0,  0,   1 ] :> [0,  0] -> [ 1,    0 ]; "up   : 2->0
026-  [ 1, 0,  1,   0 ] :> [0,  0] -> [ 1,    0 ]; "reset: 2->0
027-  [ 1, 0,  1,   1 ] :> [0,  0] -> [ 1,    0 ]; "reset: 2->0
028-  [ 1, 1,  0,   0 ] :> [0,  0] -> [ 0,    0 ]; "error->0
029-  [ 1, 1,  0,   1 ] :> [0,  0] -> [ 0,    0 ]; "error->0
030-  [ 1, 1,  1,   0 ] :> [0,  0] -> [ 0,    0 ]; "error->0
031-  [ 1, 1,  1,   1 ] :> [0,  0] -> [ 0,    0 ]; "error->0
032-
033- END CNT3
```

The VHDL code in Listing 7.2 also has special modifications so it will run on the XS40 Board. When targeting an XC4000 FPGA, the Foundation Series synthesis software tries to use special clock buffers to drive clock signals to the flip-flops. These buffers minimize the skew between the times at which the clock edges reach the individual flip-flops. The Foundation Series Implementation tools also like to have the clock input come into the XC4000 FPGA through special pins which connect directly to these clock buffers. But for our example designs we use XSPORT to drive clock signals through the parallel port interface, and none of these pins directly connects to these special clock input pins.

So we have to explicitly describe the connection of the clock input to one of the clock buffers. This is done on lines 26 and 27. On line 26, the main clock input is connected to an input buffer (IBUF) module. Then the output of the IBUF module is connected to the input of a general-purpose clock buffer (BUFG) on line 27. The output of the BUFG, buf_c, becomes the clock that drives the state updating process. (The IBUF and BUFG interfaces are described by the COMPONENT statements on lines 18–21.) You must use this clock input buffer circuitry when you are targeting the XC4000 FPGAs. For the XS95 Board and the XC9500 CPLDs, you can remove these buffers and use the clock input directly.

The simulated waveforms for the up/down counter are shown in Figure 7.10. The RESET input is activated from 0–10 ns and resets the [Q1..Q0] state vector to 0. When RESET is released, the counter begins to count down since UP=0. The counter progresses in the sequence 0, 2, 1, 0, and rolls-over back to 2 over the period 10–50 ns. Then we raise the direction control so UP=1 at 50 ns. This makes the counter increment over the 50–100 ns period. Finally, we raise RESET again at 100 ns. Since it is a synchronous reset, the counter moves from state 1 to 0 only at the next rising clock edge at 106 ns. The state machine seems to be doing what we want.

A constraint file for the CNT3 design should assign the inputs and outputs like this for the XS40 Board:

```
NET c        LOC=P44;    # B0 argument of XSPORT
NET reset    LOC=P45;    # B1 argument of XSPORT
NET up       LOC=P46;    # B2 argument of XSPORT
NET q0       LOC=P25;    # S0 LED segment
NET q1       LOC=P26;    # S1 LED segment
NET outp0    LOC=P24;    # S2 LED segment
NET outp1    LOC=P20;    # S3 LED segment
```

Or use the following constraints for the XS95 Board:

```
NET c        LOC=P46;    # B0 argument of XSPORT
NET reset    LOC=P47;    # B1 argument of XSPORT
NET up       LOC=P48;    # B2 argument of XSPORT
NET q0       LOC=P21;    # S0 LED segment
NET q1       LOC=P23;    # S1 LED segment
NET outp0    LOC=P19;    # S2 LED segment
NET outp1    LOC=P17;    # S3 LED segment
```

The clock input must be assigned to either B0 or B1. These inputs pass through a Schmitt-trigger buffer on the XS40 or XS95 Board that removes any spurious glitches caused by the slow transitions of the PC parallel port signals. Using any of the B2-B7 bits may cause the counter to decrement erratically. This will be true for any sequential design we do.

Listing 7.2 VHDL code for a 3-state up/down counter.

```
001- -- 3-state up/down counter
002-
003- LIBRARY IEEE;
004- USE IEEE.std_logic_1164.ALL;
005-
006- ENTITY cnt3 IS
007- PORT
008-    (
009-    c: IN STD_LOGIC;        -- clock input
010-    reset: IN STD_LOGIC;    -- active-high reset
011-    up: IN STD_LOGIC;       -- up/down control input
012-    outp0, outp1: OUT STD_LOGIC;-- outputs
013-    q0,q1: OUT STD_LOGIC    -- state outputs
014-    );
015- END cnt3;
016-
017- ARCHITECTURE cnt3_arch OF cnt3 IS
018- COMPONENT IBUF PORT(I: IN STD_LOGIC;
019-                        O: OUT STD_LOGIC); END COMPONENT;
020- COMPONENT BUFG PORT(I: IN STD_LOGIC;
021-                        O: OUT STD_LOGIC); END COMPONENT;
022- SIGNAL in_c, buf_c: STD_LOGIC;
023- SIGNAL curr_state, next_state: STD_LOGIC_VECTOR (1 DOWNTO 0);
024- BEGIN
025-
026- u0: IBUF PORT MAP(I=>c, O=>in_c);-- connect clock to
027- u1: BUFG PORT MAP(I=>in_c, O=>buf_c);-- clock buffer
028-
029- outp0 <= curr_state(0);  -- connect outputs to current state
030- outp1 <= curr_state(1);
031- q0 <= curr_state(0);      -- connect state outputs to current state
032- q1 <= curr_state(1);
033-
034- PROCESS(reset,up,curr_state)-- compute next state from current
035- BEGIN
036- IF reset='1' THEN
037-    next_state <= "00";
038- ELSIF curr_state="00" THEN
039-    IF up='1' THEN next_state <= "01";
040-    ELSE next_state <= "10";
041-    END IF;
042- ELSIF curr_state="01" THEN
```

Listing 7.2   VHDL code for a 3-state up/down counter. (Cont'd.)

```
043-    IF up='1' THEN next_state <= "10";
044-    ELSE next_state <= "00";
045-    END IF;
046- ELSIF curr_state="10" THEN
047-    IF up='1' THEN next_state <= "00";
048-    ELSE next_state <= "01";
049-    END IF;
050- ELSE
051-    next_state <= "00";
052- END IF;
053- END PROCESS;
054-
055- PROCESS(buf_c,next_state)-- update the current state vector
056- BEGIN
057- IF (buf_c'event AND buf_c='1') THEN
058-    curr_state <= next_state;
059- END IF;
060- END PROCESS;
061-
062- END cnt3_arch;
```

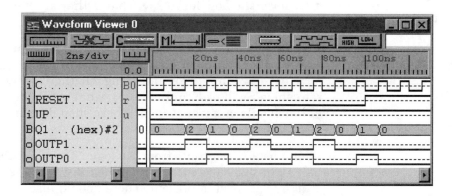

Figure 7.10   Simulated waveforms for the 3-state up/down counter.

After downloading your circuit to the XS40 or XS95 Board, you can use the following command sequences to move the state machine between its various states. Each set of commands activates a particular input and then sends a rising clock edge to the state machine. You can read off the current state by observing the S0 and S1 LED segments.

Action	XSPORT Command Sequence
Reset the counter	XSPORT 010 XSPORT 011 XSPORT 000
Increment by one	XSPORT 100 XSPORT 101 XSPORT 000
Decrement by one	XSPORT 000 XSPORT 001 XSPORT 000

A Drink Dispenser Controller

This example will show you how to design a state machine to solve an actual problem: dispensing soft-drinks.

You have been contracted to design the logic for a soft-drink machine that dispenses two brands of cola: Swill and Krunk. The dispenser should accept nickels and dimes and should allow the buyer to select one of the soft-drinks after he or she has put in 25¢. That is the extent of the requirements.

The first step is to get some idea of the inputs and outputs of the logic circuitry. From the requirements, you know that the logic must receive inputs that indicate whether the buyer has put in a nickel or a dime. You also know that there have to be two inputs for indicating whether the buyer wants to enjoy a can of refreshing Swill or Krunk. Finally, what if the buyer changes his or her mind in the middle of a transaction? There should be an input to return the buyer's money.

What about the outputs from the logic? There should be outputs to control the individual dispensers for Swill and Krunk so that a can is dispensed whenever a high logic level is present. There should also be two outputs to control the return of change: one that causes a nickel to be dropped into the change return slot, and another that drops a dime.

OK, you seem to have an initial idea of what the I/O looks like. The next step is to determine the states of your state machine (i.e., what your logic has to remember). The obvious thing to remember is how much money the buyer has put into the dispenser. This machine dispenses a soft-drink after it receives 25¢ in either nickels or dimes, so the possible states for the machine are that it has received 0¢, 5¢, 10¢, 15¢, 20¢, or 25¢. That is a total of 6 states. At this point, the drink dispenser controller looks like Figure 7.11.

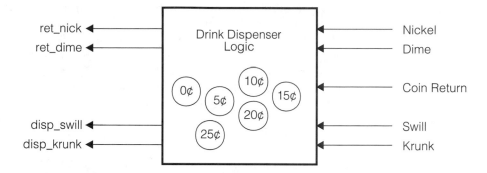

Figure 7.11 **Figure 7.11**: I/O and internal states for the drink dispenser controller.

Once you have the states and I/O, you have to decide how the logic will move from state to state under the influence of the inputs and what the outputs will do in each case. That is listed in the state transition table of Table 7.3.

Table 7.3   State transition table for the drink dispenser controller.

Current State	Input					
	Nickel	**Dime**	**Coin Return**	**Swill**	**Krunk**	**Nothing**
0¢	→5¢	→10¢	→0¢	→0¢	→0¢	→0¢
5¢	→10¢	→15¢	→0¢, ret_nick = 1	→5¢	→5¢	→5¢
10¢	→15¢	→20¢	→0¢, ret_dime = 1	→10¢	→10¢	→10¢
15¢	→20¢	→25¢	→5¢, ret_dime = 1	→15¢	→15¢	→15¢
20¢	→25¢	→25¢, ret_nick = 1	→10¢, ret_dime = 1	→20¢	→20¢	→20¢
25¢	→25¢, ret_nick = 1	→25¢, ret_dime = 1	→15¢, ret_dime = 1	→0¢, disp_swill = 1	→0¢, disp_krunk = 1	→25¢

The entries in the table are pretty simple. For example, when the state machine is in state 5¢ and the buyer puts in a dime, the state machine moves to the state 15¢. A more complicated entry occurs when the machine is in state 20¢ and the buyer puts in a dime. In this case, the state machine moves to the 25¢ state and also activates the output which returns a nickel to the buyer, thus giving change for the dime that was put in. After the state machine is in the 25¢ state, either the Swill or Krunk input will cause the machine to dispense a can of the selected drink and the state machine will return to the 0¢ state because the money has been exchanged for the soft-drink. At any point, the buyer can activate the coin-return to retrieve any money that has been put in. For instance, if the state machine is in the 20¢ state and the coin-return input is activated, the state machine will give back a dime and move into the 10¢ state. If the buyer activates the coin-return button again, the state machine will give back another dime and move into the 0¢ state, after which the coin-return input will have no more effect.

Note that in Table 7.3 only one input is active at any time. For example, we assume the buyer never presses the buttons for both Swill and Krunk simultaneously. We also had to add a new input condition called Nothing to indicate when there are no inputs currently at high logic levels. In the state table, the state machine always stays in its current state whenever Nothing happens.

Now we can assign bit patterns to each state. In this case, we will use the one-hot encoding shown in Table 7.4.

Table 7.4 One-hot state encoding for the drink dispenser controller.

State	Q5	Q4	Q3	Q2	Q1	Q0
0¢	0	0	0	0	0	1
5¢	0	0	0	0	1	0
10¢	0	0	0	1	0	0
15¢	0	0	1	0	0	0
20¢	0	1	0	0	0	0
25¢	1	0	0	0	0	0

Once you have the state encoding, you can write down the truth-table for the combinatorial logic that determines the next state and the output values given the current state and the input values. Do the following steps for each row of the state transition table in Table 7.3:

1. Find the encoding of the current state given in column 1 of the current row. In our example, the current state in the first row is 0¢ which has an encoding of 000001. Then, do the following steps for the remaining columns in the row:

2. For each of the remaining columns in the row, find the encoding for the action taken by the buyer as indicated by the label at the top of each column. For the second column, for example, the buyer is putting in a nickel. The encodings for possible inputs are shown in Table 7.5.

Table 7.5 Input encoding for the drink dispenser controller.

Buyer's Action	Input to Drink Dispenser Logic				
	Nickel	**Dime**	**Coin Return**	**Swill**	**Krunk**
Put in 5¢	1	0	0	0	0
Put in 10¢	0	1	0	0	0
Push coin return	0	0	1	0	0
Select Swill	0	0	0	1	0
Select Krunk	0	0	0	0	1
Do nothing	0	0	0	0	0

For the action of putting in a nickel, Table 7.5 indicates that the nickel input should be at a logic 1 and all the other inputs should be at logic 0. Thus, the binary code for this action is 10000.

3. Find the encoding of the next state listed in the box at the current row and column of the state transition table. For example, in the first row and second column, the buyer has put in a nickel and the current state is 0¢, so the next state is 5¢ which has an encoding of 000010.

4. Find the values of the outputs given the current state and inputs. To continue the example, none of the 4 outputs are activated by putting in a nickel to the machine if its current state is 0¢. So the output encoding is 0000.

5. For the current line of the truth-table, write down the current state encoding and the input encoding on the input side of the current line of the truth-table, and write down the next state encoding and the output encoding on the output side of the truth table. Finalizing the example we started, this would add the following line to the truth-table:

```
0 0 0 0 0 1    1 0 0 0 0    :>    0 0 0 0 1 0  ->  0 0 0 0
```

Table 7.6 shows the completed drink dispenser truth-table that was generated by this procedure. The complete state machine is shown in Figure 7.12.

Listing 7.2 shows the ABEL code for the drink dispenser state machine. It is really not much different from the previous state machine descriptions. There are a few more flip-flops and a larger truth-table, but nothing we have not seen before. Since one-hot state encoding is being used, I named each flip-flop according to the current state that exists when the flip-flop stores a logic 1. That makes it easier to tell what is going on.

A problem with the state machine as designed is that it could power-up in any random state, even the 25¢ state. If this happened, unscrupulous customers could unplug the drink machine, plug it in again, and the machine would start

Table 7.6   The truth table for the drink dispenser logic.

Current State	Nickel	Dime	Coin Return	Swill	Krunk	Next State	ret_nick	ret_dime	disp_swill	disp_krunk
000001	1	0	0	0	0	000010	0	0	0	0
000001	0	1	0	0	0	000100	0	0	0	0
000001	0	0	1	0	0	000001	0	0	0	0
000001	0	0	0	1	0	000001	0	0	0	0
000001	0	0	0	0	1	000001	0	0	0	0
000001	0	0	0	0	0	000001	0	0	0	0
000010	1	0	0	0	0	000100	0	0	0	0
000010	0	1	0	0	0	001000	0	0	0	0
000010	0	0	1	0	0	000001	1	0	0	0
000010	0	0	0	1	0	000010	0	0	0	0
000010	0	0	0	0	1	000010	0	0	0	0
000010	0	0	0	0	0	000010	0	0	0	0
000100	1	0	0	0	0	001000	0	0	0	0
000100	0	1	0	0	0	010000	0	0	0	0
000100	0	0	1	0	0	000001	0	1	0	0
000100	0	0	0	1	0	000100	0	0	0	0
000100	0	0	0	0	1	000100	0	0	0	0
000100	0	0	0	0	0	000100	0	0	0	0
001000	1	0	0	0	0	010000	0	0	0	0
001000	0	1	0	0	0	100000	0	0	0	0
001000	0	0	1	0	0	000010	0	1	0	0
001000	0	0	0	1	0	001000	0	0	0	0
001000	0	0	0	0	1	001000	0	0	0	0
001000	0	0	0	0	0	001000	0	0	0	0
010000	1	0	0	0	0	100000	0	0	0	0
010000	0	1	0	0	0	100000	1	0	0	0
010000	0	0	1	0	0	000100	0	1	0	0
010000	0	0	0	1	0	010000	0	0	0	0
010000	0	0	0	0	1	010000	0	0	0	0
010000	0	0	0	0	0	010000	0	0	0	0

Table 7.6   The truth table for the drink dispenser logic. (Cont'd.)

Current State	Nickel	Dime	Coin Return	Swill	Krunk	Next State	ret_nick	ret_dime	disp_swill	disp_krunk
100000	1	0	0	0	0	100000	1	0	0	0
100000	0	1	0	0	0	100000	0	1	0	0
100000	0	0	1	0	0	001000	0	1	0	0
100000	0	0	0	1	0	000001	0	0	1	0
100000	0	0	0	0	1	000001	0	0	0	1
100000	0	0	0	0	0	100000	0	0	0	0

off in the 25¢ state and dispense a soft-drink without receiving any money. This can be avoided by providing a reset signal which starts the state machine in the 0¢ state whenever power is first applied. A dedicated synchronous reset input could be implemented just as in the up-counter example. In this example, however, we do an asynchronous reset that is triggered by a specific combination of the existing inputs. Line 21 of Listing 7.3 specifies that the reset will be triggered when all the inputs are forced to logic 1. The reset is applied to the asynchronous preset and reset inputs of the state flip-flops on lines 29 and 30. Line 29 uses the ACLR signal extension to asynchronously clear the flip-flops for the 5¢, 10¢, 15¢, 20¢ and 25¢ states. Line 30 uses the asynchronous preset signal extension, ASET, to force a logic 1 into the 0¢ state flip-flop when the same reset trigger condition occurs. Thus, the reset forces the state machine into the 0¢ state.

The VHDL version of the drink dispenser controller is shown in Listing 7.4. It is lengthy (mainly because we are trying to maintain a close correspondence with the truth table in Table 7.6) but relatively straight-forward. The ENTITY block on lines 6 through 26 declares the same inputs and outputs as the ABEL version.

The ARCHITECTURE section has two major blocks: a large WHEN...ELSE statement that computes the outputs (lines 72–116), and a process that controls the transitions from state to state on the rising edge of the clock (122–211). Each of these blocks uses the current state and input arrays to determine which statements get executed. These arrays as well as an output array are declared on lines 36, 45, and 46 respectively. It is easier to assign values to and test the values of these arrays than it is when using the individual state flip-flops and I/O signals. The input array is connected to an aggregate of the individual inputs on line 59 and the output array is connected to the individual outputs on lines 60–63. The outputs that indicate the current state of the drink dispenser are connected to the drink_state array on lines 64–69.

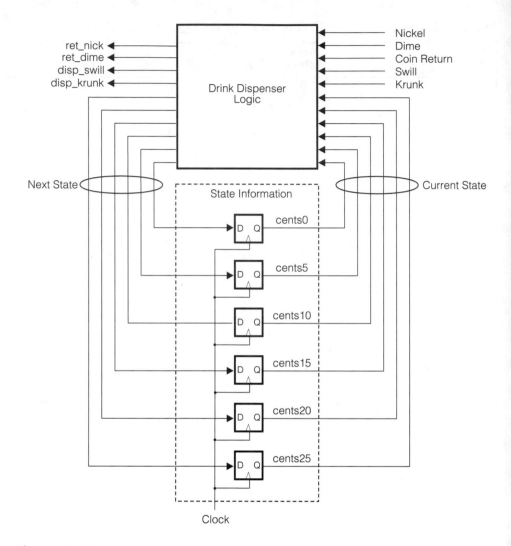

Figure 7.12   Drink dispenser state machine.

To make the meaning of the VHDL statements more obvious, the CONSTANT keyword is used to define the bit patterns for the different states (lines 38–43) and input combinations (48–54). Declaring a constant is identical to declaring a signal except you use the CONSTANT keyword and you can initialize the constant's value using the := operator.

As an example of how the VHDL code works, consider the case when the current state is 20¢. Then the cents20 flip-flop should be set so the drink_state array should be 010000. This matches the st_cents20 constant. If a dime is inserted, then line 59 indicates the inputs array should be 01000 (provided no

Listing 7.3 ABEL code for the drink dispenser controller.

```
001- MODULE DRINK1;
002- TITLE 'drink dispenser controller (1st version)'
003-
004- DECLARATIONS
005- C           PIN;                "clock input
006- NICKEL      PIN;                "nickel inserted input
007- DIME        PIN;                "dime inserted input
008- COIN_RET    PIN;                "coin-return input
009- SWILL       PIN;                "Swill selection input
010- KRUNK       PIN;                "Krunk selection input
011- RET_NICK    PIN ISTYPE 'COM';  "return nickel output
012- RET_DIME    PIN ISTYPE 'COM';  "return dime output
013- DISP_SWILL  PIN ISTYPE 'COM';  "dispense swill output
014- DISP_KRUNK  PIN ISTYPE 'COM';  "dispense Krunk output
015- CENTS0      PIN ISTYPE 'REG';  "DFF for 0-cents state
016- CENTS5      PIN ISTYPE 'REG';  "DFF for 5-cents state
017- CENTS10     PIN ISTYPE 'REG';  "DFF for 10-cents state
018- CENTS15     PIN ISTYPE 'REG';  "DFF for 15-cents state
019- CENTS20     PIN ISTYPE 'REG';  "DFF for 20-cents state
020- CENTS25     PIN ISTYPE 'REG';  "DFF for 25-cents state
021- RESET = NICKEL & DIME & COIN_RET & SWILL & KRUNK; "reset trigger
022-
023- EQUATIONS
024-
025- "connect clock input to controller state DFFs
026- [CENTS0,CENTS5,CENTS10,CENTS15,CENTS20,CENTS25].CLK = C;
027- "next two statements asynchronously reset controller
028- "to the cents0 state (000001)
029- [CENTS5,CENTS10,CENTS15,CENTS20,CENTS25].ACLR = RESET; "clear
030- CENTS0.ASET = RESET; "set cents0 flip-flop to 1
031-
032- TRUTH_TABLE
033- ([CENTS25,CENTS20,CENTS15,CENTS10,CENTS5,CENTS0,
034-               NICKEL,DIME,COIN_RET,SWILL,KRUNK]
035-        :> [CENTS25,CENTS20,CENTS15,CENTS10,CENTS5,CENTS0]
036-        -> [RET_NICK,RET_DIME,DISP_SWILL,DISP_KRUNK])
037- "state: 0 cents
038- [0,0,0,0,0,1, 1,0,0,0,0] :> [0,0,0,0,1,0] -> [0,0,0,0];
039- [0,0,0,0,0,1, 0,1,0,0,0] :> [0,0,0,1,0,0] -> [0,0,0,0];
040- [0,0,0,0,0,1, 0,0,1,0,0] :> [0,0,0,0,0,1] -> [0,0,0,0];
041- [0,0,0,0,0,1, 0,0,0,1,0] :> [0,0,0,0,0,1] -> [0,0,0,0];
042- [0,0,0,0,0,1, 0,0,0,0,1] :> [0,0,0,0,0,1] -> [0,0,0,0];
```

Listing 7.3   ABEL code for the drink dispenser controller.

```
043-    [0,0,0,0,0,1, 0,0,0,0,0] :> [0,0,0,0,0,1] -> [0,0,0,0];
044-    "state: 5 cents
045-    [0,0,0,0,1,0, 1,0,0,0,0] :> [0,0,0,1,0,0] -> [0,0,0,0];
046-    [0,0,0,0,1,0, 0,1,0,0,0] :> [0,0,1,0,0,0] -> [0,0,0,0];
047-    [0,0,0,0,1,0, 0,0,1,0,0] :> [0,0,0,0,0,1] -> [1,0,0,0];
048-    [0,0,0,0,1,0, 0,0,0,1,0] :> [0,0,0,0,1,0] -> [0,0,0,0];
049-    [0,0,0,0,1,0, 0,0,0,0,1] :> [0,0,0,0,1,0] -> [0,0,0,0];
050-    [0,0,0,0,1,0, 0,0,0,0,0] :> [0,0,0,0,1,0] -> [0,0,0,0];
051-    "state: 10 cents
052-    [0,0,0,1,0,0, 1,0,0,0,0] :> [0,0,1,0,0,0] -> [0,0,0,0];
053-    [0,0,0,1,0,0, 0,1,0,0,0] :> [0,1,0,0,0,0] -> [0,0,0,0];
054-    [0,0,0,1,0,0, 0,0,1,0,0] :> [0,0,0,0,0,1] -> [0,1,0,0];
055-    [0,0,0,1,0,0, 0,0,0,1,0] :> [0,0,0,1,0,0] -> [0,0,0,0];
056-    [0,0,0,1,0,0, 0,0,0,0,1] :> [0,0,0,1,0,0] -> [0,0,0,0];
057-    [0,0,0,1,0,0, 0,0,0,0,0] :> [0,0,0,1,0,0] -> [0,0,0,0];
058-    "state: 15 cents
059-    [0,0,1,0,0,0, 1,0,0,0,0] :> [0,1,0,0,0,0] -> [0,0,0,0];
060-    [0,0,1,0,0,0, 0,1,0,0,0] :> [1,0,0,0,0,0] -> [0,0,0,0];
061-    [0,0,1,0,0,0, 0,0,1,0,0] :> [0,0,0,0,1,0] -> [0,1,0,0];
062-    [0,0,1,0,0,0, 0,0,0,1,0] :> [0,0,1,0,0,0] -> [0,0,0,0];
063-    [0,0,1,0,0,0, 0,0,0,0,1] :> [0,0,1,0,0,0] -> [0,0,0,0];
064-    [0,0,1,0,0,0, 0,0,0,0,0] :> [0,0,1,0,0,0] -> [0,0,0,0];
065-    "state: 20 cents
066-    [0,1,0,0,0,0, 1,0,0,0,0] :> [1,0,0,0,0,0] -> [0,0,0,0];
067-    [0,1,0,0,0,0, 0,1,0,0,0] :> [1,0,0,0,0,0] -> [1,0,0,0];
068-    [0,1,0,0,0,0, 0,0,1,0,0] :> [0,0,0,1,0,0] -> [0,1,0,0];
069-    [0,1,0,0,0,0, 0,0,0,1,0] :> [0,1,0,0,0,0] -> [0,0,0,0];
070-    [0,1,0,0,0,0, 0,0,0,0,1] :> [0,1,0,0,0,0] -> [0,0,0,0];
071-    [0,1,0,0,0,0, 0,0,0,0,0] :> [0,1,0,0,0,0] -> [0,0,0,0];
072-    "state: 25 cents
073-    [1,0,0,0,0,0, 1,0,0,0,0] :> [1,0,0,0,0,0] -> [1,0,0,0];
074-    [1,0,0,0,0,0, 0,1,0,0,0] :> [1,0,0,0,0,0] -> [0,1,0,0];
075-    [1,0,0,0,0,0, 0,0,1,0,0] :> [0,0,1,0,0,0] -> [0,1,0,0];
076-    [1,0,0,0,0,0, 0,0,0,1,0] :> [0,0,0,0,0,1] -> [0,0,1,0];
077-    [1,0,0,0,0,0, 0,0,0,0,1] :> [0,0,0,0,0,1] -> [0,0,0,1];
078-    [1,0,0,0,0,0, 0,0,0,0,0] :> [1,0,0,0,0,0] -> [0,0,0,0];
079-
080-    END DRINK1
```

other inputs are active). This matches the insert_dime constant. Based on these matches, we can see that statement on line 103 will assign the value 1000 to the outputs array. This sets the outputs(3) array element. From line 60 we know that the outputs(3) element is connected to the ret_nickel output. So a nickel will be returned to the customer. Meanwhile, the drink_state array is loaded with st_cents25 by line 184. The st_cents25 constant is 100000, so the drink_state(5) element is set and the other elements are cleared. The drink_state(5) element is connected to the cents25 state output on line 69. So the drink dispenser moves into the 25¢ state. The net result is that if a dime is inserted when the dispenser has already received 20¢, then the dispenser will return a nickel and hold a total of 25¢. This is exactly the behavior we want.

As we did with the 3-state counter in the previous section, this VHDL code has been augmented with IBUF and BUFG modules to provide a buffered, low-skew clock to the state machine flip-flops. This is necessary when targeting the design to the XS40 Board. For the XS95 Board, you can remove these components and drive the state flip-flops directly with the input clock.

The simulated waveforms for the ABEL or VHDL version of the drink dispenser control circuit are shown in Figure 7.13. Here is the "instant-by-instant" description of what happens:

0–10 ns: All the inputs are raised to trigger a reset so the controller moves to the CENTS0 state. (Just look for the flip-flop with a 1 stored in it and that will tell you what state the controller is in.)

10–20 ns: A nickel is inserted (NICKEL=1) and the controller moves to the CENTS5 state.

20–40 ns: Two dimes are added (DIME=1 for two clock cycles) and the controller moves through the CENTS15 to the CENTS25 state.

40–50 ns: Swill is selected as the soft drink of choice. The controller signals the dispenser mechanism to let out a can of Swill (DISP_SWILL=1). Then the controller moves to the CENTS0 state.

50–60 ns: Nothing happens. The controller just sits quietly in the CENTS0 state, as it should.

60–90 ns: Three nickels are inserted to move the state machine through the CENTS5 and CENTS10 states to the CENTS15 state.

90–100 ns: The coin return is activated (COIN_RET=1). This makes the controller request that a dime should be released from the coin changer (RET_DIME=1). The controller also transitions to the CENTS5 state.

100–110 ns: The coin-return input is still active, so the controller releases another nickel (RET_NICK=1) and moves into the CENTS0 state.

110–120 ns: The coin-return is still active. However, the controller has returned all the money that was put in so it sits in the CENTS0 state and does nothing.

Listing 7.4   VHDL code for the drink dispenser controller.

```
001- -- Drink Dispenser Controller
002-
003- LIBRARY IEEE;
004- USE IEEE.std_logic_1164.ALL;
005-
006- ENTITY drink1 IS
007- PORT
008- (
009-   c: IN STD_LOGIC;          -- clock
010-   nickel: IN STD_LOGIC;        -- nickel is inserted
011-   dime: IN STD_LOGIC;          -- dime is inserted
012-   coin_ret: IN STD_LOGIC;      -- coin-return is pressed
013-   swill: IN STD_LOGIC;         -- swill is selected
014-   krunk: IN STD_LOGIC;         -- krunk is selected
015-   ret_nick: OUT STD_LOGIC;     -- nickel is returned
016-   ret_dime: OUT STD_LOGIC;     -- dime is returned
017-   disp_swill: OUT STD_LOGIC;  -- swill is dispensed
018-   disp_krunk: OUT STD_LOGIC;  -- krunk is dispensed
019-   cents0: OUT STD_LOGIC;       -- no money has been inserted
020-   cents5: OUT STD_LOGIC;       -- 5 cents has been inserted
021-   cents10: OUT STD_LOGIC;      -- 10 cents has been inserted
022-   cents15: OUT STD_LOGIC;      -- 15 cents has been inserted
023-   cents20: OUT STD_LOGIC;      -- 20 cents has been inserted
024-   cents25: OUT STD_LOGIC       -- 25 cents has been inserted
025- );
026- END drink1;
027-
028-
029- ARCHITECTURE drink1_arch OF drink1 IS
030-
031- COMPONENT IBUF PORT(I: IN STD_LOGIC; O: OUT STD_LOGIC); END COMPONENT;
032- COMPONENT BUFG PORT(I: IN STD_LOGIC; O: OUT STD_LOGIC); END COMPONENT;
033- SIGNAL in_c: STD_LOGIC; -- input-buffered clock
034- SIGNAL buf_c: STD_LOGIC;-- global-buffered clock signal
035-
036- SIGNAL drink_state: STD_LOGIC_VECTOR (5 DOWNTO 0);   -- state DFFs
037- -- the one-hot state definitions for the state machine
038- CONSTANT st_cents0:  STD_LOGIC_VECTOR (5 DOWNTO 0) := "000001";
039- CONSTANT st_cents5:  STD_LOGIC_VECTOR (5 DOWNTO 0) := "000010";
040- CONSTANT st_cents10: STD_LOGIC_VECTOR (5 DOWNTO 0) := "000100";
041- CONSTANT st_cents15: STD_LOGIC_VECTOR (5 DOWNTO 0) := "001000";
042- CONSTANT st_cents20: STD_LOGIC_VECTOR (5 DOWNTO 0) := "010000";
043- CONSTANT st_cents25: STD_LOGIC_VECTOR (5 DOWNTO 0) := "100000";
```

Listing 7.4   VHDL code for the drink dispenser controller.

```
044-
045- SIGNAL inputs: STD_LOGIC_VECTOR (4 DOWNTO 0);-- vector of inputs
046- SIGNAL outputs: STD_LOGIC_VECTOR (3 DOWNTO 0);-- vector of outputs
047- -- definitions for different input vectors
048- CONSTANT no_input:      STD_LOGIC_VECTOR (0 TO 4) := "00000";
049- CONSTANT insert_nickel: STD_LOGIC_VECTOR (0 TO 4) := "10000";
050- CONSTANT insert_dime:   STD_LOGIC_VECTOR (0 TO 4) := "01000";
051- CONSTANT press_return:  STD_LOGIC_VECTOR (0 TO 4) := "00100";
052- CONSTANT press_swill:   STD_LOGIC_VECTOR (0 TO 4) := "00010";
053- CONSTANT press_krunk:   STD_LOGIC_VECTOR (0 TO 4) := "00001";
054- CONSTANT reset:         STD_LOGIC_VECTOR (0 TO 4) := "11111";
055-
056- BEGIN
057-
058- -- collect all the individual inputs into a single input vector
059- inputs <= (nickel,dime,coin_ret,swill,krunk);  -- aggregate of inputs
060- ret_nick   <= outputs(3);   -- connect the components of
061- ret_dime   <= outputs(2);   -- the output vector to the
062- disp_swill <= outputs(1);   -- individual outputs
063- disp_krunk <= outputs(0);
064- cents0  <= drink_state(0);  -- connect the components of
065- cents5  <= drink_state(1);  -- the state vector to the
066- cents10 <= drink_state(2);  -- individual state outputs
067- cents15 <= drink_state(3);
068- cents20 <= drink_state(4);
069- cents25 <= drink_state(5);
070-
071- -- set the outputs according to the current drink machine state and inputs
072- outputs <=
073-            -- current state = 0 cents
074- "0000" WHEN ((drink_state=st_cents0) AND (inputs=insert_nickel)) ELSE
075- "0000" WHEN ((drink_state=st_cents0) AND (inputs=insert_dime)) ELSE
076- "0000" WHEN ((drink_state=st_cents0) AND (inputs=press_return)) ELSE
077- "0000" WHEN ((drink_state=st_cents0) AND (inputs=press_swill)) ELSE
078- "0000" WHEN ((drink_state=st_cents0) AND (inputs=press_krunk)) ELSE
079- "0000" WHEN ((drink_state=st_cents0) AND (inputs=no_input)) ELSE
080-            -- current state = 5 cents
081- "0000" WHEN ((drink_state=st_cents5) AND (inputs=insert_nickel)) ELSE
082- "0000" WHEN ((drink_state=st_cents5) AND (inputs=insert_dime)) ELSE
083- "1000" WHEN ((drink_state=st_cents5) AND (inputs=press_return)) ELSE
084- "0000" WHEN ((drink_state=st_cents5) AND (inputs=press_swill)) ELSE
085- "0000" WHEN ((drink_state=st_cents5) AND (inputs=press_krunk)) ELSE
086- "0000" WHEN ((drink_state=st_cents5) AND (inputs=no_input)) ELSE
```

Listing 7.4   VHDL code for the drink dispenser controller.

```
087-              -- current state = 10 cents
088- "0000" WHEN ((drink_state=st_cents10) AND (inputs=insert_nickel)) ELSE
089- "0000" WHEN ((drink_state=st_cents10) AND (inputs=insert_dime)) ELSE
090- "0100" WHEN ((drink_state=st_cents10) AND (inputs=press_return)) ELSE
091- "0000" WHEN ((drink_state=st_cents10) AND (inputs=press_swill)) ELSE
092- "0000" WHEN ((drink_state=st_cents10) AND (inputs=press_krunk)) ELSE
093- "0000" WHEN ((drink_state=st_cents10) AND (inputs=no_input)) ELSE
094-              -- current state = 15 cents
095- "0000" WHEN ((drink_state=st_cents15) AND (inputs=insert_nickel)) ELSE
096- "0000" WHEN ((drink_state=st_cents15) AND (inputs=insert_dime)) ELSE
097- "0100" WHEN ((drink_state=st_cents15) AND (inputs=press_return)) ELSE
098- "0000" WHEN ((drink_state=st_cents15) AND (inputs=press_swill)) ELSE
099- "0000" WHEN ((drink_state=st_cents15) AND (inputs=press_krunk)) ELSE
100- "0000" WHEN ((drink_state=st_cents15) AND (inputs=no_input)) ELSE
101-              -- current state = 20 cents
102- "0000" WHEN ((drink_state=st_cents20) AND (inputs=insert_nickel)) ELSE
103- "1000" WHEN ((drink_state=st_cents20) AND (inputs=insert_dime)) ELSE
104- "0100" WHEN ((drink_state=st_cents20) AND (inputs=press_return)) ELSE
105- "0000" WHEN ((drink_state=st_cents20) AND (inputs=press_swill)) ELSE
106- "0000" WHEN ((drink_state=st_cents20) AND (inputs=press_krunk)) ELSE
107- "0000" WHEN ((drink_state=st_cents20) AND (inputs=no_input)) ELSE
108-              -- current state = 25 cents
109- "1000" WHEN ((drink_state=st_cents25) AND (inputs=insert_nickel)) ELSE
110- "0100" WHEN ((drink_state=st_cents25) AND (inputs=insert_dime)) ELSE
111- "0100" WHEN ((drink_state=st_cents25) AND (inputs=press_return)) ELSE
112- "0010" WHEN ((drink_state=st_cents25) AND (inputs=press_swill)) ELSE
113- "0001" WHEN ((drink_state=st_cents25) AND (inputs=press_krunk)) ELSE
114- "0000" WHEN ((drink_state=st_cents25) AND (inputs=no_input)) ELSE
115-              -- default output
116- "0000";
117-
118- u0: IBUF PORT MAP(I=>c, O=>in_c);        -- buffer the clock
119- u1: BUFG PORT MAP(I=>in_c, O=>buf_c);
120-
121- -- this process changes the drink machine state vector
122- PROCESS (buf_c,nickel,dime,coin_ret,swill,krunk)
123- BEGIN
124- IF inputs=reset THEN   -- asynchronous reset
125-    drink_state <= st_cents0;
126- -- otherwise, change the state on each rising clock edge
127- ELSIF (buf_c'event AND buf_c='1') THEN
128-    -- current state = 0 cents
129-    IF ((drink_state=st_cents0) AND (inputs=insert_nickel)) THEN
```

Listing 7.4   VHDL code for the drink dispenser controller.

```
130-                drink_state <= st_cents5;
131-    ELSIF ((drink_state=st_cents0) AND (inputs=insert_dime)) THEN
132-                drink_state <= st_cents10;
133-    ELSIF ((drink_state=st_cents0) AND (inputs=press_return)) THEN
134-                drink_state <= st_cents0;
135-    ELSIF ((drink_state=st_cents0) AND (inputs=press_swill)) THEN
136-                drink_state <= st_cents0;
137-    ELSIF ((drink_state=st_cents0) AND (inputs=press_krunk)) THEN
138-                drink_state <= st_cents0;
139-    ELSIF ((drink_state=st_cents0) AND (inputs=no_input)) THEN
140-                drink_state <= st_cents0;
141-       -- current state = 5 cents
142-    ELSIF ((drink_state=st_cents5) AND (inputs=insert_nickel)) THEN
143-                drink_state <= st_cents10;
144-    ELSIF ((drink_state=st_cents5) AND (inputs=insert_dime)) THEN
145-                drink_state <= st_cents15;
146-    ELSIF ((drink_state=st_cents5) AND (inputs=press_return)) THEN
147-                drink_state <= st_cents0;
148-    ELSIF ((drink_state=st_cents5) AND (inputs=press_swill)) THEN
149-                drink_state <= st_cents5;
150-    ELSIF ((drink_state=st_cents5) AND (inputs=press_krunk)) THEN
151-                drink_state <= st_cents5;
152-    ELSIF ((drink_state=st_cents5) AND (inputs=no_input)) THEN
153-                drink_state <= st_cents5;
154-       -- current state = 10 cents
155-    ELSIF ((drink_state=st_cents10) AND (inputs=insert_nickel)) THEN
156-                drink_state <= st_cents15;
157-    ELSIF ((drink_state=st_cents10) AND (inputs=insert_dime)) THEN
158-                drink_state <= st_cents20;
159-    ELSIF ((drink_state=st_cents10) AND (inputs=press_return)) THEN
160-                drink_state <= st_cents0;
161-    ELSIF ((drink_state=st_cents10) AND (inputs=press_swill)) THEN
162-                drink_state <= st_cents10;
163-    ELSIF ((drink_state=st_cents10) AND (inputs=press_krunk)) THEN
164-                drink_state <= st_cents10;
165-    ELSIF ((drink_state=st_cents10) AND (inputs=no_input)) THEN
166-                drink_state <= st_cents10;
167-       -- current state = 15 cents
168-    ELSIF ((drink_state=st_cents15) AND (inputs=insert_nickel)) THEN
169-                drink_state <= st_cents20;
170-    ELSIF ((drink_state=st_cents15) AND (inputs=insert_dime)) THEN
171-                drink_state <= st_cents25;
172-    ELSIF ((drink_state=st_cents15) AND (inputs=press_return)) THEN
```

Listing 7.4   VHDL code for the drink dispenser controller.

```
173-              drink_state <= st_cents5;
174-     ELSIF ((drink_state=st_cents15) AND (inputs=press_swill)) THEN
175-              drink_state <= st_cents15;
176-     ELSIF ((drink_state=st_cents15) AND (inputs=press_krunk)) THEN
177-              drink_state <= st_cents15;
178-     ELSIF ((drink_state=st_cents15) AND (inputs=no_input)) THEN
179-              drink_state <= st_cents15;
180-        -- current state = 20 cents
181-     ELSIF ((drink_state=st_cents20) AND (inputs=insert_nickel)) THEN
182-              drink_state <= st_cents25;
183-     ELSIF ((drink_state=st_cents20) AND (inputs=insert_dime)) THEN
184-              drink_state <= st_cents25;
185-     ELSIF ((drink_state=st_cents20) AND (inputs=press_return)) THEN
186-              drink_state <= st_cents10;
187-     ELSIF ((drink_state=st_cents20) AND (inputs=press_swill)) THEN
188-              drink_state <= st_cents20;
189-     ELSIF ((drink_state=st_cents20) AND (inputs=press_krunk)) THEN
190-              drink_state <= st_cents20;
191-     ELSIF ((drink_state=st_cents20) AND (inputs=no_input)) THEN
192-              drink_state <= st_cents20;
193-        -- current state = 25 cents
194-     ELSIF ((drink_state=st_cents25) AND (inputs=insert_nickel)) THEN
195-              drink_state <= st_cents25;
196-     ELSIF ((drink_state=st_cents25) AND (inputs=insert_dime)) THEN
197-              drink_state <= st_cents25;
198-     ELSIF ((drink_state=st_cents25) AND (inputs=press_return)) THEN
199-              drink_state <= st_cents15;
200-     ELSIF ((drink_state=st_cents25) AND (inputs=press_swill)) THEN
201-              drink_state <= st_cents0;
202-     ELSIF ((drink_state=st_cents25) AND (inputs=press_krunk)) THEN
203-              drink_state <= st_cents0;
204-     ELSIF ((drink_state=st_cents25) AND (inputs=no_input)) THEN
205-              drink_state <= st_cents25;
206-        -- default transition
207-     ELSE
208-              drink_state <= st_cents0;
209-     END IF;
210- END IF;
211- END PROCESS;
212-
213- END drink1_arch;
214-
```

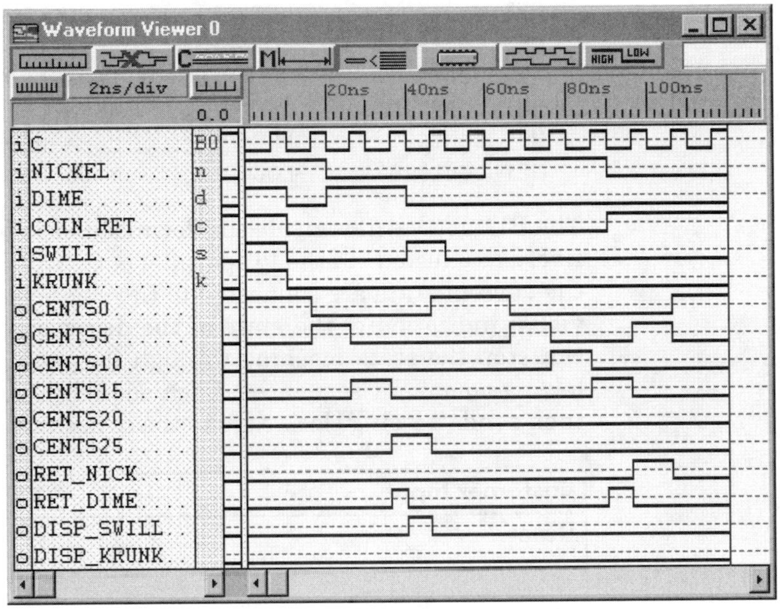

Figure 7.13   Simulated waveforms for the drink dispenser controller.

A constraint file for the DRINK1 design should assign the inputs and outputs like this for the XS40 Board:

```
NET c              LOC=P44;    # B0 argument of XSPORT
NET nickel         LOC=P45;    # B1 argument of XSPORT
NET dime           LOC=P46;    # B2 argument of XSPORT
NET coin_ret       LOC=P47;    # B3 argument of XSPORT
NET swill          LOC=P48;    # B4 argument of XSPORT
NET krunk          LOC=P49;    # B5 argument of XSPORT
NET cents0         LOC=P25;    # S0 LED segment
NET cents5         LOC=P26;    # S1 LED segment
NET cents10        LOC=P24;    # S2 LED segment
NET cents15        LOC=P20;    # S3 LED segment
NET cents20        LOC=P23;    # S4 LED segment
NET cents25        LOC=P18;    # S5 LED segment
```

Or use the following constraints for the XS95 Board:

```
NET c              LOC=P46;    # B0 argument of XSPORT
NET nickel         LOC=P47;    # B1 argument of XSPORT
NET dime           LOC=P48;    # B2 argument of XSPORT
NET coin_ret       LOC=P50;    # B3 argument of XSPORT
NET swill          LOC=P51;    # B4 argument of XSPORT
NET krunk          LOC=P52;    # B5 argument of XSPORT
```

```
NET  cents0        LOC=P21;      # S0 LED segment
NET  cents5        LOC=P23;      # S1 LED segment
NET  cents10       LOC=P19;      # S2 LED segment
NET  cents15       LOC=P17;      # S3 LED segment
NET  cents20       LOC=P18;      # S4 LED segment
NET  cents25       LOC=P14;      # S5 LED segment
```

The BIT or SVF file can also be downloaded to the XS40 or XS95 Board, respectively, and tested via the PC parallel port. The outputs of the DFFs which store the current state are connected to the segments of the LED digit. This gives a visual indication of the state of the drink machine after each operation. You can step the state machine through its paces using combinations of the following command sequences. Each sequence activates a particular input and then sends a rising clock edge to the drink dispenser controller.

Customer Input	XSPORT Command Sequence
Reset	XSPORT 111110 XSPORT 111111 XSPORT 111110
Put in a nickel	XSPORT 000010 XSPORT 000011 XSPORT 000000
Put in a dime	XSPORT 000100 XSPORT 000101 XSPORT 000000
Get money returned	XSPORT 001000 XSPORT 001001 XSPORT 000000
Get Swill	XSPORT 010000 XSPORT 010001 XSPORT 000000
Get Krunk	XSPORT 100000 XSPORT 100001 XSPORT 000000

The Drink Dispenser Controller Revisited

If you were looking carefully, you would have noticed a problem in the simulation waveforms shown in Figure 7.13: the RET_DIME output that causes a dime to be returned is active during the 36–40 ns period. This occurs when a dime is entered by the rising clock edge at 36 ns and this causes the machine to move immediately from the 15¢ state to the 25¢ state with the DIME input still active until the 40 ns mark. The state machine sees the logic 1 is still present on the DIME input and believes the customer is putting in a new dime. Since the state machine is now in the 25¢ state and no more money is needed to get a drink, the state machine dutifully returns a dime. The end result is the customer can get a can of Swill or Krunk for only 15¢.

The problem just described can be solved by adding a new state called "Armed" to the state machine and changing the actions that occur in the 25¢ state, as shown in Table 7.7.

Table 7.7   The revised truth table for the drink dispenser logic.

Current State	Input					
	Nickel	**Dime**	**Coin Return**	**Swill**	**Krunk**	**Nothing**
0¢	→5¢	→10¢	→0¢	→0¢	→0¢	→0¢
5¢	→10¢	→15¢	→0¢, ret_nick = 1	→5¢	→5¢	→5¢
10¢	→15¢	→20¢	→0¢, ret_dime = 1	→10¢	→10¢	→10¢
15¢	→20¢	→25¢	→5¢, ret_dime = 1	→15¢	→15¢	→15¢
20¢	→25¢	→25¢, ret_nick = 1	→10¢, ret_dime = 1	→20¢	→20¢	→20¢
25¢	→25¢	→25¢	→25¢	→25¢	→25¢	→Armed
Armed	→Armed, ret_nick = 1	→Armed, ret_dime = 1	→15¢, ret_dime = 1	→0¢, disp_swill = 1	→0¢, disp_krunk = 1	→Armed

In the revised transition table, the state machine remains in the 25¢ state as long as any input is at a logic 1 level. Also, no output signals are generated for the coin-return or drink dispenser mechanisms in the 25¢ state. The transition from the 25¢ state to the Armed state can only occur when all the inputs are at logic 0 levels (i.e., this is the Nothing input). In the Armed state, the state machine will once again respond to inputs for selecting the desired beverage and accepting and returning money.

At this time you should be thinking: "Oh no! I have to go through my entire truth-table and insert this new state and correct all my transitions and outputs!" That thought would make anybody hesitant to make a correction. For that reason, ABEL provides language constructs like those in VHDL to make it easy to describe a state machine (at least, easier than what we have been doing). Listing 7.5 shows the ABEL code for our revised drink dispenser controller. Now we can look at how this differs from the previous ABEL design.

First, we added a new state flip-flop on line 21 called ARMED. This flip-flop will be on when the controller is in the Armed state. On lines 26–32 we define a set of state names and assign a particular pattern of bits to each one. Note that this is just the one-hot encoding we used previously, with the addition of the ARMED flip-flop. (The names of the flip-flops are listed in the comment on line 25 so you can see which flip-flop is active for each state.) Lines 36–41 connect the clock, preset, and reset inputs of the flip-flops just like in the previous design. The ARMED flip-flop has been added to the statements.

The STATE_DIAGRAM keyword on line 43 begins the description of our drink dispenser controller's behavior. The keyword is followed by a bracket-enclosed list of the flip-flops that store the state for this state machine. For correct operation, the order of these flip-flop names must match with how you ordered the flip-flops in the state assignments on lines 26–32.

The remainder of the state machine description consists of 7 sets of statements - one set for each state in the controller. Each set begins with a colon-terminated label of the state to which the following statements will apply. On line 46, for example, the ST_CENTS0 label indicates that the statements on lines 47, 48, and 49 will describe how the state machine should act when it is in the 0¢ state.

Lines 47–49 are a simple IF-THEN-ELSE construct. The basic syntax is as follows:

```
IF <boolean expression A == TRUE> THEN <next state B>
ELSE IF <boolean expression C == TRUE> THEN <next state D>
....
ELSE <next state Z>;
```

The IF-THEN-ELSE statement generates a logic circuit that evaluates a set of Boolean expressions involving the inputs to the state machine. The first expression that evaluates to a logic 1 (TRUE) will cause the state machine to move into the next state listed after the corresponding THEN keyword. If none of the expressions are TRUE, the state machine transitions to the next state listed in the final ELSE clause. (Be careful to always terminate an IF-THEN-ELSE statement with a semicolon.)

To make this more concrete, examine lines 46–49. Given that the state machine is in the 0¢ state, line 47 says that the controller will move to the 5¢ state if a nickel is inserted. If no nickel is inserted but a dime is, then the controller moves to the 10¢ state. Finally, if neither a nickel nor a dime is inserted, the controller remains in the 0¢ state. This behavior *almost* matches what is described on the first row of our revised state transition table. (We will discuss the slight difference later).

The IF-THEN-ELSE statement is a very readable way to describe the state transitions of a state machine, but how do we describe the actions of the outputs that control the dispenser and coin-return machanisms? This is accomplished with the WITH keyword which lets us assign a Boolean expression to a particular output. WITH is used after the next-state label in the IF-THEN-ELSE statement. The revised syntax is:

```
IF <boolean expression A == TRUE> THEN <next state B>
                    WITH <output E = boolean expression F>
ELSE IF <boolean expression C == TRUE> THEN <next state D>
                    WITH <output G = boolean expression H>
....
ELSE <next state Z> WITH <output W = boolean expression X>;
```

Basically, the IF-THEN-WITH-ELSE statement not only generates the next-state logic, but it also causes the values of the Boolean expressions to appear on the specified outputs when certain conditions apply. As before, the circuit evaluates a set of Boolean expressions involving the inputs to the state machine. The first expression that evaluates to a logic 1 (TRUE) will cause the value of the Boolean equation listed after the WITH keyword to appear immediately on the associated output. If none of the expressions are TRUE, the WITH clause in the final ELSE clause is activated.

Take lines 83–89 as an example of the use of the WITH keyword. Given that the controller is in the Armed state, line 84 states that the insertion of a nickel will cause the controller to activate the coin-return and return a nickel (RET_NICK=1). Lines 85 and 86 require that a dime be returned if a dime is inserted or the coin return is pressed, respectively. Lines 87 and 88 activate the outputs for dispensing cans of Swill or Krunk if the button for Swill or Krunk is activated, respectively. No outputs are activated at all if none of the previous inputs are TRUE (line 89).

It is possible to control more than one output at a time using the WITH clause and a brace-enclosed group of Boolean equations. For instance, if the coin return input was pressed in the Armed state, the controller could return both a dime and a nickel simultaneously (provided the coin return mechanism could handle it). The modified IF-THEN-WITH-ELSE statement looks like this:

```
STATE ST_ARMED:
  IF NICKEL THEN ST_ARMED WITH RET_NICK=1
  ELSE IF DIME THEN ST_ARMED WITH RET_DIME=1
  ELSE IF COIN_RET THEN ST_CENTS10 WITH {RET_DIME=1; RET_NICK=1;}
  ELSE IF SWILL THEN ST_CENTSO WITH DISP_SWILL=1
  ELSE IF KRUNK THEN ST_CENTSO WITH DISP_KRUNK=1
  ELSE ST_ARMED;
```

With your new knowledge of the ABEL state machine syntax, you should be able to interpret each line of Listing 7.5 and relate it back to the original state transition table. In fact, the ABEL compiler which you use to synthesize netlists does the same thing. The compiler parses the STATE_DIAGRAM statements and constructs its own state transition table. Then the compiler can derive the circuitry for generating the next state and the outputs just as you did in the previous sections. So the software tools are not doing anything magical.

We can also re-do the VHDL version of the drink dispenser state machine as shown in Listing 7.6. Like the 3-state counter VHDL example (Listing 7.2), this code uses two processes: one for computing the outputs and the next state based upon the values of the current state and inpus (lines 59–113), and another process which just updates the current state with the new next state value (lines 119–127). For this reason, we declare two arrays for holding the current and next state on lines 37 and 38, respectively.

The new armed output is added to the interface description on line 25, and the current and next state arrays and state definitions are expanded to handle this extra bit on lines 37 through 46. The state outputs for this design are attached to the elements of the current state array on lines 49 through 55.

The first process starts by assigning a low logic level to all the combinational outputs of the state machine (lines 61–64). This ensures that these outputs are set to some value if none of the following conditional statements does so. If we did not perform this initialization step, then these outputs would become implied latches. The next state array is also initialized to the reset state (with 0¢) on line 65. This will cause the state machine to restart if it ever gets into an illegal state.

A large IF...THEN...ELSE statement follows on lines 67–112 that uses the current state and inputs to compute the next state and the outputs. As an example of how the VHDL code works, consider the case when the current state is 20¢. Then the cents20 flip-flop should be on so the curr_drink_state array should be 0010000. This matches the st_cents20 constant so the statement on line 90 is executed. This activates the nested IF...THEN...ELSE statement on lines 91–95. If a dime is inserted, then execution falls through to line 92. The statement on line 92 assigns a logic 1 to the ret_nick output. So a nickel will be returned to the customer. At the same time, the next_drink_state array is loaded with st_cents25. The st_cents25 constant is 0100000, so the next_drink_state(5) element is set and the other elements are cleared.

Continuing the example, the second process transfers the next_drink_state array into the curr_drink_state array when a rising edge occur on the clock input. This loads curr_drink_state with st_cents25 which is 0100000. This sets the curr_drink_state(5) element and resets all the others. The curr_drink_state(5) element is connected to the cents25 state output on line

Listing 7.5   ABEL code for the revised drink dispenser controller.

```
001- MODULE DRINK2;
002- TITLE 'drink dispenser controller (2nd version)'
003-
004- DECLARATIONS
005- C          PIN;                "clock input
006- NICKEL     PIN;                "nickel inserted input
007- DIME       PIN;                "dime inserted input
008- COIN_RET   PIN;                "coin-return input
009- SWILL      PIN;                "Swill selection
010- KRUNK      PIN;                "Krunk selection
011- RET_NICK   PIN ISTYPE 'COM';   "return nickel output
012- RET_DIME   PIN ISTYPE 'COM';   "return dime output
013- DISP_SWILL PIN ISTYPE 'COM';   "dispense swill output
014- DISP_KRUNK PIN ISTYPE 'COM';   "dispense Krunk output
015- CENTS0     PIN ISTYPE 'REG';   "DFF for 0-cents state
016- CENTS5     PIN ISTYPE 'REG';   "DFF for 5-cents state
017- CENTS10    PIN ISTYPE 'REG';   "DFF for 10-cents state
018- CENTS15    PIN ISTYPE 'REG';   "DFF for 15-cents state
019- CENTS20    PIN ISTYPE 'REG';   "DFF for 20-cents state
020- CENTS25    PIN ISTYPE 'REG';   "DFF for 25-cents state
021- ARMED      PIN ISTYPE 'REG';   "DFF for armed state
022- RESET = NICKEL & DIME & COIN_RET & SWILL & KRUNK; "reset
trigger
023-
024- "now make some state assignments
025- "          ARMED CENTS25 CENTS20 CENTS15 CENTS10 CENTS5 CENTS0
026- ST_CENTS0  = [0,    0,      0,      0,      0,      0,     1];
027- ST_CENTS5  = [0,    0,      0,      0,      0,      1,     0];
028- ST_CENTS10 = [0,    0,      0,      0,      1,      0,     0];
029- ST_CENTS15 = [0,    0,      0,      1,      0,      0,     0];
030- ST_CENTS20 = [0,    0,      1,      0,      0,      0,     0];
031- ST_CENTS25 = [0,    1,      0,      0,      0,      0,     0];
032- ST_ARMED   = [1,    0,      0,      0,      0,      0,     0];
033-
034- EQUATIONS
035-
036- "clock controller state DFFs
037- [CENTS0,CENTS5,CENTS10,CENTS15,CENTS20,CENTS25,ARMED].CLK = C;
038- "next two statements asynchronously reset controller
039- "to the cents0 state (0000001)
040- CENTS0.ASET = RESET;
041- [CENTS5,CENTS10,CENTS15,CENTS20,CENTS25,ARMED].ACLR = RESET;
042-
```

Listing 7.5   ABEL code for the revised drink dispenser controller. (Cont'd.)

```
043-  STATE_DIAGRAM [ARMED, CENTS25, CENTS20, CENTS15,
044-                  CENTS10, CENTS5, CENTS0]
045-
046-    STATE ST_CENTS0:
047-      IF NICKEL THEN ST_CENTS5
048-      ELSE IF DIME THEN ST_CENTS10
049-      ELSE ST_CENTS0;
050-
051-    STATE ST_CENTS5:
052-      IF NICKEL THEN ST_CENTS10
053-      ELSE IF DIME THEN ST_CENTS15
054-      ELSE IF COIN_RET THEN ST_CENTS0 WITH RET_NICK=1
055-      ELSE ST_CENTS5;
056-
057-    STATE ST_CENTS10:
058-      IF NICKEL THEN ST_CENTS15
059-      ELSE IF DIME THEN ST_CENTS20
060-      ELSE IF COIN_RET THEN ST_CENTS0 WITH RET_DIME=1
061-      ELSE ST_CENTS10;
062-
063-    STATE ST_CENTS15:
064-      IF NICKEL THEN ST_CENTS20
065-      ELSE IF DIME THEN ST_CENTS25
066-      ELSE IF COIN_RET THEN ST_CENTS5 WITH RET_DIME=1
067-      ELSE ST_CENTS15;
068-
069-    STATE ST_CENTS20:
070-      IF NICKEL THEN ST_CENTS25
071-      ELSE IF DIME THEN ST_CENTS25 WITH RET_NICK=1
072-      ELSE IF COIN_RET THEN ST_CENTS10 WITH RET_DIME=1
073-      ELSE ST_CENTS20;
074-
075-    STATE ST_CENTS25:
076-      IF NICKEL THEN ST_CENTS25
077-      ELSE IF DIME THEN ST_CENTS25
078-      ELSE IF COIN_RET THEN ST_CENTS25
079-      ELSE IF SWILL THEN ST_CENTS25
080-      ELSE IF KRUNK THEN ST_CENTS25
081-      ELSE ST_ARMED;
082-
083-    STATE ST_ARMED:
084-      IF NICKEL THEN ST_ARMED WITH RET_NICK=1
085-      ELSE IF DIME THEN ST_ARMED WITH RET_DIME=1
```

Listing 7.5   ABEL code for the revised drink dispenser controller. (Cont'd.)

```
086-      ELSE IF COIN_RET THEN ST_CENTS15 WITH RET_DIME=1
087-      ELSE IF SWILL THEN ST_CENTSO WITH DISP_SWILL=1
088-      ELSE IF KRUNK THEN ST_CENTSO WITH DISP_KRUNK=1
089-      ELSE ST_ARMED;
090-
091- END DRINK2
```

54. So the drink dispenser moves into the 25¢ state. The net result is that if a dime is inserted when the dispenser has already received 20¢, then the dispenser will return a nickel and hold a total of 25¢. This replicates the behavior of our previous state machine described by Listing 7.4.

The simulated waveforms for the revised ABEL or VHDL drink dispenser control circuit are shown in Figure 7.14. Notice that our "problem pulse" on the RET_DIME output is no longer present during the 36–40 ns period.

The revised drink dispenser controller circuit can be downloaded and tested in exactly the same manner as the controller from the previous section.

We mentioned earlier that the behavior of our revised controller *almost* matches the behavior described in Table 7.3. Remember when we created Table 7.3, we specified that none of the inputs could be active simultaneously. For example, if a customer pressed both the Swill and Krunk buttons at the same time, the result would be undefined because there is no listing in the table for such an occurrence. (With 5 inputs, the table would need 32 columns to cover every possibility. I am not going to try writing *that* out!)

But the circuit as described in Listings 7.5 or 7.6 *does* handle the cases where multiple inputs are active simultaneously. For example, the IF clause on line 47 of Listing 7.5 (or line 68 of Listing 7.6) is active if NICKEL=1 regardless of the state of the other inputs. So the same clause would also be active if NICKEL=1 *and* DIME=1. Interpreted in this way, we can see that all 32 possible input combinations are covered for all 7 states in the revised drink dispenser controller. There is no longer any unspecified behavior. Of course, this does not mean that the specified behavior is *what we want!* To be sure of that, we would have to examine each response to each combination of inputs in each state and ask if that is the most desireable behavior. But that is a job for another day.

Using the State Editor

At this point we have described state machines using ABEL and VHDL. These are text-based design methods. For those who want a more graphical approach, the Foundation Series software includes a state editor that lets you design a

Listing 7.6   VHDL code for the revised drink dispenser controller.

```
001- -- Drink Dispenser Controller
002-
003- LIBRARY IEEE;
004- USE IEEE.std_logic_1164.ALL;
005-
006- ENTITY drink2 IS
007- PORT
008- (
009-         c: IN STD_LOGIC;                 -- clock
010-         nickel: IN STD_LOGIC;            -- nickel is inserted
011-         dime: IN STD_LOGIC;              -- dime is inserted
012-         coin_ret: IN STD_LOGIC;          -- coin-return is pressed
013-         swill: IN STD_LOGIC;             -- swill is selected
014-         krunk: IN STD_LOGIC;             -- krunk is selected
015-         ret_nick: OUT STD_LOGIC;         -- nickel is returned
016-         ret_dime: OUT STD_LOGIC;         -- dime is returned
017-         disp_swill: OUT STD_LOGIC;       -- swill is dispensed
018-         disp_krunk: OUT STD_LOGIC;       -- krunk is dispensed
019-         cents0: OUT STD_LOGIC;           -- no money has been inserted
020-         cents5: OUT STD_LOGIC;           -- 5 cents has been inserted
021-         cents10: OUT STD_LOGIC;          -- 10 cents has been inserted
022-         cents15: OUT STD_LOGIC;          -- 15 cents has been inserted
023-         cents20: OUT STD_LOGIC;          -- 20 cents has been inserted
024-         cents25: OUT STD_LOGIC;          -- 25 cents has been inserted
025-         armed: OUT STD_LOGIC
026- );
027- END drink2;
028-
029- ARCHITECTURE drink2_arch OF drink2 IS
030-
031- COMPONENT IBUF PORT(I: IN STD_LOGIC; O: OUT STD_LOGIC); END COMPONENT;
032- COMPONENT BUFG PORT(I: IN STD_LOGIC; O: OUT STD_LOGIC); END COMPONENT;
033- SIGNAL in_c: STD_LOGIC; -- input-buffered clock
034- SIGNAL buf_c: STD_LOGIC;-- global-buffered clock
035-
036- -- declare current and next state vectors for drink machine state
037- SIGNAL curr_drink_state: STD_LOGIC_VECTOR (6 DOWNTO 0);
038- SIGNAL next_drink_state: STD_LOGIC_VECTOR (6 DOWNTO 0);
039- -- the one-hot state definitions for the state machine
040- CONSTANT st_cents0: STD_LOGIC_VECTOR (6 DOWNTO 0) := "0000001";
041- CONSTANT st_cents5: STD_LOGIC_VECTOR (6 DOWNTO 0) := "0000010";
042- CONSTANT st_cents10: STD_LOGIC_VECTOR (6 DOWNTO 0) := "0000100";
043- CONSTANT st_cents15: STD_LOGIC_VECTOR (6 DOWNTO 0) := "0001000";
```

```
044- CONSTANT st_cents20: STD_LOGIC_VECTOR (6 DOWNTO 0) := "0010000";
045- CONSTANT st_cents25: STD_LOGIC_VECTOR (6 DOWNTO 0) := "0100000";
046- CONSTANT st_armed:  STD_LOGIC_VECTOR (6 DOWNTO 0) := "1000000";
047-
048- BEGIN
049- cents0  <= curr_drink_state(0); -- connect the components of
050- cents5  <= curr_drink_state(1); -- the current state vector
051- cents10 <= curr_drink_state(2); -- to the state outputs
052- cents15 <= curr_drink_state(3);
053- cents20 <= curr_drink_state(4);
054- cents25 <= curr_drink_state(5);
055- armed   <= curr_drink_state(6);
056-
057- -- this process sets both the outputs and the next state
058- -- given the current state and inputs
059- PROCESS (curr_drink_state,nickel,dime,coin_ret,swill,krunk)
060- BEGIN
061- ret_nick   <= '0';    -- first set the default values
062- ret_dime   <= '0';    -- for the combinational outputs
063- disp_swill <= '0';
064- disp_krunk <= '0';
065- next_drink_state <= st_cents0; -- set default next state
066-
067- IF curr_drink_state=st_cents0 THEN
068-    IF nickel='1'       THEN next_drink_state <= st_cents5;
069-    ELSIF dime='1'      THEN next_drink_state <= st_cents10;
070-    ELSE                     next_drink_state <= st_cents0;
071-    END IF;
072- ELSIF curr_drink_state=st_cents5 THEN
073-    IF nickel='1'        THEN next_drink_state <= st_cents10;
074-    ELSIF dime='1'      THEN next_drink_state <= st_cents15;
075-    ELSIF coin_ret='1' THEN next_drink_state<=st_cents0; ret_nick<='1';
076-    ELSE                     next_drink_state <= st_cents5;
077-    END IF;
078- ELSIF curr_drink_state=st_cents10 THEN
079-    IF nickel='1'       THEN next_drink_state <= st_cents15;
080-    ELSIF dime='1'      THEN next_drink_state <= st_cents20;
081-    ELSIF coin_ret='1' THEN next_drink_state<=st_cents0; ret_dime<='1';
082-    ELSE                     next_drink_state <= st_cents10;
083-    END IF;
084- ELSIF curr_drink_state=st_cents15 THEN
085-    IF nickel='1'       THEN next_drink_state <= st_cents20;
086-    ELSIF dime='1'      THEN next_drink_state <= st_cents25;
```

```
087-     ELSIF coin_ret='1' THEN next_drink_state<=st_cents5; ret_dime<='1';
088-     ELSE                    next_drink_state <= st_cents15;
089-     END IF;
090- ELSIF curr_drink_state=st_cents20 THEN
091-     IF nickel='1'      THEN next_drink_state <= st_cents25;
092-     ELSIF dime='1'     THEN next_drink_state<=st_cents25; ret_nick<='1';
093-     ELSIF coin_ret='1' THEN next_drink_state<=st_cents10; ret_dime<='1';
094-     ELSE                    next_drink_state <= st_cents20;
095-     END IF;
096- ELSIF curr_drink_state=st_cents25 THEN
097-     IF nickel='1'      THEN next_drink_state <= st_cents25;
098-     ELSIF dime='1'     THEN next_drink_state <= st_cents25;
099-     ELSIF coin_ret='1' THEN next_drink_state <= st_cents25;
100-     ELSIF swill='1'    THEN next_drink_state <= st_cents25;
101-     ELSIF krunk='1'    THEN next_drink_state <= st_cents25;
102-     ELSE                    next_drink_state <= st_armed;
103-     END IF;
104- ELSIF curr_drink_state=st_armed THEN
105-     IF nickel='1'      THEN next_drink_state<=st_armed; ret_nick<='1';
106-     ELSIF dime='1'     THEN next_drink_state<=st_armed; ret_dime<='1';
107-     ELSIF coin_ret='1' THEN next_drink_state<=st_cents15; ret_dime<='1';
108-     ELSIF swill='1'    THEN next_drink_state<=st_cents0; disp_swill<='1';
109-     ELSIF krunk='1'    THEN next_drink_state<=st_cents0; disp_krunk<='1';
110-     ELSE                    next_drink_state <= st_armed;
111-     END IF;
112- END IF;
113- END PROCESS;
114-
115- u0: IBUF PORT MAP(I=>c, O=>in_c);
116- u1: BUFG PORT MAP(I=>in_c, O=>buf_c);
117-
118- -- this process the current state with the next state
119- PROCESS (buf_c,nickel,dime,coin_ret,swill,krunk)
120- BEGIN
121-     -- asynchronously reset the state when all inputs are high
122- IF (nickel AND dime AND coin_ret AND swill AND krunk)='1' THEN
123-   curr_drink_state <= st_cents0;
124- ELSIF (buf_c'event AND buf_c='1') THEN
125-   curr_drink_state <= next_drink_state;
126- END IF;
127- END PROCESS;
128- END drink2_arch;
129-
```

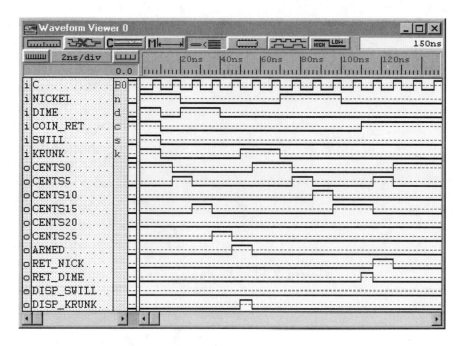

Figure 7.14 Simulated waveforms for the revised drink dispenser controller.

state machine using labeled circles for states and labeled arrows for conditional transitions.

We will go through an example of using the state editor to design the 3-state up/down counter (Figure 7.6). Start by creating a new schematic mode project called CNT3SE_4 for the XC4005XL FPGA (or CNT3SE_9 for the XC95108 CPLD) in the XCPROJ directory. Then select the **Tools → Design Entry → State Editor** menu item (or click on the ⚙ button in the **Flow** tab) in the **Project Manager** window. This will bring up the initial window for the state editor (Figure 7.15). Check the **Use HDL Design Wizard** in the window so that we can be led through the initial stages of creating a state machine. Then click on **OK**.

Click **Next** on the Design Wizard window that appears, after which a window appears that asks us to select the type of HDL that will be used to describe the state machine (Figure 7.16). Check the radio-button labeled **ABEL** so the state machine compiler will output the design in ABEL. You could also select the **VHDL** button to output VHDL code. Then click the **Next** button.

Next, we are asked for the name of the state machine design file (Figure 7.17). Enter CNT3SE and click on the **Next** button.

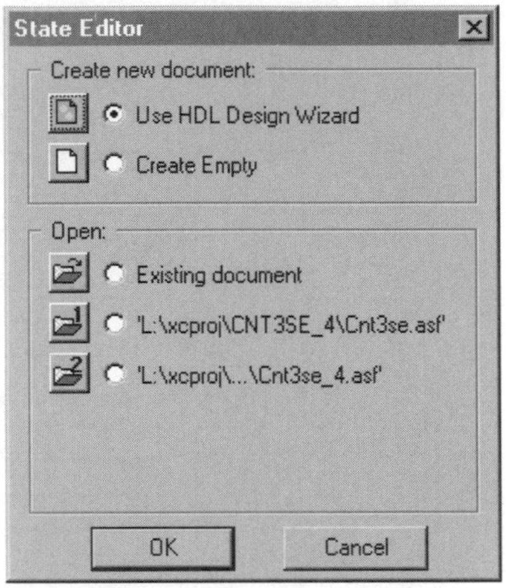

Figure 7.15 Beginning window for using the state editor.

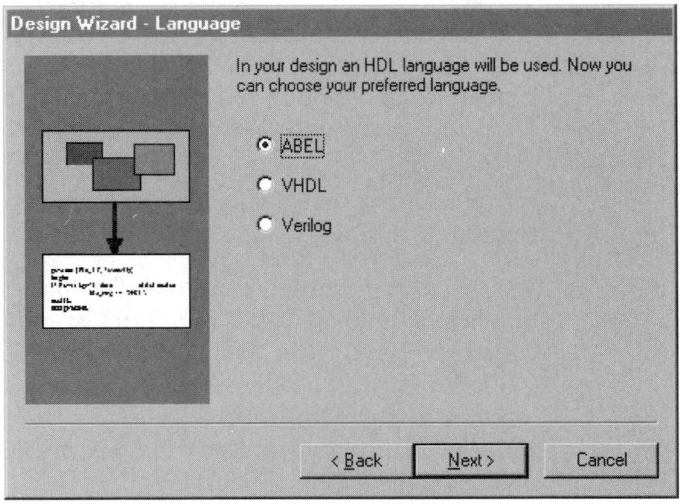

Figure 7.16 Selecting the type of HDL code generated by the state editor.

The next window lets us enter the names of the inputs and outputs for our state machine (Figure 7.18). From Figure 7.6, we can see we need an input called UP that sets the direction of the counter. We also need a RESET input to initialize the state machine. Two combinational ports (OUTPUT0 and OUTPUT1) are also needed to output the current count in the state machine. Finally, a clock input (CLK) is needed to step the counter through its states. After entering these inputs and outputs, click **Next** to advance to the next window.

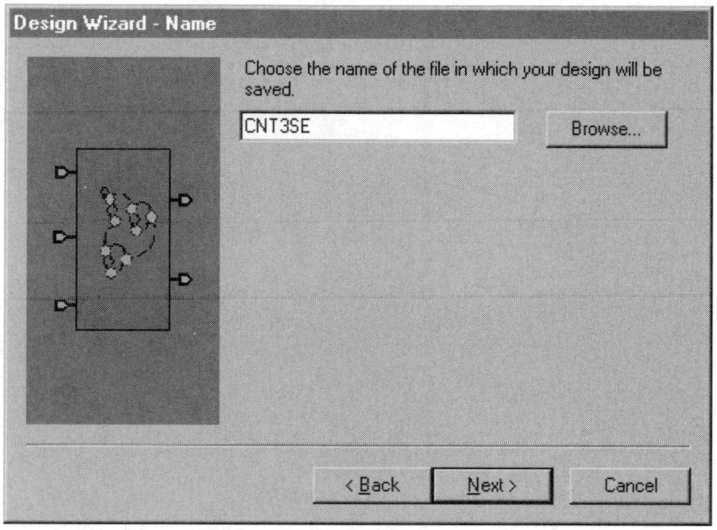

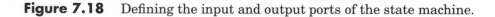

Figure 7.17 Setting the name of the state editor design file.

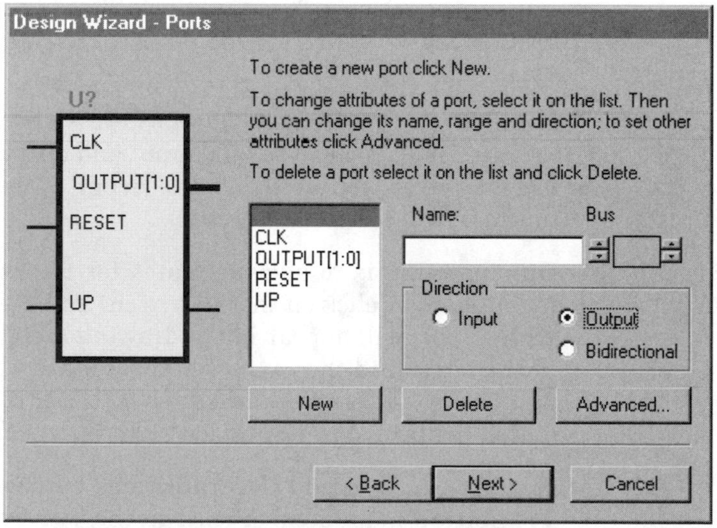

Figure 7.18 Defining the input and output ports of the state machine.

The next window lets us select the number of concurrent state machines that will be running in our design (Figure 7.19). Since this is a simple counter, we only need a single state machine, so click the radio button labeled **One**. (More complicated designs may have multiple state machines running in parallel and communicating information back and forth. This is somewhat analogous to modern computer operating systems which can have several independent processes running at the same time.) Then click on **Finish**.

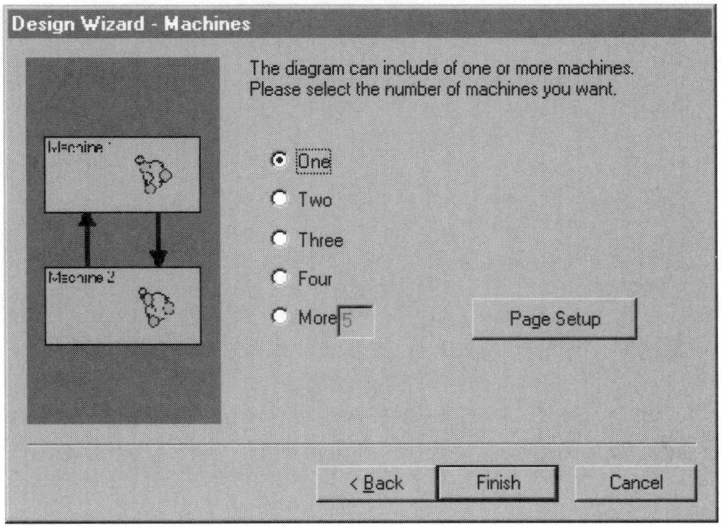

Figure 7.19   Selecting the number of state machines in the design.

Finally we arrive in the window where we can graphically describe our counter (Figure 7.20). You can see the input and output ports arranged at the top of the drawing area. An area labeled **Sreg0** appears below the ports. Our state machine will be entered here.

Our first step is to add the states for the counter. Select the **FSM → State** menu item (or click on the $\boxed{s}$ button) and then click in the Sreg0 area. A circle with a label will appear in the drawing area. You can double-click on the label in the circle and then type whatever state name you want. In this example, I will use S0, S1, and S2 for states 0, 1, and 2 in Figure 7.6. After adding all 3 states, your window should look like Figure 7.21.

Next, we need to add the transitions between the states. Select the **FSM → Transition** menu item or click on the $\boxed{\nearrow}$ button. Then click on the originating state for a transition. After this, click on the destination state for the transition. At this point an arrow will appear between the two states. Then you can use the mouse to pull the arrow into any shape you want. Figure 7.22 shows our state machine after all the transitions are added.

The transitions will do us no good unless we can specify the conditions under which each transition can occur. To do this, select the **FSM → Condition** menu item or click on the $\boxed{=?}$ button. Then click on one of the transition arrows

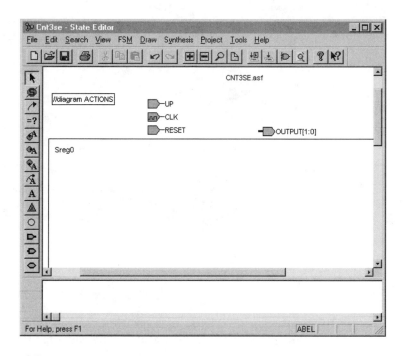

Figure 7.20 Beginning screen for the CNT3SE_4 design.

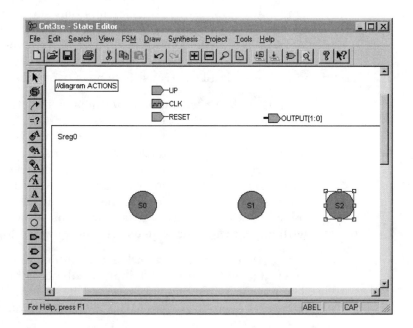

Figure 7.21 Adding the 3 states for the CNT3SE_4 design.

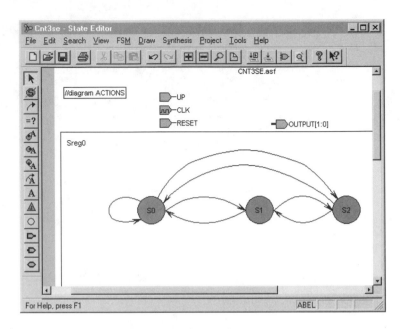

Figure 7.22 Adding state transitions for the CNT3SE_4 design.

in the drawing area. A box will appear in which you can type the Boolean equation that expresses when this transition should be taken. (You have to write the expressions using ABEL or VHDL operators depending upon which HDL you selected.) After typing in the expression, you can use the mouse to move the text to a convenient spot near the arrow. Figure 7.23 shows our state machine after all the transitions have been labeled with conditions.

Even though the transition conditions include the RESET input, it is always a good idea to include an explicit reset condition. Select the **FSM → Reset** menu item (or click on the ⚠ button) and drop the triangular reset symbol into the drawing area. Then connect it with a transition arrow to the state that should be entered when a reset occurs (this is the reset state). Label the transition with a Boolean expression which, if true, will cause the state machine to enter the reset state. You can click the right mouse button on the reset symbol to select whether the reset is asynchronous or synchronous. The state machine with an explicit, synchronous reset is shown in Figure 7.24.

(Given that an explicit reset has been added to the state machine, do we really need to include the RESET input in all the other state transition conditions? The answer is no. The only reason I have kept the more complicated expressions is to keep this example close to the state machine diagram in Figure 7.6.)

Finally, we have to set the actions that will occur in each state. In this example, the only action is to set the value that will appear on the outputs. Select the

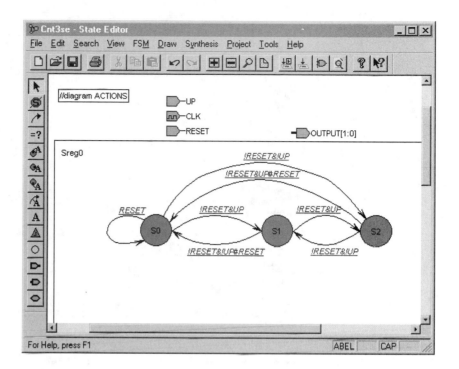

Figure 7.23 Labeling the state transitions with conditions.

FSM → **Action** → **State** menu item (or use the button) and then click on one of the states. A text box connected to the state circle will appear. You can type an expression for the two outputs in this box. (For example, in state S0 the outputs should both be zero so I entered the ABEL statement OUTPUT=^b00.) Then you can move the box with the text to a convenient location. Figure 7.25 shows our state machine after all the states have been labeled with actions.

After the state machine is drawn, select the **File** → **Save** menu item. Now we can generate the HDL code that corresponds to our drawing. Select the **Synthesis** → **HDL Code Generation** menu item to create the ABEL code shown in Listing 7.7. (By now you should be familiar enough with the ABEL's STATE_DIAGRAM construct to detect the correspondence between the code and the graphical state machine.) Then select the **Synthesis** → **Synthesize** menu item to generate the netlist from the ABEL code.

At this point, you can use the **Document** → **Add** menu item in the **Project Manager** window to add either the CNT3SE.ABL file or the CNT3SE.ASF file (which holds the graphical rendition of the counter) to the project. Then you can use the Foundation Series Implementation Tools to compile it for the XC4005XL FPGA or XC95108 CPLD.

Experimental

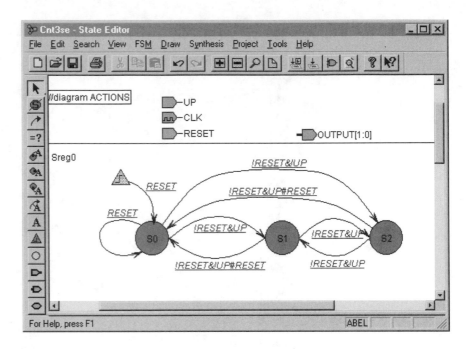

Figure 7.24 Adding a reset to the state machine.

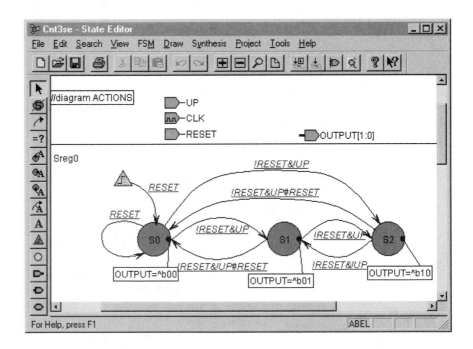

Figure 7.25 Labeling the states with output actions.

Listing 7.7   ABEL code generated for the CNT3SE state machine.

```
001- "
002- "   File:    L:\xcproj\CNT3SE_9\Cnt3se.abl
003- "   created:09/23/98 14:17:30
004- "   from:     'L:\xcproj\CNT3SE_9\Cnt3se.asf'
005- "   by:      fsm2hdl - version: 2.0.1.49
006- "
007- module cnt3se
008- Title 'cnt3se'
009-
010- Declarations
011-
012- "clocks
013- CLK PIN;
014-
015- "input ports
016- RESET PIN;
017- UP PIN;
018-
019- "output ports
020- OUTPUT1..OUTPUT0 PIN;
021- OUTPUT = [OUTPUT1..OUTPUT0];
022-
023- "******** SYMBOLIC state machine: Sreg0 ******
024- Sreg0 STATE_REGISTER;
025- S0, S1, S2 STATE;
026-
027-
028- Equations
029-
030- "diagram ACTIONS
031-
032- "************* state machine: Sreg0 *************
033- " clock signals definitions
034- Sreg0.clk = CLK;
035-
036- State_diagram Sreg0
037- SYNC_RESET S0 : RESET;
038-
039-
040- State S0:
041-
```

Listing 7.7   ABEL code generated for the CNT3SE state machine. (Cont'd.)

```
042-       OUTPUT=^b00;
043-        IF (!RESET&!UP) THEN
044-       S2
045-        ELSE  IF (RESET) THEN
046-            S0
047-        ELSE  IF (!RESET&UP) THEN
048-               S1;
049-
050-
051- State S1:
052-
053-       OUTPUT=^b01;
054-        IF (!RESET&!UP#RESET) THEN
055-       S0
056-        ELSE  IF (!RESET&UP) THEN
057-           S2;
058-
059-
060- State S2:
061-
062-       OUTPUT=^b10;
063-        IF (!RESET&UP#RESET) THEN
064-       S0
065-        ELSE  IF (!RESET&!UP) THEN
066-           S1;
067-
068- " end of state machine - Sreg0
069-
070-
071- end cnt3se
072-
```

A user-constraint file for the CNT3SE design should assign the inputs and outputs like this for the XS40 Board:

```
NET CLK          LOC=P44;    # B0 argument of XSPORT
NET RESET        LOC=P45;    # B1 argument of XSPORT
NET UP           LOC=P46;    # B2 argument of XSPORT
NET OUTPUT0      LOC=P25;    # S0 LED segment
NET OUTPUT1      LOC=P26;    # S1 LED segment
```

Or use the following constraints for the XS95 Board:

```
NET CLK          LOC=P46;    # B0 argument of XSPORT
NET RESET        LOC=P47;    # B1 argument of XSPORT
NET UP           LOC=P48;    # B2 argument of XSPORT
NET OUTPUT0      LOC=P21;    # S0 LED segment
NET OUTPUT1      LOC=P23;    # S1 LED segment
```

After downloading your circuit to the XS40 or XS95 Board, you can use the following command sequences to move the state machine between its various states. Each set of commands activates a particular input and then sends a rising clock edge to the state machine. You can see the output for each state by observing the S0 and S1 LED segments.

Action	XSPORT Command Sequence
Reset the counter	XSPORT 010 XSPORT 011 XSPORT 000
Increment by one	XSPORT 100 XSPORT 101 XSPORT 000
Decrement by one	XSPORT 000 XSPORT 001 XSPORT 000

Projects

1. Design a circuit that counts in the following sequence:

   7,6,5,4,3,4,5,6,7,6,5,4,3,4,5,6,7,....

   The circuit should change states on the falling edge of the clock. The circuit should have a reset input that forces it into the 3 state on the next falling clock edge. The output of the circuit should drive an LED digit and display the appropriate digit as it operates.

2. Design a 4-bit counter that will count up when a control signal is at logic 0, and will count down when the control signal is logic 1. The output of the circuit should drive an LED digit and display the appropriate digit as it operates.

3. Add a new 20¢ drink selection to the DRINK2.ABL or DRINK2.VHD design.

4. Modify DRINK2.ABL or DRINK2.VHD so that multiple simultaneous button presses are ignored.

5. Change DRINK2.ABL or DRINK2.VHD from a one-hot state encoding to a minimal encoding. Compile the design for the XC4005XL FPGA and XC95108 CPLD. Compare the device utilizations for both the one-hot and minimal encodings.

6. Redesign the drink dispenser controller using the state editor.

Memories

Objectives

- Show how various types of random access memory operate.
- Build a RAM using an FPLD.
- Show how to use the built-in RAMs of the FPGA.
- Show how to use an external RAM with an FPLD to build a FIFO.

Discussion

Simple Memory

Memory merely stores information so it can be recalled later. You already have some experience with memory since you used D flip-flops to store the current state of the state machines in the previous couple of chapters. But those examples used only a small amount of memory—just a few bits. This chapter will discuss larger memories capable of holding millions of bits. These large memories are constructed using very advanced circuit-design techniques and semiconductor technologies. However, they can be understood using only our knowledge of D flip-flops and combinational logic, and small, working memories can be built using the logic in FPLDs.

First, you must understand the function and operations of memory from a user's point of view (this is typically called a programmer's model of memory.) The programmer's model for a memory is shown in Figure 8.1. This memory consists of eight memory locations or registers with 3-bit addresses of 000, 001, 010, 011, 100, 101, 110, and 111. Each memory location is 4 bits wide, so each location can store a single number between 0 and 15, inclusive. There are three memory buses: (1) the input bus that transfers numbers to be written into the memory; (2) the output bus that reads numbers back out of the memory; and (3) the address bus that selects one of the memory locations for reading or writing.

The memory operates in one of two modes: (1) write mode which is used to enter numbers into the memory; and (2) read mode which is used to get the numbers back out. Figure 8.2 depicts these two modes. In Figure 8.2a, the user has forced a value of 13 (1101 in binary) onto the input bus and an address of 5 (101

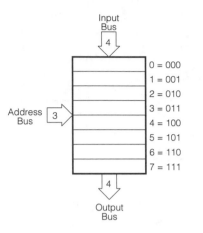

Input
Bus

4

Address
Bus

3

0 = 000
1 = 001
2 = 010
3 = 011
4 = 100
5 = 101
6 = 110
7 = 111

4

Output
Bus

Figure 8.1   Programmer's model for a random access memory.

in binary) onto the address bus. This activates memory location 5 and causes it to write the number on the input bus into its 4 bits of storage. An identical write operation occurs in Figure 8.2b, except that the address and data are changed to 2 and 8, respectively. Thus, the number 8 is now stored in location 2.

The stored numbers can be read out of the memory by placing an address on the address bus. In Figure 8.2c, the memory location at address 5 is activated, which causes it to output the number it stores onto the output bus. Thus, the previously written value of 13 is read out on the output bus.

Notice that you can read or write any memory location merely by specifying its address. This lets you access any randomly selected memory location whenever you want. That is why this type of memory is called random access memory or RAM.

Once you understand the basic functions of a RAM, you can begin to consider how to build one. The first thing you might build would be the circuitry for selecting which memory location is read or written. This is called the address decoder and is shown in Figure 8.3. The address decoder for our example accepts five inputs:

- Three address inputs (A0, A1, and A2) to specify the register address
- A Write input, which indicates that the number on the input bus should be written into the addressed memory location
- A Read input, which indicates the addressed memory location should output its stored data onto the output bus.

The address decoder has two outputs to every memory location (for a total of 16 outputs in this example). One output instructs the memory location to output

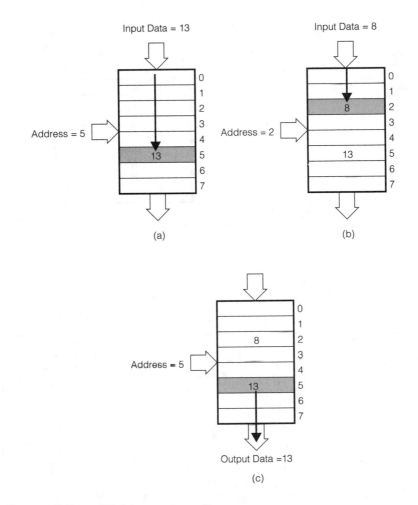

Figure 8.2 Writing and reading a memory.

its stored data, and the other output causes the memory location to write new data into itself.

The truth table of an address decoder for a four-location memory is shown in Table 8.1. The read line, rn, for memory location n goes to logic 1 when the read control input is high and the value on the address bus is n. A memory location is written in a similar manner, except the write control line and the individual write lines for each memory location are all active when they are at a low logic level. Note that none of the read or write lines are active if the user activates both the read and the write control inputs (read = 1 and write = 0) and creates a read/write conflict. Note also that only one read or write line is active during a memory operation; there is no case where two or more registers are reading or writing data simultaneously.

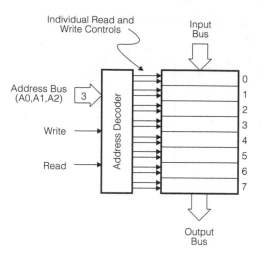

Figure 8.3   RAM address decoder.

Table 8.1   Truth table for an address decoder. (Read is active-high, and Write is active-low.)

Address				0		1		2		3		
A1	**A0**	**Read**	**Write**	**r0**	**w0**	**r1**	**w1**	**r2**	**w2**	**r3**	**w3**	**Operation**
0	0	0	0	0	0	0	1	0	1	0	1	write 0
0	0	0	1	0	1	0	1	0	1	0	1	inactive
0	0	1	0	0	1	0	1	0	1	0	1	conflict
0	0	1	1	1	1	0	1	0	1	0	1	read 0
0	1	0	0	0	1	0	0	0	1	0	1	write 1
0	1	0	1	0	1	0	1	0	1	0	1	inactive
0	1	1	0	0	1	0	1	0	1	0	1	conflict
0	1	1	1	0	1	1	1	0	1	0	1	read 1
1	0	0	0	0	1	0	1	0	0	0	1	write 2
1	0	0	1	0	1	0	1	0	1	0	1	inactive
1	0	1	0	0	1	0	1	0	1	0	1	conflict
1	0	1	1	0	1	0	1	1	1	0	1	read 2
1	1	0	0	0	1	0	1	0	1	0	0	write 3
1	1	0	1	0	1	0	1	0	1	0	1	inactive
1	1	1	0	0	1	0	1	0	1	0	1	conflict
1	1	1	1	0	1	0	1	0	1	1	1	read 3

The next issue is how to build the memory storage array that is controlled by the address decoder. Let's start by building an individual memory location or register, shown in Figure 8.4. The four D inputs of the four D flip-flops act as the input bus to the 4-bit memory location. The four Q outputs make up the 4-bit output bus. The clock inputs of all the D flip-flops are connected, and this acts as the write line, w, for this memory location. Whenever a rising edge occurs on w, the logic levels on the input bus will be latched into the D flip-flops, thus fulfilling the requirements of a write operation. Notice that there is no equivalent read line, r, because the Q outputs are active all the time. This will cause problems later, as you will see.

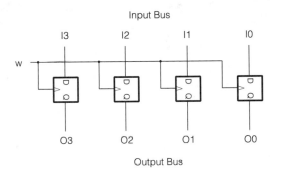

Figure 8.4   An individual memory location.

How do you combine several registers to build a larger storage array? Again, that is easy (at least when considering the write operation). Figure 8.5 shows how to do it with three registers. Merely construct three identical registers and connect their equivalent D inputs. When a write operation occurs, the value to be written will be on the input bus and goes to each register. However, only one of the registers will store the data because the address decoder will only activate one of the w0, w1, or w2 lines. For example, if the address inputs to the address decoder had the value 001, then the w1 write line would be active and the value on the input bus would be stored in memory location 1. Larger memories can be built merely by connecting more registers and building a larger address decoder.

The memory I have just described is asynchronous in that a write will occur anytime a rising edge occurs on one of the w lines. Each w line is affected by the address inputs as well as the write control input. No global clock signal determines when the writes will take place. Because of the way the address decoder works, a sequence of RAM addresses can be written by applying different addresses while holding the write line low. The changing address lines will cause the rising edges on the w lines as long as write = 0. But this is not a very reliable way to write to memory since it is hard to make sure all the

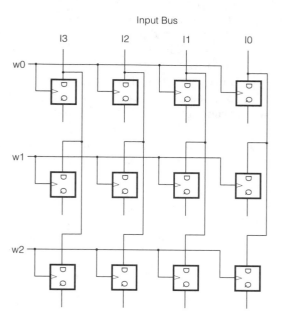

Figure 8.5   Three writable memory locations.

address lines change at the same time. If you applied address [A1..A0] = 11 and then 00, it is possible that addresses 10 or 01 may also get written if A0 changes from 1 to 0 slightly before or after A1 does, respectively. It is much safer to write to memory using the following procedure:

1. Raise the write control input to logic 1. This disables writing to the RAM.

2. Apply the address you want to write to the address lines. You can simultaneously apply a value to the input data bus. This ensures that the address and data lines are stable before the actual write occurs.

3. Lower the write control input to logic 0 and then raise it back to logic 1. This writes the input data value into the memory location selected by the address.

The next piece of the memory circuit we need to design allows data to be read from the registers. We cannot simply connect the Q outputs of each register to the output bus because the outputs would interfere with one another and erroneous data would be read out of the memory (see Figure 8.6). Instead, you have to build a multiplexer. A multiplexer selects one binary value from several inputs and places that value on its output. A simple 2-to-1 multiplexer is used in Figure 8.7 to build a memory with two registers. The multiplexer consists of AND gates that feed OR gates and operates as follows (see Figure 8.8). If r0 is high and r1 is low (meaning memory location 0 has been selected), then the

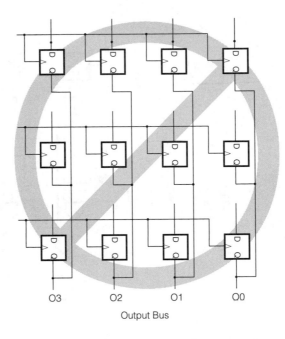

Figure 8.6   The incorrect way to hook up the output bus of a memory.

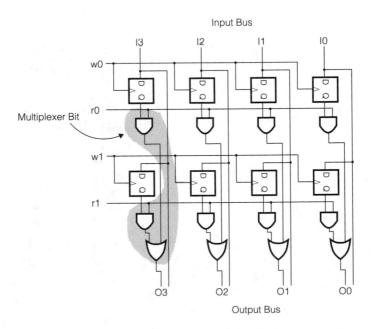

Figure 8.7   Using a 2-to-1 multiplexer to build a readable memory.

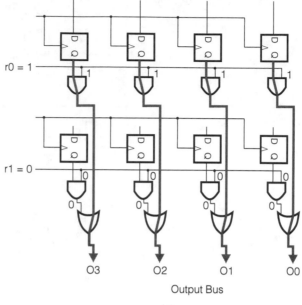

r0 = 1

O3 O2 O1 O0

Output Bus

(a)

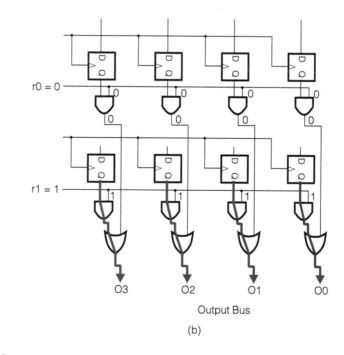

r0 = 0

r1 = 1

O3 O2 O1 O0

Output Bus

(b)

Figure 8.8   Reading the memory.

Memories   Chapter 8

outputs of the lower set of AND gates are all zero, while the values of the Q outputs in memory location 0 show up on the outputs of the upper set of AND gates. The OR gates logically OR the contents of memory location 0 (from the upper set of AND gates) with a set of zeros (from the lower set of AND gates). The result is that the OR gates output the value stored in memory location 0. Similarly, if r0 = 0 and r1 = 1, then the OR gates output the value stored in memory location 1 (Figure 8.8b). The combination of the AND gates and the read lines serves to block all but one register from putting their values on the output bus. The memory size can be increased merely by adding more memory locations and concatenating the multiplexers. A RAM with four memory locations is shown in Figure 8.9.

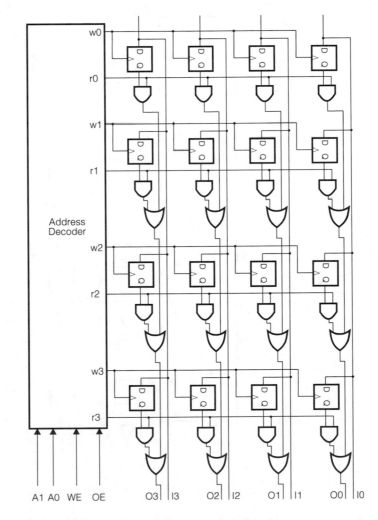

Figure 8.9   A RAM with four memory locations.

By this time you should realize that real memories are not built this way. For example, a memory containing 32,000 memory locations would require data from a register to travel through as many as 32,000 gates in order to exit from the chip. If a gate delay is 5 ns, the cumulative delay would be 160 μs. (In contrast, you can buy cheap memory for your PC with access times of 70 ns—more than 2000 times faster.) In addition, a separate input and output bus are needed and this is wasteful since the buses are not active at the same time. In real memories, both these problems are solved using tristate gates (which we discussed in Chapter 5). Figure 8.10 shows the memory with the multiplexers replaced with tristate gates and both the input and output buses merged into a single memory data bus. The tristate gates allow the value in a register to be driven onto the

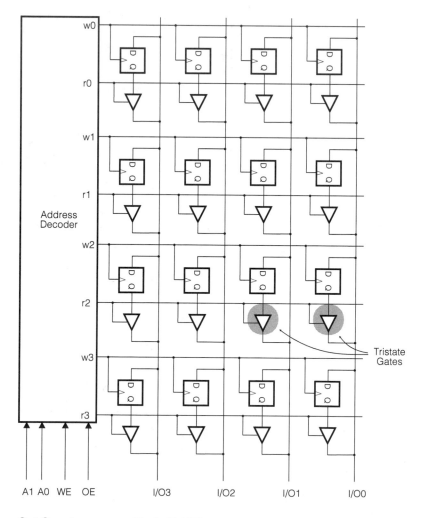

Figure 8.10 A more realistic RAM.

output bus if the associated read line is high. Otherwise, the tristate drivers are disabled and the contents of the register have no affect on the bus. Since no more than one register is ever selected at any time (because of the way the decoder circuitry is built), there is never any contention on the memory bus.

When writing a memory with a single bus, all of the individual read lines are disabled so that none of the registers is driving the memory bus. This leaves the bus free for use as an input bus that can bring a value into the memory array. This value can be stored in a register by activating the associated write line. The single bus method works because read lines and write lines are never active simultaneously.

Specialized Memories

Synchronous Memory

The memory described in the previous subsection was asynchronous because new data could be written to it just by changing the address lines and/or the write control input. As we discussed, reliable writing to an asynchronous memory requires that the address lines be stable before the write pulse occurs. And as we have seen in the previous chapter, synchronous systems in which all components change their state on a global clock edge are easier to analyze for setup and hold-time violations. A synchronous RAM component lets us maintain this advantage.

Figure 8.11 shows a RAM modified for synchronous write operations. Each D flip-flop has been modified to include a 2-to-1 multiplexer on its input. One input to the multiplexer is the data bus and the other input is driven by the output of the flip-flop. The w lines now connect to the multiplexer instead of the DFF clock. If w = 0, the value on the data bus is passed to the DFF's D input. Otherwise, the DFF's Q output is recycled back to its input.

The entire set of flip-flops in the synchronous RAM all receive the same global clock. When a rising edge occurs on the clock line, the DFFs in row i will be written with new data from the bus if $w_i = 0$. All the other flip-flops will reload themselves with their current contents.

Dual-port Memory

The RAM architectures we have examined so far are single-port types: There is a single bus on which data enters and exits. In some applications it is necessary for two different systems to access a RAM at the same time. Dual-port memories are used in these applications.

A simple dual-read/single-write RAM is shown in Figure 8.12. This RAM has one I/O bus for reading and writing to the memory locations, and a second

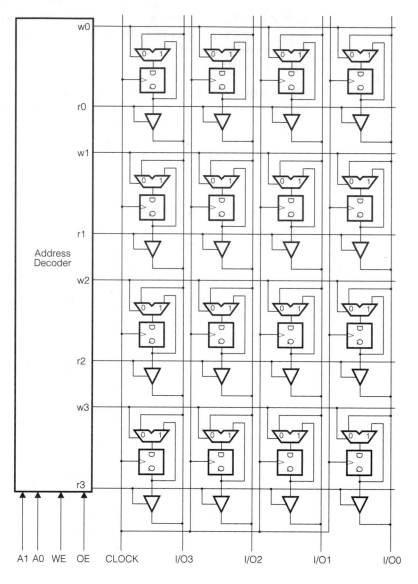

Figure 8.11 A synchronous RAM.

output-only bus for reading the memory contents. A second memory address bus and output-enable are provided so the second bus can output a value while the first bus is performing an independent read or write operation.

We could also build the circuitry to allow writing of memory locations via the second bus. But we would also have to build some control circuitry to detect when both ports are trying to write data to the same register.

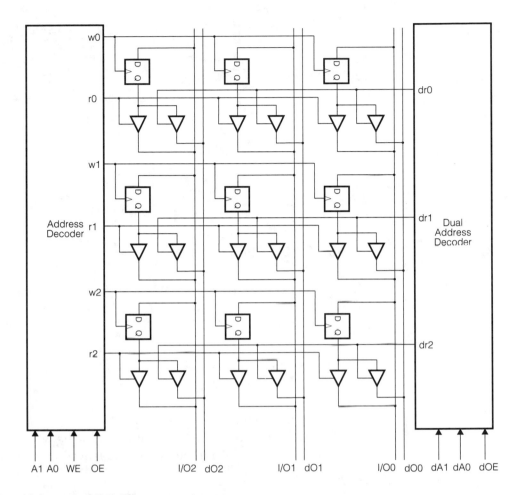

Figure 8.12   A dual-port RAM.

First-In, First-Out Buffers

In a communication system, a transmitter may send data in bursts faster than the receiver can handle it. Buffers are needed which can accept short bursts of high-speed data and then allow the data to be read out as needed. In addition, the bytes in the data stream must remain in order so that the first byte entered into the buffer is the first byte read out from the buffer. A device that performs these operations is called a first-in, first-out buffer or FIFO.

An example of a FIFO in operation is shown in Figure 8.13. The FIFO is built from a RAM with room for four data bytes. Two memory pointers called rdptr and wrptr are used to address the next location in the RAM that will be read or written, respectively. There are also two flags called empty and full to indicate how much data is in the FIFO. When all four RAM locations are written

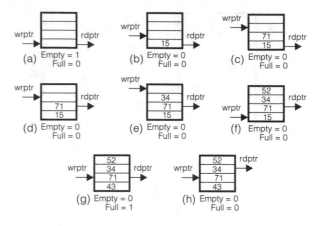

Figure 8.13 FIFO operations.

with data, then full is set to 1. When all four locations are empty, then empty is set to 1. Otherwise, both empty and full are reset to 0. An example of the FIFO in operation goes as follows (the letters in parentheses refer to parts of Figure 8.13):

(a) The FIFO has no data in it so both wrptr and rdptr point to the same RAM address and empty is set to 1.

(b) A write operation occurred. The value 15 was written into the RAM location addressed by the wrptr. Then the wrptr was incremented so it points to the next RAM location. Now the write and read pointers are no longer equal and the empty flag is zero since the FIFO contains some data.

(c) Another write operation has occurred. The value 71 was written to the FIFO and the wrptr was incremented again.

(d) A read operation occurred. The value 15 is read out of the RAM location pointed to by the read pointer. Then rdptr is incremented. Notice that the first value stored in the FIFO was also the first value read out of it. Notice also that the value stored in the FIFO is not removed after the read operation.

(e) A value of 34 was written to the FIFO.

(f) A value of 52 was written to the FIFO. Notice that wrptr has rolled over so that it now points to the first RAM location again.

(g) A value of 43 has overwritten the value of 15 that was stored in the first RAM location. Then the wrptr is incremented. This makes the read and write pointers equal, and all four locations of the RAM are loaded with data that has not been read yet. In this condition, the full flag is set to 1.

(h) Another read operation was performed. The value of 71 that was loaded in step (c) was finally read and the rdptr was incremented. The FIFO is no longer completely full, so the full flag is reset to zero.

From the example, you can see that the reader of the FIFO gets the data in the same order as it was entered by the writer. The advantage of using the FIFO is that it decouples the reader and writer so they do not have to do their operations in lock-step. For instance, a sensor might provide four bytes of data at 1 μs intervals, but each burst occurs only once a second. Without a FIFO, each data byte would have to be transmitted as soon as it arrived at a rate of 8 bits/byte × 1 byte/μs = 8 bits/μs = 8,000,000 bits/sec. Instead, a FIFO could be used to accept the high-speed sensor bursts and then dole them out slowly to a transmitter that operates at a rate of 8 bits/byte × 4 bytes/s = 32 bits/s. The decoupling of the sensor and the transmitter by the FIFO allows you to build a transmitter based on the average data rate from the sensor rather than the peak rate. And the equipment needed to transmit at 32 bits/s is much cheaper than what is needed to do 8 megabytes/s.

Now that you know how a FIFO operates and the advantages it provides, you have to return to the question, "How do I build one?" First, you should define the interface as shown in Figure 8.14. There are two ports:

Write port: The write port consists of an 8-bit in bus for entering the data bytes. A low pulse on the wr control line writes the data into the FIFO. The full status output informs the data supplier if the FIFO is full and cannot accept any more data.

Read port: The read port has an 8-bit out bus for sending out the data bytes. A low level on the rd line places the next byte of data from the FIFO onto the out bus. The empty status output informs the data consumer if the FIFO is empty and no data is available.

The final signal in the interface is the reset, which has the obvious function of initializing the FIFO to the empty state.

From the interface description, you can work inward to create a simple block diagram like the one in Figure 8.15. The components you will need are as follows:

RAM: Obviously, you need static RAM to store the data.

rdptr: The read pointer is a counter that increments in the wrap-around sequence $0, 1, ..., N-1, 0, 1$ so that all N memory locations can be addressed. A read operation must cause the pointer to be incremented, but not if the FIFO is empty. If that happened, then the write pointer would be one behind the read pointer and the FIFO would interpret this to mean the FIFO had $N-1$ data bytes in it. Then these $N-1$ bytes of old data would be erroneously read out as if they were new.

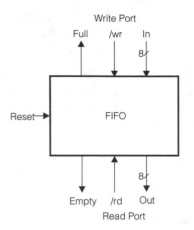

Figure 8.14 FIFO interface signals.

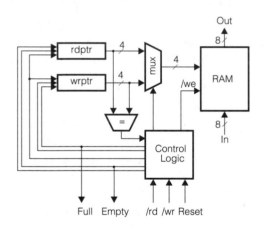

Figure 8.15 FIFO components.

wrptr: The write pointer also increments in the wrap-around sequence 0, 1, ..., $N - 1$, 0, 1. A write operation increments the pointer, but not if the FIFO is full. If that happened, then the read pointer would be one behind the write pointer and the FIFO would interpret this to mean that the FIFO contained only one data byte. Thus, $N - 1$ bytes of data would be lost.

Multiplexer: This component directs either the read or write pointer to the address inputs of the RAM depending on the operation being performed.

Comparator: The read and write pointer values must be compared for equality to determine when the FIFO is full or empty.

Control logic: This is the block where you throw all the stuff you do not know how to do yet. It controls the write enable line of the RAM and the address multiplexer. It also controls the incrementing of both the read and write pointers depending upon whether the FIFO is empty or full.

The circuitry for a FIFO pointer is shown in Figure 8.16. The pointer has clk, clr, and stop inputs, which cause the pointer to increment, asynchronously clear to zero, or stop incrementing, respectively. The register is made of D flip-flops, which hold the current address in the pointer. The pointer changes its value on the rising edge of the clk signal. If stop = 0, then the multiplexer steers the incremented value of the register back into the register so that an increment occurs. If stop = 1, the register just reloads its current value.

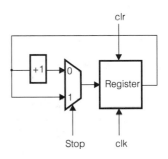

Figure 8.16   FIFO pointer module.

The FIFO pointer module is used twice to build the read and write pointers. The clocks for these pointers are the rd and wr inputs. Both rd and wr are normally high and are pulsed low to cause a read or write operation, respectively. The rising edge at the end of either a read or a write pulse causes the respective pointer to change state. The empty or full outputs are used to stop the increment of rdptr or wrptr, respectively. The global reset line is also hooked to the clr input of each pointer so that they can both be reset to zero upon startup.

The outputs of both pointers go to the equality comparator and the multiplexer. The multiplexer selects which pointer addresses the RAM using the rd input. If rd is low (i.e., a read operation), then the read pointer addresses the RAM. Otherwise, the write pointer addresses the RAM. It is important that the write pointer should address the RAM whenever there is not a read operation. If you built the opposite arrangement so that the read pointer normally addressed the RAM except when a write occurred, there is the possibility of corrupting data in the FIFO. This occurs because the write address does not have enough time to propagate through the multiplexer and enable the correct RAM location for writing before the write pulse gets to the RAM. Thus, the location addressed by the read pointer may get written into instead. You can avoid this problem by always addressing the RAM with the write pointer except when a read operation occurs. (This does mean that you have to wait a bit longer to read the output of the FIFO so that the address has time to settle and the RAM has time to access the memory location. But no erroneous data can slip into the FIFO during a read operation.)

The control logic to determine whether the FIFO is empty or full is described next. The FIFO is definitely neither empty nor full if the read and write pointers store different addresses. But how do you determine whether the FIFO is actually empty or full when the read and write pointers have the same value? (Look at Figures 8.13a and 8.13g to see that pointer equality is not enough to distinguish empty from full.) The answer lies in recording the type of the previous operation on the FIFO. If the previous operation was a write, then the FIFO cannot be empty because something was just written into it. Conversely, if the previous operation was a read, then the FIFO cannot be full because something was just read from it. So we can state the logic as follows:

> If the read and write pointers are equal and the last operation was a write, then the FIFO is full.

> If the read and write pointers are equal and the last operation was a read, then the FIFO is empty.

The logic is simple so all that remains is finding a way to record the type of the previous operation. Obviously, you need a flip-flop, but where does its clock come from? Figure 8.17 shows the circuitry for this. The prevop0 flip-flop can

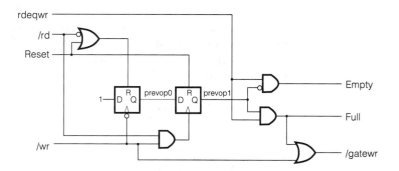

Figure 8.17   FIFO control logic.

be immediately set to logic 1 to indicate a write operation if we tie the D input high and clock the flip-flop with the falling edge of the write line. Conversely, we can drive the asynchronous clear with the read line to reset the flip-flop when a read occurs. Otherwise, the flip-flop retains its value.

The output of prevop0 cannot be used directly because it changes value as soon as a read or write operation begins. The current operation must complete before the flip-flop is updated. Therefore a second flip-flop, prevop1, is used to delay the output from prevop0 until after the current operation is complete. The clock for the prevop1 flip-flop is formed from the logical AND of the read and write control lines so that a low pulse on either control line updates the prevop1 flip-flop with the value stored in prevop0.

A final piece of control logic is needed to prevent the RAM from being corrupted by a write to a full FIFO. The active-low write-enable signal to the RAM, gatewr, is driven by an OR gate that has the write control line of the FIFO and the full status signal as inputs. If the FIFO is full, then gatewr is held high so that no data can be written into the RAM. If the FIFO is not full, then the write pulse from the FIFO write control input passes through to the RAM write-enable unchanged.

Experimental

Building a Small Asynchronous Memory

The ABEL code for the 4-register memory with 4 bits per register (similar to the one in Figure 8.9 which we will call a 4×4 memory) is shown in Listing 8.1.

Lines 6–9 declare pins for the write and output-enable control inputs, the 2-bit address bus, and the 4-bit input data bus, respectively. Internal signals are declared on lines 10–14. Line 10 defines a 4-bit register composed of 4 D flip-flops. The other three registers are defined on lines 11–13. The signals that control writing into these registers are declared on line 14.

The 4-bit memory output bus is declared on line 15. In keeping with Figure 8.9, we are using separate input and output buses. A combined I/O bus (as in Figure 8.10) can be used if the design is targeted only toward the XC4000 FPGAs. An I/O bus requires internal tristate buses, which are not available in the XC9500 CPLD.

We also add another set of 7 outputs on line 16 to display the output from the RAM on the LED digit.

Lines 18–25 just define some shorter names for some of the signals that we declared earlier. The construction of the memory begins by connecting the input data bus to the 4 registers on lines 30–33. Note the use of the := operator to signify that the value on the input data bus is being clocked into the registers.

Lines 35–48 describe the actual signals that clock the registers. The operation of the write line for each register is described on lines 38–41 using the ABEL WHEN...THEN...ELSE construct. If the Boolean expression following the

Listing 8.1 ABEL version of a 4-bit wide asynchronous memory built from 4 register modules.

```
001- MODULE MEM1
002- TITLE 'A 4x4 asynchronous memory built from DFFs'
003-
004- DECLARATIONS
005-
006- WEB          PIN;          "active-low write-enable input
007- OE           PIN;               "active-high output enable
008- A1..A0       PIN;               "2-bit address inputs
009- INP3..INP0   PIN;               "4-bit input data bus
010- RGA3..RGA0   NODE ISTYPE 'REG'; "first 4-bit register
011- RGB3..RGB0   NODE ISTYPE 'REG'; "second 4-bit register
012- RGC3..RGC0   NODE ISTYPE 'REG'; "third 4-bit register
013- RGD3..RGD0   NODE ISTYPE 'REG'; "fourth 4-bit register
014- W3..W0       NODE ISTYPE 'COM'; "register write control lines
015- OUTP3..OUTP0 PIN  ISTYPE 'COM'; "4-bit output data bus
016- S6..S0       PIN  ISTYPE 'COM'; "LED driver outputs
017-
018- A    = [A1..A0];         "synonym for the address bus
019- INP  = [INP3..INP0];     "synonym for input bus
020- OUTP = [OUTP3..OUTP0];   "synonym for output bus
021- RGA  = [RGA3..RGA0];     "synonym for first register
022- RGB  = [RGB3..RGB0];     "synonym for second register
023- RGC  = [RGC3..RGC0];     "synonym for third register
024- RGD  = [RGD3..RGD0];     "synonym for fourth register
025- S    = [S6..S0];         "synonym for LED drivers
026-
027- EQUATIONS
028-
029- "each register receives its input from the input bus
030- RGA := INP;
031- RGB := INP;
032- RGC := INP;
033- RGD := INP;
034-
035- "a simple address decoder that raises the write-line
036- "to the selected register when the input address
037- "matches its address and WEB=0
038- WHEN      ((A==0) & !WEB) THEN W0=1;
039- ELSE WHEN ((A==1) & !WEB) THEN W1=1;
040- ELSE WHEN ((A==2) & !WEB) THEN W2=1;
041- ELSE WHEN ((A==3) & !WEB) THEN W3=1;
```

```
042-
043-  "clock the registers with the inverse of the write
044-  "control lines so we get a rising edge when WEB rises
045-  RGA.CLK = !WO;
046-  RGB.CLK = !W1;
047-  RGC.CLK = !W2;
048-  RGD.CLK = !W3;
049-
050-  "a simple address decoder that passes the value from
051-  "one of the registers to the output bus
052-  WHEN      (A==0) THEN OUTP=RGA;
053-  ELSE WHEN (A==1) THEN OUTP=RGB;
054-  ELSE WHEN (A==2) THEN OUTP=RGC;
055-  ELSE                  OUTP=RGD;
056-
057-  "enable the memory output drivers when OE=1
058-  OUTP.OE = OE;
059-
060-  "drive LED digit to display memory output bus
061-  TRUTH_TABLE (OUTP->S)
062-     0 -> ^b1110111; "0 (we've used binary number format
063-     1 -> ^b0010010; "1  to express the pattern of ones
064-     2 -> ^b1011101; "2  and zeroes on the LED segments
065-     3 -> ^b1011011; "3  for each numeral)
066-     4 -> ^b0111010; "4
067-     5 -> ^b1101011; "5
068-     6 -> ^b1101111; "6
069-     7 -> ^b1010010; "7
070-     8 -> ^b1111111; "8
071-     9 -> ^b1111011; "9
072-    10 -> ^b1111110; "A
073-    11 -> ^b0101111; "b
074-    12 -> ^b1100101; "C
075-    13 -> ^b0011111; "d
076-    14 -> ^b1101101; "E
077-    15 -> ^b1101100; "F
078-
079-  END MEM1
```

WHEN keyword evaluates to a logic 1, then the equations following the THEN
keyword are activated. If the expression evaluates to a logic 0 instead, then the
equations following the ELSE keyword are active. In this case, all the write

lines, W0-W3, are usually at logic 0. However, on line 38, if the input address equals 0 (A==0) and the write input is low (!WEB), then the W0 line goes to logic 1. If the conditions for activating W0 are not met, but instead A==1 and the WEB==0, then the W1 line goes to logic 1 (Line 39). The logic to control the W2 and W3 write lines is similar and is described on lines 40 and 41. If none of the Boolean expressions evaluates to a logic 1, then none of the equations is activated and the write lines all stay at logic 0.

Lines 45–48 attach the write control lines to the clock inputs of their respective registers. Note that the clock inputs are inverted so they are normally high except when the register is being written. Thus, a rising edge will occur on the clock inputs at the end of the write operation, causing the D flip-flops to be loaded with new values. This will clock the value on the input bus into the selected register. The memory is asynchronous since each of the registers receives its own individual clock, which is derived from the main write control input to the RAM.

Lines 52–55 use the WHEN...THEN...ELSE construct to describe a multiplexer circuit. The multiplexer just examines the input address and gates the output of one of the registers onto the output bus. Line 58 attaches the OE input to the output enable control of the output data bus buffers. Thus, the buffers will be in a high impedance state if OE is low. The value of a memory location can only be read out when OE is at a logic 1.

The output from the memory drives an LED decoder circuit described by the truth-table on lines 61–77.

The VHDL version of the 4×4 memory is shown in Listing 8.2. The ENTITY block on lines 7–17 declares the interface to the memory. The following ARCHITECTURE section begins with a declaration of the four 4-bit signal arrays that will store the register contents (lines 21–24). This is followed by declarations for the write-enable controls for the registers (lines 26 and 27).

The memory address decoder is described on lines 33 through 41. The write control line for a register is lowered if its address matches the address inputs and the main write-enable input is low. If the write-enable input is not low or none of the addresses match, then the write control lines all remain at logic 1.

Each write control line serves as a clock for one of the individual register processes on lines 43–48, 50–55, 57–62, or 64–69. In these processes, the value on the input data bus is written into the register on the rising edge of the write control signal.

The output multiplexer on lines 71 through 74 uses the input address to select one of the registers and pass its contents onto the intermediate output bus, outbus. Then the value on the intermediate output bus is passed onto the main output bus if the output enable is high (line 75). If the output enable is low, then the output bus is placed in a high-impedance state.

Listing 8.2 VHDL version of a 4-bit wide asynchronous memory built from 4 register modules.

```
001- -- 4x4 asynchronous memory
002-
003- LIBRARY IEEE,xse;          -- access IEEE and XSE libraries
004- USE IEEE.std_logic_1164.ALL;
005- USE xse.led.ALL;           -- get access to the LED decoder module
006-
007- ENTITY mem1 IS
008- PORT
009- (
010-   web: IN STD_LOGIC;       -- active-low write-enable
011-   oe: IN STD_LOGIC;        -- active-high output-enable
012-   a: IN STD_LOGIC_VECTOR (1 DOWNTO 0); -- 2-bit address
013-   inp: IN STD_LOGIC_VECTOR (3 DOWNTO 0); -- 4-bit input bus
014-   outp: BUFFER STD_LOGIC_VECTOR (3 DOWNTO 0);-- 4-bit out bus
015-   s: OUT STD_LOGIC_VECTOR (6 DOWNTO 0) -- LED display outputs
016- );
017- END mem1;
018-
019- ARCHITECTURE mem1_arch OF mem1 IS
020-
021- SIGNAL rga: STD_LOGIC_VECTOR (3 DOWNTO 0);-- register a
022- SIGNAL rgb: STD_LOGIC_VECTOR (3 DOWNTO 0);-- register b
023- SIGNAL rgc: STD_LOGIC_VECTOR (3 DOWNTO 0);-- register c
024- SIGNAL rgd: STD_LOGIC_VECTOR (3 DOWNTO 0);-- register d
025- -- the write enable array and individual signals
026- SIGNAL w: STD_LOGIC_VECTOR (3 DOWNTO 0);
027- SIGNAL w0,w1,w2,w3: STD_LOGIC;
028- -- the intermediate output bus
029- SIGNAL outbus: STD_LOGIC_VECTOR (3 DOWNTO 0);
030-
031- BEGIN
032- -- activate register write-enable when web=0 and address matches
033- w <= "1110" WHEN ((a="00") AND (web='0')) ELSE
034-      "1101" WHEN ((a="01") AND (web='0')) ELSE
035-      "1011" WHEN ((a="10") AND (web='0')) ELSE
036-      "0111" WHEN ((a="11") AND (web='0')) ELSE
037-      "1111";                -- don't activate a write-enable otherwise
038- w0 <= w(0);                -- connect the elements of the
039- w1 <= w(1);                -- write enable register to the
040- w2 <= w(2);                -- individual registers
```

```
041- w3 <= w(3);
042-
043- PROCESS (w0,inp)          -- register a
044- BEGIN
045-    IF(w0'event AND w0='1') THEN
046-            rga <= inp;  -- write register a with inp bus on
047-    END IF;              -- a rising edge of the w0 write-enable
048- END PROCESS;
049-
050- PROCESS (w1,inp)          -- register b
051- BEGIN
052-    IF(w1'event AND w1='1') THEN
053-            rgb <= inp;  -- write register b with inp bus on
054-    END IF;              -- a rising edge of the w1 write-enable
055- END PROCESS;
056-
057- PROCESS (w2,inp)          -- register c
058- BEGIN
059-    IF(w2'event AND w2='1') THEN
060-            rgc <= inp;  -- write register c with inp bus on
061-    END IF;              -- a rising edge of the w2 write-enable
062- END PROCESS;
063-
064- PROCESS (w3,inp)          -- register d
065- BEGIN
066-    IF(w3'event AND w3='1') THEN
067-            rgd <= inp;  -- write register d with inp bus on
068-    END IF;              -- a rising edge of the w3 write-enable
069- END PROCESS;
070-
071- outbus <=                 rga WHEN a="00" ELSE-- output addressed
072-                           rgb WHEN a="01" ELSE-- register value
073-                           rgc WHEN a="10" ELSE-- onto intermediate
bus
074-                           rgd;
075- outp <= outbus WHEN oe='1' ELSE "ZZZZ"; -- tristate bus when oe=0
076-
077- u0: leddcd PORT MAP (d=>outp,s=>s);-- display digit on LEDs
078-
079- END mem1_arch;
080-
```

Finally, the value on the output bus is passd to an LED decoder module on line 77. This allows the display of the 4-bit output value as a hexadecimal digit on the LED display.

Simulated waveforms of the small memory's operation are shown in Figure 8.18. The design was targeted for an XC95108 CPLD. Four writes were made to the memory by applying an address and data value and then pulsing the WEB input low. During the interval 0–110 ns the hex values of 0xC, 0x3, 0x5, and 0xA were written to memory locations 0, 1, 2, and 3, respectively. Then the values were read back out of these locations between 120 ns and 160 ns. Note that the value read from each location matches the data that was written to it. Then the output enable is lowered at 160 ns and the output bus tristates to the Z state.

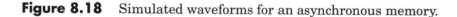

Figure 8.18   Simulated waveforms for an asynchronous memory.

The following constraints are used to map the I/O of the ABEL version of the MEM1 design to the pins of the XC95108 CPLD on the XS95 Board:

```
NET WEB      LOC=P46;    # B0 argument of XSPORT
NET OE       LOC=P47;    # B1 argument of XSPORT
NET A0       LOC=P48;    # B2 argument of XSPORT
NET A1       LOC=P50;    # B3 argument of XSPORT
NET INP0     LOC=P51;    # B4 argument of XSPORT
NET INP1     LOC=P52;    # B5 argument of XSPORT
NET INP2     LOC=P81;    # B6 argument of XSPORT
NET INP3     LOC=P80;    # B7 argument of XSPORT
NET S0       LOC=P21;    # S0 LED segment
NET S1       LOC=P23;    # S1 LED segment
NET S2       LOC=P19;    # S2 LED segment
NET S3       LOC=P17;    # S3 LED segment
NET S4       LOC=P18;    # S4 LED segment
NET S5       LOC=P14;    # S5 LED segment
NET S6       LOC=P15;    # S6 LED segment
```

And these constraints are used to map the VHDL version of the MEM1 design to the XS95 Board:

```
NET web        LOC=P46;      # B0 argument of XSPORT
NET oe         LOC=P47;      # B1 argument of XSPORT
NET a<0>       LOC=P48;      # B2 argument of XSPORT
NET a<1>       LOC=P50;      # B3 argument of XSPORT
NET inp<0>     LOC=P51;      # B4 argument of XSPORT
NET inp<1>     LOC=P52;      # B5 argument of XSPORT
NET inp<2>     LOC=P81;      # B6 argument of XSPORT
NET inp<3>     LOC=P80;      # B7 argument of XSPORT
NET s<0>       LOC=P21;      # S0 LED segment
NET s<1>       LOC=P23;      # S1 LED segment
NET s<2>       LOC=P19;      # S2 LED segment
NET s<3>       LOC=P17;      # S3 LED segment
NET s<4>       LOC=P18;      # S4 LED segment
NET s<5>       LOC=P14;      # S5 LED segment
NET s<6>       LOC=P15;      # S6 LED segment
```

Some examples of the actions that can be performed with the memory are as follows:

Action	XSPORT Command Sequence
Write location 1 (01) with 7 (111)	XSPORT 01110111 XSPORT 01110110 XSPORT 01110111
Write location 3 (11) with 10 (1010)	XSPORT 10101111 XSPORT 10101110 XSPORT 10101111
Read location 0 (00)	XSPORT 0011
Read location 2 (10)	XSPORT 1011

Using command sequences like these, you can recreate the sequence of writes and reads shown in the simulation of Figure 8.18. You will see the data written to each register on the LED digit.

You have to make some modifications to compile the MEM1.ABL or MEM1.VHD design for the XS40 Board. MEM1 has 8 inputs so arguments B6 and B7 of the XSPORT command are necessary. But B6 and B7 map to the special-purpose M0 and M2 pins of the XC4005XL FPGA and can not be specified in a user-constraint file (as we saw back in Chapter 4). So in this case we will compile MEM1.ABL into a macro and then hook the macro into a schematic as follows:

1. Create an MEM1_40 project targeted toward the XC4005XL FPGA. Once the project directory exists, use the standard file copying tools of your oper-

ating system to bring the MEM1.ABL file from the C:\XCPROJ\MEM1_95 directory to the C:\XCPROJ\MEM1_40 directory. (Do not use the **Document → Add...** menu item to do this because we do not want these ABEL files actually added to the project.)

2. Open the MEM1.ABL file in an **HDL Editor** window. Remove lines 57–58 of Listing 8.1 because the tristate outputs will serve no purpose within this macro. Then select the **Synthesis → Options** menu item. Click on the **Macro** radio-button in the **ABEL6** window. This directs the synthesizer to create a netlist for a module that will be included within a larger design. Click **OK** to remove the **ABEL6** window. Then select **Project → Create Macro** to synthesize the netlist for the 4×4 memory macro and add it to the project library. Then close the **HDL Editor** window.

3. Open a **Schematic Editor** window and select the **Mode → Symbols** menu item to bring up the **SC Symbols** window. Scroll through the list of symbols until you find the MEM1 part. Highlight this part, move your cursor into the drawing area of the window, and then click to drop the memory module into the schematic.

4. Select IBUF symbols from the **SC Symbols** window and connect their outputs to the [A1..A0], [INP3..INP0], OE, and WEB inputs of the MEM1 module. Then create input terminals and connect them to the inputs of all the IBUFs except for those connected to the INP2 and INP3 inputs.

5. . Select OBUF symbols from the **SC Symbols** window and connect their inputs to the [S6..S0] outputs of the MEM1 module. Then create output terminals and connect them to the outputs of the OBUFs.

6. . Select the MD0 and MD2 symbols from the **SC Symbols** window and connect them to the inputs of the IBUFs that drive the INP2 and INP3 inputs of the MEM1 module, respectively. At this point, your schematic should look like the one in Figure 8.19.

7. Save the schematic and then use the **Options → Create Netlist**, **Options → Integrity Test**, and **Options → Export Netlist...** menu items to create a netlist.

Now you can assign the pins of your memory design to the pins of the XC4005XL FPGA using the following constraint file:

```
NET  WEB        LOC=P44;     # B0 argument of XSPORT
NET  OE         LOC=P45;     # B1 argument of XSPORT
NET  A0         LOC=P46;     # B2 argument of XSPORT
NET  A1         LOC=P47;     # B3 argument of XSPORT
NET  INP0       LOC=P48;     # B4 argument of XSPORT
NET  INP1       LOC=P49;     # B5 argument of XSPORT
# NET INP2      LOC=P32;     # B6 argument of XSPORT
# NET INP3      LOC=P34;     # B7 argument of XSPORT
```

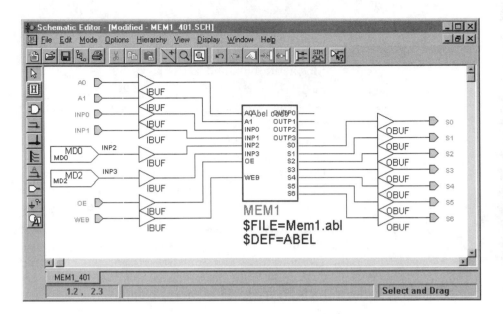

Figure 8.19 Connecting the MEM1 module for the XS40 Board.

```
NET S0        LOC=P25;      # S0 LED segment
NET S1        LOC=P26;      # S1 LED segment
NET S2        LOC=P24;      # S2 LED segment
NET S3        LOC=P20;      # S3 LED segment
NET S4        LOC=P23;      # S4 LED segment
NET S5        LOC=P18;      # S5 LED segment
NET S6        LOC=P19;      # S6 LED segment
```

Note that the INP2 and INP3 inputs are commented so they have no effect. (I have left them in only to document which arguments of XSPORT affect them.) Their pin assignments have already been handled by the attachment of the MD0 and MD2 symbols in the schematic. Unfortunately, this spreads some of your design information between two different files in two different formats. But that is life.

Now you can compile and download the MEM1_40 project and download it to the XS40 Board. The same command sequences listed for the XS95 Board will work for the XS40 Board, and you should see the same results.

Building a Small Synchronous Memory

The synchronous memory described in Listing 8.3 differs from the asynchronous memory of Listing 8.1 in the following ways:

Line 7: The output enable signal was replaced by a clock signal. (There is no reason a synchronous memory cannot have an output enable. I just needed the pin for the clock input since only 8 inputs can be driven from the PC into the XS95 and XS40 Boards.)

Lines 25–28: The equations for the register write control signals are specified here rather than using a WHEN-THEN-ELSE construct in the EQUATIONS section. This is done to condense the description of the individual registers and their multiplexers on lines 34–37.

Lines 34–37: The input to each register is driven by a multiplexer that selects either the input data bus or the register's output depending upon the value of the write control signal.

Lines 39–42: The registers are all clocked by the main clock input instead of by their individual write control lines.

The VHDL version of the synchronous memory is shown in Listing 8.4. This design is targeted toward the XC4005XL FPGA and the XS40 Board as can be seen by the presence of the clock buffer and mode input components and signals on lines 23–25 and 28–30, respectively. The mode inputs are instantiated on lines 43 through 46. On line 47 the outputs of the mode pin modules are concatenated with the 2-bit input bus to form a complete 4-bit input bus. The circuitry which generates the write control signals for the 4 registers is described on lines 49–57.

The connection of the input clock to the low-skew clock buffer occurs on lines 60 and 61. The buffered clock and the write control lines are used by the process on lines 64–80 to select which register is loaded with the value on the input bus on the rising clock edge.

The output multiplexer on lines 82–85 selects one of the registers based on the current address inputs and then places its value on the output bus. The output value also goes to an LED decoder on line 88 so the hexadecimal representation of the 4-bit value will appear on the LED display.

Simulated waveforms of the synchronous memory's operation are shown in Figure 8.20. The same test sequence that was used for the asynchronous memory is used here: the hex values of 0xC, 0x3, 0x5, and 0xA are written to memory locations 0, 1, 2, and 3, respectively, and then read back out. The actual write to memory occurs whenever the write control input is low and a rising edge occurs on the clock input. With the synchronous memory we only need to keep the address and data values stable during the rising clock edge (taking account of setup and hold times, of course), whereas the asynchronous memory requires the address and data be held steady for the entire time the write control line is low. This makes it a bit easier to work with synchronous memories.

The ABEL version of the MEM2 design can be compiled for the XS95 Board with these constraints:

```
NET CLK        LOC=P46;    # B0 argument of XSPORT
NET WEB        LOC=P47;    # B1 argument of XSPORT
NET A0         LOC=P48;    # B2 argument of XSPORT
NET A1         LOC=P50;    # B3 argument of XSPORT
NET INP0       LOC=P51;    # B4 argument of XSPORT
NET INP1       LOC=P52;    # B5 argument of XSPORT
NET INP2       LOC=P81;    # B6 argument of XSPORT
NET INP3       LOC=P80;    # B7 argument of XSPORT
NET S0         LOC=P21;    # S0 LED segment
NET S1         LOC=P23;    # S1 LED segment
NET S2         LOC=P19;    # S2 LED segment
NET S3         LOC=P17;    # S3 LED segment
NET S4         LOC=P18;    # S4 LED segment
NET S5         LOC=P14;    # S5 LED segment
NET S6         LOC=P15;    # S6 LED segment
```

For the VHDL version of MEM2, use these constraints when targeting the XS40 Board:

```
NET clk        LOC=P44;    # B0 argument of XSPORT
NET web        LOC=P45;    # B1 argument of XSPORT
NET a<0>       LOC=P46;    # B2 argument of XSPORT
NET a<1>       LOC=P47;    # B3 argument of XSPORT
NET inp<0>     LOC=P48;    # B4 argument of XSPORT
NET inp<1>     LOC=P49;    # B5 argument of XSPORT
# NET inp<2>   LOC=P32;    # B6 argument of XSPORT mapped to M0
# NET inp<3>   LOC=P34;    # B7 argument of XSPORT mapped to M2
NET s<0>       LOC=P25;    # S0 LED segment
NET s<1>       LOC=P26;    # S1 LED segment
NET s<2>       LOC=P24;    # S2 LED segment
NET s<3>       LOC=P20;    # S3 LED segment
NET s<4>       LOC=P23;    # S4 LED segment
NET s<5>       LOC=P18;    # S5 LED segment
NET s<6>       LOC=P19;    # S6 LED segment
```

Listing 8.3   ABEL version of a 4-bit wide synchronous memory built from 4 register modules.

```
001- MODULE MEM2
002- TITLE 'A 4x4 synchronous memory built from DFFs'
003-
004- DECLARATIONS
005-
006- WEB          PIN;                    "active-low write-enable input
007- CLK          PIN;                    "clock input
008- A1..A0       PIN;                    "2-bit address inputs
009- INP3..INP0   PIN;                    "4-bit input data bus
010- RGA3..RGA0   NODE ISTYPE 'REG'; "first 4-bit register
011- RGB3..RGB0   NODE ISTYPE 'REG'; "second 4-bit register
012- RGC3..RGC0   NODE ISTYPE 'REG'; "third 4-bit register
013- RGD3..RGD0   NODE ISTYPE 'REG'; "fourth 4-bit register
014- OUTP3..OUTP0 PIN ISTYPE 'COM';  "4-bit output data bus
015- S6..S0       PIN ISTYPE 'COM';  "LED driver outputs
016-
017- A    = [A1..A0];      "synonym for the address bus
018- INP  = [INP3..INP0];  "synonym for input bus
019- OUTP = [OUTP3..OUTP0]; "synonym for output bus
020- RGA  = [RGA3..RGA0];  "synonym for first register
021- RGB  = [RGB3..RGB0];  "synonym for second register
022- RGC  = [RGC3..RGC0];  "synonym for third register
023- RGD  = [RGD3..RGD0];  "synonym for fourth register
024- S    = [S6..S0];      "synonym for LED drivers
025- W0   = ((A==0) & !WEB); "write signal for first register
026- W1   = ((A==1) & !WEB); "write signal for second register
027- W2   = ((A==2) & !WEB); "write signal for third register
028- W3   = ((A==3) & !WEB); "write signal for fourth register
029-
030- EQUATIONS
031-
032- "write signal for each register controls multiplexer which
033- "directs input bus or register contents back into register
034- RGA := W0 & INP # !W0 & RGA;
035- RGB := W1 & INP # !W1 & RGB;
036- RGC := W2 & INP # !W2 & RGC;
037- RGD := W3 & INP # !W3 & RGD;
038-
039- RGA.CLK = CLK; "all registers get the same clock
040- RGB.CLK = CLK; "so the entire memory is synchronous
041- RGC.CLK = CLK;
042- RGD.CLK = CLK;
043-
044- "a simple address decoder that passes the value from
045- "one of the registers to the output bus
046- WHEN       (A==0) THEN OUTP=RGA;
047- ELSE WHEN (A==1) THEN OUTP=RGB;
048- ELSE WHEN (A==2) THEN OUTP=RGC;
```

Listing 8.3 ABEL version of a 4-bit wide synchronous memory built from 4 register modules. (Cont'd.)

```
049- ELSE                   OUTP=RGD;
050-
051- "drive the LED digit to display the memory output
052- TRUTH_TABLE (OUTP->S)
053-    0 -> ^b1110111;  "0
054-    1 -> ^b0010010;  "1
055-    2 -> ^b1011101;  "2
056-    3 -> ^b1011011;  "3
057-    4 -> ^b0111010;  "4
058-    5 -> ^b1101011;  "5
059-    6 -> ^b1101111;  "6
060-    7 -> ^b1010010;  "7
061-    8 -> ^b1111111;  "8
062-    9 -> ^b1111011;  "9
063-   10 -> ^b1111110;  "A
064-   11 -> ^b0101111;  "b
065-   12 -> ^b1100101;  "C
066-   13 -> ^b0011111;  "d
067-   14 -> ^b1101101;  "E
068-   15 -> ^b1101100;  "F
069-
070- END MEM2
```

```
001- -- 4x4 synchronous memory
002-
003- LIBRARY IEEE,xse;
004- USE IEEE.std_logic_1164.ALL;
005- USE xse.led.ALL;            -- access the LED decoder module
006-
007- ENTITY mem2 IS
008- PORT
009- (
010-   web: IN STD_LOGIC;        -- active-low write-enable
011-   clk: IN STD_LOGIC;        -- master clock
012-   a: IN STD_LOGIC_VECTOR (1 DOWNTO 0); -- 2-bit address bus
013-   inp: IN STD_LOGIC_VECTOR (1 DOWNTO 0); -- 4-bit input bus
014-   outp: BUFFER STD_LOGIC_VECTOR (3 DOWNTO 0); -- 4-bit output bus
015-   s: OUT STD_LOGIC_VECTOR (6 DOWNTO 0) -- LED outputs
016- );
017- END mem2;
018-
019-
020- ARCHITECTURE mem2_arch OF mem2 IS
021-
022- -- components and signals for buffering clock
023- COMPONENT IBUF PORT(I: IN STD_LOGIC; O: OUT STD_LOGIC); END COMPONENT;
024- COMPONENT BUFG PORT(I: IN STD_LOGIC; O: OUT STD_LOGIC); END COMPONENT;
025- SIGNAL in_clk, buf_clk: STD_LOGIC;
026-
027- -- components/signals for bringing in the special-purpose mode pin signals
028- COMPONENT md0 PORT(MD0: OUT STD_LOGIC); END COMPONENT;
029- COMPONENT md2 PORT(MD2: OUT STD_LOGIC); END COMPONENT;
030- SIGNAL MD0in, MD2in, MD0out, MD2out: STD_LOGIC;
031- SIGNAL inbus: STD_LOGIC_VECTOR (3 DOWNTO 0);
032-
033- -- signals for the individual 4-bit registers that are in the memory
034- SIGNAL rga: STD_LOGIC_VECTOR (3 DOWNTO 0);-- register a
035- SIGNAL rgb: STD_LOGIC_VECTOR (3 DOWNTO 0);-- register b
036- SIGNAL rgc: STD_LOGIC_VECTOR (3 DOWNTO 0);-- register c
037- SIGNAL rgd: STD_LOGIC_VECTOR (3 DOWNTO 0);-- register d
038- SIGNAL w: STD_LOGIC_VECTOR (3 DOWNTO 0);-- active-low write-enables
039- SIGNAL w0,w1,w2,w3: STD_LOGIC; -- individual active-low write-enables
040-
041- BEGIN
042-
043- u0: md0 PORT MAP(MD0=>MD0in);    -- connect the mode inputs to
044- u1: md2 PORT MAP(MD2=>MD2in);    -- the bus feeding the registers
045- u2: IBUF PORT MAP(I=>MD0in, O=>MD0out);
046- u3: IBUF PORT MAP(I=>MD2in, O=>MD2out);
047- inbus <= MD2out & MD0out & inp(1 DOWNTO 0); -- group signals into inbus
048-
```

```
049- w <="1110" WHEN ((a="00") AND (web='0')) ELSE
050-    "1101" WHEN ((a="01") AND (web='0')) ELSE
051-    "1011" WHEN ((a="10") AND (web='0')) ELSE
052-    "0111" WHEN ((a="11") AND (web='0')) ELSE
053-    "1111";
054- w0 <= w(0);
055- w1 <= w(1);
056- w2 <= w(2);
057- w3 <= w(3);
058-
059- -- buffer the input clock
060- u4: IBUF PORT MAP(I=>clk, O=>in_clk);
061- u5: BUFG PORT MAP(I=>in_clk, O=>buf_clk);
062-
063- -- this process loads the selected reg on the rising clock edge
064- PROCESS (buf_clk,w0,w1,w2,w3,inp)
065- BEGIN
066- IF(buf_clk'event AND buf_clk='1') THEN
067-    IF w0='0' THEN
068-             rga <= inbus;
069-    END IF;
070-    IF w1='0' THEN
071-             rgb <= inbus;
072-    END IF;
073-    IF w2='0' THEN
074-             rgc <= inbus;
075-    END IF;
076-    IF w3='0' THEN
077-             rgd <= inbus;
078-    END IF;
079- END IF;
080- END PROCESS;
081-
082- outp <= rga WHEN a="00" ELSE    -- select one of the four
083-         rgb WHEN a="01" ELSE    -- registers and output
084-         rgc WHEN a="10" ELSE    -- its value on the output bus
085-    rgd;
086-
087- -- display the output value on the LED display
088- u6: leddcd PORT MAP (d=>outp,s=>s);
089-
090- END mem2_arch;
091-
```

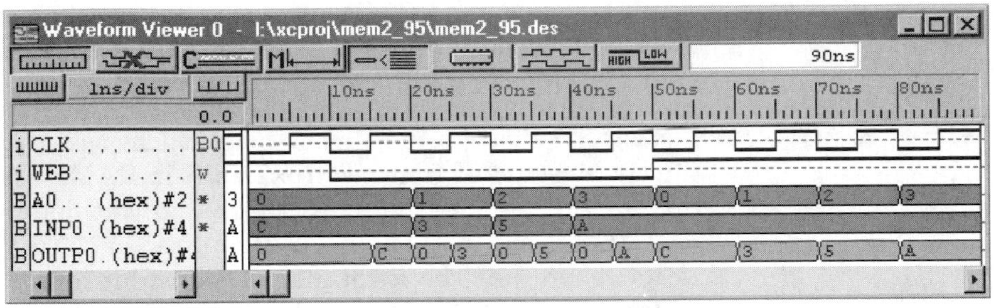

Figure 8.20   Simulated waveforms for a synchronous memory.

Both the XS95 and XS40 Board implementations will work the same way. Some examples of the actions that can be performed with the synchronous memory are as follows:

Action	XSPORT Command Sequence
Write location 1 (01) with 7 (0111)	XSPORT 01110100 XSPORT 01110101
Write location 3 (11) with 10 (1010)	XSPORT 10101100 XSPORT 10101101
Read location 0 (00)	XSPORT 0011
Read location 2 (10)	XSPORT 1011

The MEM2 design uses 51 macrocells of an XC95108 CPLD. Since the LED decoder uses 7 macrocells, the 4×4 memory must use the other $51 - 7 = 44$ macrocells. This is a large fraction of the total resources of the CPLD. Obviously, you cannot build a very large memory using the techniques shown in this or the previous section.

Using Built-In Memory

To allow designers to use larger amounts of memory in their designs, the XC4000 series of FPGAs is equipped with internal RAMs. Since the LUTs in FPGAs are essentially RAMs which store the logic truth-tables (refer back to Figure 2.6), it is reasonable to allow the designer to use these RAMs for general-purpose data storage. Each CLB in an XC4000 FPGA has 2 LUTs that can be used as RAMs (Figure 2.7). Each 4-input LUT has room to store 2^4=16 bits, so a single CLB can store 32 bits of data. The data can be organized as a 16×2 RAM, a 32×1 RAM, or dual 16×1 RAMs.

(Unfortunately, the XC9500 series of CPLDs does not offer the option of using macrocells as either logic or memory. So the material in this section is irrelevant to those using the XS95 Board. Sorry.)

Listing 8.5 shows some VHDL code which uses the RAM16X4 RAM module that is constructed from two 16×2 built-in RAMs. The COMPONENT declaration for the module is on lines 20–28 and was determined by looking into the RAM16X4.XNF file found in the directory Xilinx\Synth\xilinx\macros\Xc4000e\v6_xnf. The RAM module has 4-bit input and output buses, a 4-bit address bus, and an active-high write-enable input.

The RAM module is instantiated on lines 52–58. Since we are only building a 4×4 memory, we only need two address bits. Therefore, on line 54 the lower two address inputs of the module are connected to the address bus. As they are not needed, the upper two address bits are connected to the gnd signal which is held at logic 0 on line 49.

Since we are targeting the XS40 Board with this design, we need to get some of our inputs from the mode pins of the XC4005XL FPGA. This is done on lines 43–46. The inputs from the mode pins are concatenated with the 2-bit input bus to create the 4-bit inb bus. The inb bus connects to the RAM module on line 55. The remaining output, and write-enable ports of the RAM module are connected to their counterpart signals in the mem3 design (lines 56 and 57).

Figure 8.21 shows the simulated waveforms for the MEM3 design. Four writes were made to the memory by applying an address and data value and then placing a positive pulse on the WE input. During the interval 0–110 ns the hex values of 0xC, 0x3, 0x5, and 0xA were written to memory locations 0, 1, 2, and 3, respectively. Then the values were read back out of these locations between 120 ns and 160 ns. The value read from each location matches the data that was written to it, so the memory seems to be working.

The user-constraint file for the MEM3 design is as follows:

```
NET  we         LOC=P44;     # B0 argument of XSPORT
# NET oe        LOC=P45;     # OE is not used with the internal memories
NET  a<0>       LOC=P46;     # B2 argument of XSPORT
NET  a<1>       LOC=P47;     # B3 argument of XSPORT
NET  inp<0>     LOC=P48;     # B4 argument of XSPORT
NET  inp<1>     LOC=P49;     # B5 argument of XSPORT
# NET inp<2>    LOC=P32;     # B6 argument of XSPORT
# NET inp<3>    LOC=P34;     # B7 argument of XSPORT
NET  s<0>       LOC=P25;     # S0 LED segment
NET  s<1>       LOC=P26;     # S1 LED segment
NET  s<2>       LOC=P24;     # S2 LED segment
NET  s<3>       LOC=P20;     # S3 LED segment
NET  s<4>       LOC=P23;     # S4 LED segment
```

Listing 8.5   VHDL version of a 4-bit wide asynchronous memory constructed using a built-in FPGA memory.

```
001- -- Using Built-in Memory of the XC4000 FPGAs
002-
003- LIBRARY IEEE,xse;
004- USE IEEE.std_logic_1164.ALL;
005- USE xse.led.ALL;
006-
007- ENTITY mem3 IS
008- PORT
009- (
010-    we: IN STD_LOGIC;      -- active-high write-enable control input
011-    a: IN STD_LOGIC_VECTOR (1 DOWNTO 0);-- 2-bit address bus
012-    inp: IN STD_LOGIC_VECTOR (1 DOWNTO 0);-- 2-bit input data bus
013-    s: OUT STD_LOGIC_VECTOR (6 DOWNTO 0)-- drivers for LED
014- );
015- END mem3;
016-
017-
018- ARCHITECTURE mem3_arch OF mem3 IS
019-
020- COMPONENT ram16x4   -- interface for the RAM16X4.XNF component
021- PORT
022- (
023-    A0,A1,A2,A3: IN STD_LOGIC;-- 4-bit address bus
024-    D0,D1,D2,D3: IN STD_LOGIC;-- 4-bit input data bus
025-    WE: IN STD_LOGIC;        -- active-high write-enable
026-    O0,O1,O2,O3: OUT STD_LOGIC-- 4-bit output data bus
027- );
028- END COMPONENT;
029-
030- SIGNAL o: STD_LOGIC_VECTOR(3 DOWNTO 0); -- intermediate output bus
031-
032- -- components/signals for input from the special-purpose mode pins
033- COMPONENT md0 PORT(MD0: OUT STD_LOGIC); END COMPONENT;
034- COMPONENT md2 PORT(MD2: OUT STD_LOGIC); END COMPONENT;
035- COMPONENT IBUF PORT(I: IN STD_LOGIC; O: OUT STD_LOGIC); END COMPONENT;
036- SIGNAL MD0in, MD2in, MD0out, MD2out: STD_LOGIC;
037- SIGNAL inb: STD_LOGIC_VECTOR (3 DOWNTO 0);
038-
039- SIGNAL gnd: STD_LOGIC;  -- signal held at logic 0
040-
041- BEGIN
```

```
042-
043- u0: md0 PORT MAP(MD0=>MD0in);-- connect the mode inputs to
044- u1: md2 PORT MAP(MD2=>MD2in);-- the bus feeding the RAM module
045- u2: IBUF PORT MAP(I=>MD0in, O=>MD0out);
046- u3: IBUF PORT MAP(I=>MD2in, O=>MD2out);
047- inb <= MD2out & MD0out & inp(1 DOWNTO 0); -- concatenate bits into bus
048-
049- gnd <= '0'; -- connect signal to logic 0;
050-
051- -- connect the RAM16X4 component I/O to the I/O of this module
052- u4: ram16x4 PORT MAP
053-       (
054-       A0=>a(0),A1=>a(1),A2=>gnd,A3=>gnd,       -- connect address bus
055-       D0=>inb(0),D1=>inb(1),D2=>inb(2),D3=>inb(3), -- connect input bus
056-       O0=>o(0),O1=>o(1),O2=>o(2),O3=>o(3),     -- connect output bus
057-       WE=>we                                   -- connect write-enable
058-       );
059-
060- u5: leddcd PORT MAP(d=>o,s=>s); -- show output hex digit on LED
061- END mem3_arch;
062-
```

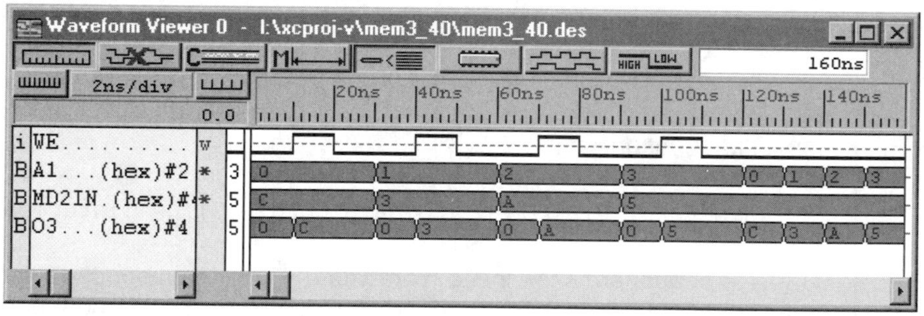

Figure 8.21   A simulation of the VHDL design that uses the built-in
asynchronous memory of the XC4005XL FPGA.

```
NET s<5>        LOC=P18;    # S5 LED segment
NET s<6>        LOC=P19;    # S6 LED segment
```

Examples of the actions that can be performed with this memory are quite similar to what we had with the asynchronous memory we built from D flip-flops (except that the write-enable is inverted):

Action	XSPORT Command Sequence
Write location 1 (01) with 7 (0111)	XSPORT 01110110 XSPORT 01110111 XSPORT 01110110
Write location 3 (11) with 10 (1010)	XSPORT 10101110 XSPORT 10101111 XSPORT 10101110
Read location 0 (00)	XSPORT 0010
Read location 2 (10)	XSPORT 1010

In addition to asynchronous RAMs, the XC4000 series also offers synchronous RAMs and dual-port RAMs. Figure 8.22 gives an example of using an internal 16×2 dual-port synchronous RAM in a schematic. The RAM was instantiated by selecting the **RAM16x2D** symbol from the **SC Symbols** window of the schematic editor. It is equipped with a normal address bus (A3..A0) that controls which memory location will be written or read using the INP1..INP0 input bus and SPO1..SPO0 output buses, respectively. In addition, a second read-only port is provided that lets you enter an address through DPRA3..DPRA0 and read the contents of the memory location through output bus DPO1..DPO0. (There are not enough pins on the PC parallel port to drive all these address bits, so the upper two bits of each address were attached to ground in the schematic.)

The constraint file for the MEM4 design follows. Note that each 2-bit output port of the memory is applied to unique pair of LED segments so that we can observe the values output by both independent ports.

```
NET CLK         LOC=P44;    # B0 argument of XSPORT
NET WE          LOC=P45;    # B1 argument of XSPORT
NET INP0        LOC=P46;    # B2 argument of XSPORT
NET INP1        LOC=P47;    # B3 argument of XSPORT
NET A0          LOC=P48;    # B4 argument of XSPORT
NET A1          LOC=P49;    # B5 argument of XSPORT
# NET DPRA0     LOC=P32;    # B6 argument of XSPORT
# NET DPRA1     LOC=P34;    # B7 argument of XSPORT
NET S1          LOC=P26;    # S1 LED segment (for SP00)
NET S2          LOC=P24;    # S2 LED segment (for SP01)
```

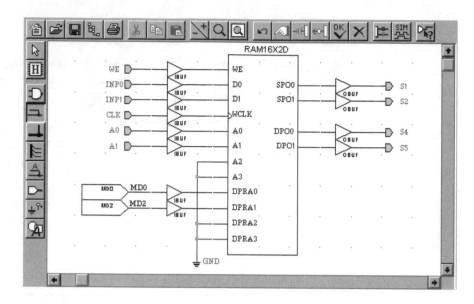

Figure 8.22 Using the built-in synchronous dual-port memory of the XC4005XL FPGA.

```
NET S4          LOC=P23;     # S4 LED segment (for DPO0)
NET S5          LOC=P18;     # S5 LED segment (for DPO1)
```

The following command sequences perform the indicated operations:

Action	XSPORT Command Sequence
Write location 1 (01) with 3 (11) while reading from location 2 (10)	XSPORT 10011110 XSPORT 10011111
Write location 0 (00) with 2 (10) while reading from location 1 (01)	XSPORT 01001010 XSPORT 01001011
Read locations 0 (00) and 1 (01)	XSPORT 00010001
Read location 3 (11) and 2 (10)	XSPORT 11100001

Using the internal RAMs, the asynchronous memory consumes only 13 LUTs (most of which are used in the LED decoder) while the dual-port synchronous design takes up 4 LUTs. This is quite a reduction in LUT usage when compared to the flip-flop-based designs of the previous sections.

Using a Memory Chip

Even the internal RAMs of the XC4000 FPGAs will not suffice when you need a large amount of memory. Connecting external memory chips to the pins of an

FPLD is the only solution when you need more than a few kilobytes of RAM. In this section, we will give an example of using an external RAM chip with an FPLD to build a FIFO.

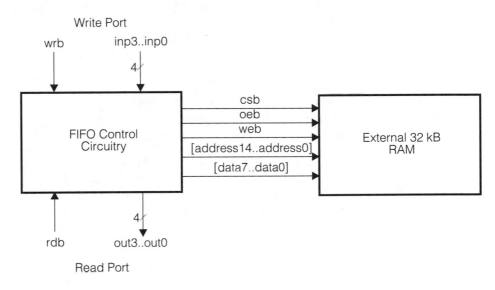

Figure 8.23 Connections between the FPLD and the external RAM for a FIFO.

The high-level block diagram in Figure 8.23 shows the signals that connect the FIFO control circuitry in the FPLD to the external RAM. The interface to the external RAM is very similar to those of the RAMs we have built in the previous sections. There is a single data I/O bus, an address bus, and active-low output-enable and write-enable control lines. In the external RAM, the write-enable input controls both reading and writing of the RAM: when web=0 an addressed memory location is overwritten with the value on the data bus, and when web=1 and oeb=0 the contents of the memory location are output on the data bus. A chip-select control line is also available that disables the reading or writing of the RAM when csb=1. In large memories built from multiple RAM chips, the chip-select inputs are used so that only one RAM chip is active at a time. Our FIFO design will only use one RAM chip, so csb will be tied to logic 0 to keep it permanently enabled.

Listing 8.6 displays the ABEL code for the FIFO controller that manipulates the external RAM. Lines 7–10 declare the user-interface to the FIFO. Lines 13–17 define the FPLD pins that connect to the external RAM. Note that the RAM has address, data, output-enable, and write-enable pins just like the smaller RAMs we have worked with. For a 32 KByte RAM, there are fifteen address lines (2^{15}=32768). Note also that the RAM has an 8-bit wide data bus, but we will use only the lower 4 bits for our FIFO.

Line 20 declares the outputs which drive the LED digit. We will use these to observe the 4-bit output of the FIFO. Lines 23–29 declare internal nodes for the FIFO control circuitry. The reset signal is driven high to clear the internal flip-flops. The empty and full signals indicate the state of the FIFO. (Normally, these signals would come to external pins but we have not bothered with that in this example.) The read and write pointers are declared as 4 D flip-flops each, giving us a total FIFO size of 16 registers. (This is considerably less than the size of the external RAMs. You can increase the size of the FIFO by adding more flip-flops to each pointer.) Finally, the flag flip-flops for storing the type of the last FIFO operation are declared.

The description of the circuitry begins after line 39. Lines 45–47 and 52–54 describe the read and write pointers, respectively. The ABEL code replicates the functions shown in Figure 8.16. For example, the read-pointer, rdptr, is incremented whenever a read operation occurs as long as the FIFO is not empty. The use of the ABEL addition operator (+) lets us describe this behavior very concisely.

Line 57 generates the signal that indicates when the read and write pointers address the same memory location. The ABEL equality operator, ==, provides a logic 1 whenever the two pointers are equal. The reset signal is generated on line 61 whenever both the read and write control signals are low at the same time. Since the external RAM has only a single data port, there will never be a case in the normal operation of the FIFO when these signals should both be low. Therefore, we have used this combination of inputs to generate a reset signal. This would not be a good way to reset the circuit if a dual-port RAM were used such that a producer and consumer of data could both access the FIFO simultaneously.

Lines 64–75 describe the circuitry shown in Figure 8.17 that records the type of the last FIFO operation and generates the FIFO empty and full status signals. The description of the interface circuitry between the FPLD and the external RAM begins on line 77. The RAM must be activated, and this is done by holding its chip select line low as shown on line 78. Since we are only using one external RAM, there is no need for more sophisticated chip select logic.

Since we are only using 16 locations of the external RAM for our FIFO, the upper 11 bits of the RAM address bus are held at logic 0 on line 81. Line 84 describes a 4-bit multiplexer that connects the read-pointer to the lower 4 bits of the RAM address bus when the FIFO is being read. At all other times, the write-pointer drives the lower part of the RAM address.

Lines 87–90 control the writing of data to the external RAM. The 4-bit FIFO input bus drives the lower 4 bits of the RAM data bus while the upper 4 bits are held at logic 0. The output pins of the FPLD are enabled when the FIFO is being written to (wrb=0). This lets the FPLD drive new data into the data bus of the RAM. The flow of data is depicted in Figure 8.24a. The RAM write control line is pulsed low during a FIFO write operation as long as the FIFO is not full (line 90).

Lines 94–96 handle the situation when data is being read from the FIFO. On line 94, the output enable of the external RAM is pulled low during a FIFO read operation so that the RAM data bus drivers are activated. This lets the RAM drive a data value into the FPLD as shown in Figure 8.24b. Naturally we do not want the FPLD to try to drive data into the RAM at the same time or one of the I/O pins might burn out, so the FPLD pin drivers are disabled by holding the write control input high (see line 89). Lines 95 and 96 pass the data from the RAM to the outputs of the FIFO and enable the drivers, respectively. This lets the FIFO output bus drive whatever circuitry is requesting a read of the FIFO.

Finally, lines 99–117 define an LED decoder that takes the value output by the FIFO during a read operation and displays it on an LED digit.

The FIFO can be compiled for the XS95 Board with the following pin assignments. There are a lot more pin assignments than we normally use because we have to pass the signals through the right pins of the XC95108 CPLD that are hooked to the 15 address lines, 8 data lines, and 3 control lines of the external RAM.

```
NET wrb      LOC=P46;    # B0 argument of XSPORT
NET rdb      LOC=P47;    # B1 argument of XSPORT
NET inp0     LOC=P48;    # B2 argument of XSPORT
NET inp1     LOC=P50;    # B3 argument of XSPORT
NET inp2     LOC=P51;    # B4 argument of XSPORT
NET inp3     LOC=P52;    # B5 argument of XSPORT
NET out0     LOC=P5;
NET out1     LOC=P6;
NET out2     LOC=P7;
NET out3     LOC=P11;
NET csb      LOC=P65;
NET web      LOC=P63;
NET oeb      LOC=P62;
NET data0    LOC=P44;
NET data1    LOC=P43;
NET data2    LOC=P41;
NET data3    LOC=P40;
NET data4    LOC=P39;
NET data5    LOC=P37;
NET data6    LOC=P36;
```

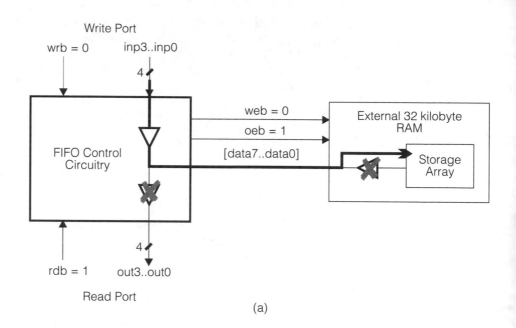

(a)

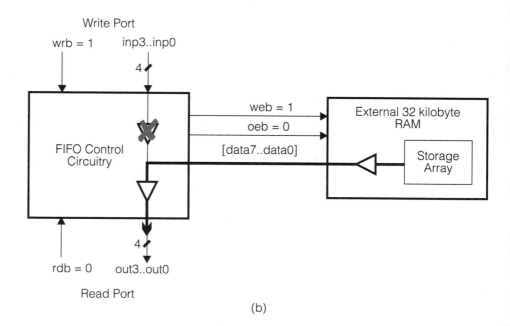

(b)

Figure 8.24 Pin driver activations during (a) FIFO write and (b) FIFO read operations.

Listing 8.6   A 4-bit wide FIFO built with an external RAM.

```
001- MODULE fifo
002- TITLE 'FIFO with 16 storage locations'
003-
004- DECLARATIONS
005-
006- "I/O pins for the user-interface of the FIFO
007- wrb          PIN; "active-low write control input
008- rdb          PIN; "active-low read control input
009- inp3..inp0   PIN; "4-bit input bus to FIFO
010- out3..out0   PIN; "4-bit output bus from FIFO
011-
012- "I/O pins to the external RAM
013- csb          PIN ISTYPE 'COM'; "active-low chip select
014- web          PIN ISTYPE 'COM'; "active low write control
015- oeb          PIN ISTYPE 'COM'; "active-low read control
016- data7..data0 PIN ISTYPE 'COM'; "RAM I/O bus
017- address14..address0 PIN ISTYPE 'COM'; "15-bit RAM address
018-
019- "output pins to the LED for diagnostic purposes
020- s6..s0       PIN ISTYPE 'COM';
021-
022- "internal nodes
023- reset            NODE;               "active-high FIFO reset
024- empty            NODE ISTYPE 'COM'; "FIFO empty signal
025- full             NODE ISTYPE 'COM'; "FIFO full signal
026- rdptr3..rdptr0   NODE ISTYPE 'REG'; "read ptr. flip-flops
027- wrptr3..wrptr0   NODE ISTYPE 'REG'; "write ptr. flip-flops
028- rdeqwr           NODE ISTYPE 'COM'; "rdptr==wrptr?
029- prevop1..prevop0 NODE ISTYPE 'REG'; "previous op flags
030-
031- "some synonyms for groups of signals
032- inp    = [inp3..inp0];       "4-bit input to FIFO
033- out= [out3..out0];           "4-bit output from FIFO
034- data= [data3..data0];    "4-bit intfc. to external RAM
035- rdptr  = [rdptr3..rdptr0]; "FIFO read pointer
036- wrptr  = [wrptr3..wrptr0]; "FIFO write pointer
037- address = [address3..address0]; "4-bit address to ext. RAM
038-
039- EQUATIONS
040-
041- "FIFO READ AND WRITE POINTERS
042- "read pointer logic. Increment pointer when reading
```

Listing 8.6 A 4-bit wide FIFO built with an external RAM. (Cont'd.)

```
043-  "non-empty FIFO. If FIFO is empty, reload read pointer
044-  "with its current contents. Clear pointer on reset.
045-  rdptr := !empty & (rdptr+1) # empty & rdptr;
046-  rdptr.CLK = rdb; "read operation updates pointer
047-  rdptr.ACLR = reset;
048-
049-  "write pointer logic. Increment pointer when writing
050-  "non-full FIFO. If FIFO is full, reload write pointer
051-  "with its current contents. Clear pointer on reset.
052-  wrptr := !full  & (wrptr+1) #  full & wrptr;
053-  wrptr.CLK = wrb; "write operation updates pointer
054-  wrptr.ACLR = reset;
055-
056-  "check read and write pointers to see if they are equal
057-  rdeqwr = (rdptr==wrptr);
058-
059-  "FIFO CONTROL LOGIC
060-  "reset FIFO if both read and write lines are low
061-  reset = !rdb & !wrb;
062-  "prevop0=0 if most recent operation was a read or reset
063-  "prevop0=1 if most recent operation was a write
064-  prevop0 := 1;
065-  prevop0.CLK = !wrb;
066-  prevop0.ACLR = !rdb; "async. clear on read
067-  "prevop1=prevop0 and gets updated whenever there is a
068-  "read or a write or a reset.
069-  prevop1 := prevop0;
070-  prevop1.CLK = wrb & rdb;
071-  prevop1.ACLR = reset; "async. clear on reset
072-  "FIFO is empty if rdptr==wrptr and most recent op was a read
073-  "FIFO is full if rdptr==wrptr and most recent op was a write
074-  empty = rdeqwr & !prevop1;
075-  full = rdeqwr & prevop1;
076-
077-  "INTERFACE TO THE EXTERNAL RAM
078-  csb = 0; "lower chip select to activate the external RAM
079-  "upper bits of RAM address are 0 since we only use
080-  "16 locations for the FIFO
081-  [address14..address4] = 0;
082-  "lower 4 bits come from read or write pointer depending
083-  "upon whether it is a read of the FIFO or not.
084-  address = !rdb & rdptr # rdb & wrptr;
085-  "input data to the FIFO is passed to the external RAM,
```

Listing 8.6   A 4-bit wide FIFO built with an external RAM. (Cont'd.)

```
086-  "but the drivers are only activated during a write operation
087-  [data3..data0] = inp.PIN;
088-  [data7..data4] = 0; "extend 4-bit input bus value with 0s
089-  [data7..data0].OE = !wrb;
090-  web = full # wrb; "write only occurs when FIFO is not full
091-  "the external RAM output drivers are only activated during
092-  "a read of the FIFO. The data from the RAM is passed to
093-  "the FIFO outputs.
094-  oeb = rdb; "external RAM drivers active when oeb=0
095-  out = [data3..data0].PIN;
096-  out.OE = !rdb;
097-
098-  "ATTACH LED DECODER TO FIFO OUTPUT
099-  [s6..s0].OE = !rdb; "activate LED only during FIFO reads
100-  TRUTH_TABLE
101-  (out -> [s6, s5, s4, s3, s2, s1, s0])
102-     0 -> [1,  1,  1,  0,  1,  1,  1 ];
103-     1 -> [0,  0,  1,  0,  0,  1,  0 ];
104-     2 -> [1,  0,  1,  1,  1,  0,  1 ];
105-     3 -> [1,  0,  1,  1,  0,  1,  1 ];
106-     4 -> [0,  1,  1,  1,  0,  1,  0 ];
107-     5 -> [1,  1,  0,  1,  0,  1,  1 ];
108-     6 -> [1,  1,  0,  1,  1,  1,  1 ];
109-     7 -> [1,  0,  1,  0,  0,  1,  0 ];
110-     8 -> [1,  1,  1,  1,  1,  1,  1 ];
111-     9 -> [1,  1,  1,  1,  0,  1,  1 ];
112-    10-> [1,  1,  1,  1,  1,  1,  0 ];
113-    11-> [0,  1,  0,  1,  1,  1,  1 ];
114-    12-> [1,  1,  0,  0,  1,  0,  1 ];
115-    13-> [0,  0,  1,  1,  1,  1,  1 ];
116-    14-> [1,  1,  0,  1,  1,  0,  1 ];
117-    15-> [1,  1,  0,  1,  1,  0,  0 ];
118-
119-  END fifo
```

```
NET data7    LOC=P35;
NET address0 LOC=P75;
NET address1 LOC=P79;
NET address2 LOC=P82;
NET address3 LOC=P84;
NET address4 LOC=P1;
NET address5 LOC=P3;
NET address6 LOC=P61;
```

Experimental

```
NET address7  LOC=P55;
NET address8  LOC=P57;
NET address9  LOC=P58;
NET address10 LOC=P83;
NET address11 LOC=P2;
NET address12 LOC=P53;
NET address13 LOC=P56;
NET address14 LOC=P54;
NET s0        LOC=P21;    # S0 LED segment
NET s1        LOC=P23;    # S1 LED segment
NET s2        LOC=P19;    # S2 LED segment
NET s3        LOC=P17;    # S3 LED segment
NET s4        LOC=P18;    # S4 LED segment
NET s5        LOC=P14;    # S5 LED segment
NET s6        LOC=P15;    # S6 LED segment
```

The user-constraint file for the XS40 Board is as follows:

```
NET wrb       LOC=P44;    # B0 argument of XSPORT
NET rdb       LOC=P45;    # B1 argument of XSPORT
NET inp0      LOC=P46;    # B2 argument of XSPORT
NET inp1      LOC=P47;    # B3 argument of XSPORT
NET inp2      LOC=P48;    # B4 argument of XSPORT
NET inp3      LOC=P49;    # B5 argument of XSPORT
NET csb       LOC=P65;
NET web       LOC=P62;
NET oeb       LOC=P61;
NET data0     LOC=P41;
NET data1     LOC=P40;
NET data2     LOC=P39;
NET data3     LOC=P38;
NET data4     LOC=P35;
NET data5     LOC=P81;
NET data6     LOC=P80;
NET data7     LOC=P10;
NET address0  LOC=P78;
NET address1  LOC=P79;
NET address2  LOC=P82;
NET address3  LOC=P84;
NET address4  LOC=P3;
NET address5  LOC=P5;
NET address6  LOC=P60;
NET address7  LOC=P56;
NET address8  LOC=P58;
NET address9  LOC=P59;
NET address10 LOC=P83;
```

```
NET address11 LOC=P4;
NET address12 LOC=P50;
NET address13 LOC=P57;
NET address14 LOC=P51;
NET s0         LOC=P25;      # S0 LED segment
NET s1         LOC=P26;      # S1 LED segment
NET s2         LOC=P24;      # S2 LED segment
NET s3         LOC=P20;      # S3 LED segment
NET s4         LOC=P23;      # S4 LED segment
NET s5         LOC=P18;      # S5 LED segment
NET s6         LOC=P19;      # S6 LED segment
```

Some examples of the actions that can be performed with the FIFO are as follows:

Action	XSPORT Command Sequence
Reset the FIFO	XSPORT 00 XSPORT 01 XSPORT 11
Write 12 (1100) into the FIFO	XSPORT 110011 XSPORT 110010 XSPORT 110011
Read the next value from the FIFO	XSPORT 11 XSPORT 01 XSPORT 11

As a simple test of the FIFO, you can reset the FIFO and write in the sequence of values 0, 1, 2, 3, 4, 5, 6, 7. Then when you read the FIFO you will see the same sequence of values displayed on the LED digit. No new numerals will be displayed once all the values have been read from the FIFO.

The VHDL version of the FIFO controller is shown in Listing 8.7. On line 3 we access the IEEE and xse libraries as usual. But on line 5 we have added a USE statement to get access to the std_logic_unsigned package in the IEEE library. This provides us with some arithmetic operators that we can use to concisely increment the read and write pointers.

Lines 8–22 declare the user-interface to the FIFO. Lines 15–19 define the FPLD pins that connect to the chip-select, write-enable, output-enable, address and data pins of the external RAM. Line 20 declares the outputs which drive the LED digit. We will use these to observe the 4-bit output of the FIFO.

Lines 27–37 declare internal signals for the FIFO control circuitry. The reset signal is driven high to clear the internal flip-flops. The empty and full signals indicate the state of the FIFO and are used to prevent over or under writing

Listing 8.7

```
001- -- Using a Memory Chip for a FIFO
002-
003- LIBRARY IEEE,xse;
004- USE IEEE.std_logic_1164.ALL;
005- USE IEEE.std_logic_unsigned.ALL;      -- access arithmetic operators
006- USE xse.led.ALL;                      -- access LED decoder module
007-
008- ENTITY fifo IS
009- PORT
010- (
011-   wrb: IN STD_LOGIC;      -- active-low FIFO write control input
012-   rdb: IN STD_LOGIC;      -- active-low FIFO read control input
013-   inp: IN STD_LOGIC_VECTOR (3 DOWNTO 0);-- 4-bit input data bus
014-   outp: OUT STD_LOGIC_VECTOR (3 DOWNTO 0);-- 4-bit output data bus
015-   csb: OUT STD_LOGIC;     -- active-low chip-select for external RAM
016-   web: OUT STD_LOGIC;     -- active-low write-enable for external RAM
017-   oeb: OUT STD_LOGIC;     -- active-low output-enable for external RAM
018-   data: INOUT STD_LOGIC_VECTOR (7 DOWNTO 0); -- data bus to/from RAM
019-   address: OUT STD_LOGIC_VECTOR (14 DOWNTO 0); -- address bus to RAM
020-   s: OUT STD_LOGIC_VECTOR (6 DOWNTO 0) -- outputs to 7-segment LED
021- );
022- END fifo;
023-
024-
025- ARCHITECTURE fifo_arch OF fifo IS
026-
027- SIGNAL reset: STD_LOGIC;-- reset signal for clearing the FIFO
028- SIGNAL empty: STD_LOGIC;-- high when the FIFO is empty
029- SIGNAL full: STD_LOGIC; -- high when the FIFO is full
030- SIGNAL rdptr: STD_LOGIC_VECTOR (3 DOWNTO 0);-- read pointer
031- SIGNAL wrptr: STD_LOGIC_VECTOR (3 DOWNTO 0);-- write pointer
032- SIGNAL rdeqwr: STD_LOGIC;-- high when read pointer = write pointer
033- SIGNAL prevop0,prevop1: STD_LOGIC;  -- previous FIFO operation flags
034- SIGNAL loc_clk: STD_LOGIC;-- this is a local clock signal
035- COMPONENT IBUF PORT( I: IN STD_LOGIC; O: OUT STD_LOGIC); END COMPONENT;
036- COMPONENT BUFG PORT( I: IN STD_LOGIC; O: OUT STD_LOGIC); END COMPONENT;
037- SIGNAL buf_rdb,bufg_rdb,buf_wrb,bufg_wrb: STD_LOGIC; -- buff. R/W signals
038-
039- BEGIN
040-
041- buf0: IBUF PORT MAP (I=>rdb,O=>buf_rdb);   -- buffer the read and write
042- buf1: IBUF PORT MAP (I=>wrb,O=>buf_wrb);   -- inputs and transfer them
043- bufg0: BUFG PORT MAP (I=>buf_rdb,O=>bufg_rdb);  -- to low-skew drivers
```

Listing 8.7   (Cont'd.)

```
044- bufg1: BUFG PORT MAP (I=>buf_wrb,O=>bufg_wrb);
045-
046- -- assert reset signal when both the read and write inputs are low
047- reset <= '1' WHEN (bufg_wrb='0' AND bufg_rdb='0') ELSE '0';
048-
049- -- next two processes describe the operation of the flags which
050- -- store the type of the last FIFO operation.
051- PROCESS (bufg_rdb,bufg_wrb)
052- BEGIN
053-   IF bufg_rdb='0' THEN
054-        prevop0 <= '0'; -- clear flag on FIFO read operation
055-   ELSIF (bufg_wrb'event AND bufg_wrb='0') THEN
056-        prevop0 <= '1'; -- set flag on initiation of a FIFO write
057-   END IF;
058- END PROCESS;
059-
060- PROCESS (bufg_rdb,bufg_wrb,prevop0)
061- BEGIN
062-    -- local clock signal pulses for either a FIFO read or write
063-   loc_clk <= bufg_wrb AND bufg_rdb;
064-   IF reset='1' THEN
065-     prevop1 <= '0';
066-     -- prevop1 loaded with prevop0 at end of read or write operation
067-   ELSIF (loc_clk'event AND loc_clk='1') THEN
068-     prevop1 <= prevop0;
069-   END IF;
070- END PROCESS;
071-
072- rdeqwr <= '1' WHEN rdptr=wrptr ELSE '0';
073- -- the FIFO is empty if the read and write pointers are equal and
074- -- the last FIFO operation was a read (prevop1=0).
075- empty <= rdeqwr AND NOT(prevop1);
076- -- the FIFO is full if the read and write pointers are equal and
077- -- the last FIFO operation was a write (prevop1=1).
078- full <= rdeqwr AND prevop1;
079-
080- PROCESS (bufg_rdb,reset) -- inc read pointer process
081- BEGIN
082-   IF reset='1' THEN    -- async reset of read pointer to start of RAM
083-     rdptr <= "0000";
084-   ELSIF (bufg_rdb'event AND bufg_rdb='1') THEN
085-          -- inc read ptr at end of read operation only if FIFO is not empty
```

Experimental

Listing 8.7 (Cont'd.)

```
086-    IF empty='0' THEN
087-            rdptr <= rdptr + 1;
088-    END IF;
089- END IF;
090- END PROCESS;
091-
092- PROCESS (bufg_wrb,reset) -- inc write pointer process
093- BEGIN
094-   IF reset='1' THEN    -- async reset of write pointer to start of RAM
095-        wrptr <= "0000";
096-   ELSIF (bufg_wrb'event AND bufg_wrb='1') THEN
097-        -- inc write ptr at end of write op only if FIFO is not full
098-   IF full='0' THEN
099-           wrptr <= wrptr + 1;
100-   END IF;
101- END IF;
102- END PROCESS;
103-
104- csb <= '0';-- keep the chip-select low so the RAM is always enabled
105- address(14 DOWNTO 4) <= "00000000000"; -- upper address bits forced to 0
106- -- lower address bits driven by rdptr during FIFO read, else by wrptr
107- address(3 DOWNTO 0) <= rdptr WHEN bufg_rdb='0' ELSE wrptr;
108- -- data on the inp bus is passed to the RAM on a write operation.
109- -- Otherwise, the drivers are disabled.
110- data(3 DOWNTO 0) <= inp     WHEN bufg_wrb='0' ELSE "ZZZZ";
111- data(7 DOWNTO 4) <= "0000" WHEN bufg_wrb='0' ELSE "ZZZZ";
112- -- RAM write-enable follows the FIFO write control if FIFO is not full.
113- web <= bufg_wrb WHEN NOT(full) ELSE '1';
114- -- enable RAM data bus drivers during a FIFO read
115- oeb <= bufg_rdb;
116- -- FIFO output is driven by the lower four bits of the RAM data bus
117- outp <= data(3 DOWNTO 0);
118-
119- u0: leddcd PORT MAP(d=>data(3 DOWNTO 0),s=>s);  -- show output on LED
120-
121- END fifo_arch;
122-
```

the RAM. The 4-bit read and write pointers will give us a total FIFO size of 16 registers. The prevop0 and prevop1 flag flip-flops for storing the type of the last FIFO operation are declared. Then the various clock signals and buffers which drive the read/write pointers and the flags are declared.

The description of the circuitry begins on line 41–44 by connecting the read and write inputs to low-skew buffers (this is only necessary when targeting the XC4005XL FPGA). The internal reset signal is generated when the FIFO read and write inputs are both low (line 47).

The two processes on lines 51–58 and 60–70 describe the circuitry shown in Figure 8.17 that records the type of the last FIFO operation. Lines 72–78 use the values in the flag flip-flops to generate the FIFO full and empty status signals. The processes on lines 80–90 and 92–102 describe the read and write pointers, respectively. The VHDL code replicates the functions shown in Figure 8.16. For example, the read-pointer is incremented whenever a read operation occurs as long as the FIFO is not empty. The use of the addition operator (+) provided by the std_logic_unsigned package lets us describe this behavior very concisely.

The description of the interface circuitry between the FPLD and the external RAM follows. The external RAM is enabled by holding its chip select line low as shown on line 104. Since we are only using 16 locations of the external RAM for our FIFO, the upper 11 bits of the RAM address bus are held at logic 0 on line 105. Line 107 describes a 4-bit multiplexer that connects the read-pointer to the lower 4 bits of the RAM address bus when the FIFO is being read. At all other times, the write-pointer drives the lower part of the RAM address.

Lines 110–111 setup the data bus to the external RAM. The 4-bit input bus to the FIFO drives the lower 4 bits of the RAM data bus and the upper 4 bits are held at logic 0 during a FIFO write operation. This lets the FPLD drive new data into the data bus of the RAM. Also, the RAM write control line is pulsed low as long as the FIFO is not full (line 113). The flow of data is depicted in Figure 8.24a.

When data is being read from the FIFO, line 115 lowers the output enable of the external RAM so that the RAM data bus drivers are activated. This lets the RAM drive a data value into the FPLD as shown in Figure 8.24b. To prevent interfering with the data coming from the RAM, the FPLD pin drivers are tristated (lines 110 and 111 with the write control input at logic 1). Line 117 passes the data coming from the RAM to the outputs of the FIFO. This lets the FIFO output bus drive whatever circuitry is requesting a read of the FIFO.

Finally, line 119 instantiates an LED decoder module and connects it to the FIFO output bus so we can see what value is coming out of the FIFO.

The VHDL in Listing 8.7 is targeted at an XC4005XL FPGA, but it can also be used with an XC95108 CPLD by removing the clock buffers on the read and write control inputs. The constraint files used for the ABEL version can also be used with the VHDL version, but all the bus indices have to be surrounded with "<" and ">" (e.g., address4 should be changed to address<4> in the VHDL constraint file).

Once it is compiled, the VHDL version of the FIFO will respond to the same XSPORT command sequences as the ABEL version does.

Projects

1. Design the largest 4-bit-wide memory that you can in a single FPLD using just flip-flops and gates.

2. Design the largest 4-bit-wide memory that you can with the internal RAM of the XC4005XL.

3. Design a 7-segment LED decoder using only the internal RAM and the fact that the RAM can be loaded when the XC4005XL is configured.

4. Modify the FIFO design so that the prevop0 flip-flop output is used in place of the prevop1 output. Compile, download, and test the design. What happens?

5. Modify the FIFO design so that the multiplexer will output the rdptr instead of the wrptr when no operation is occurring. Compile, download, and test the design. What happens?

6. Increase the size of the FIFO from 16 to 32768. How much of the FPLD is used now?

7. Replace the external RAM with the internal memories of the XC4005XL. How do you handle the interface to the RAM data input and output buses? Does the new design perform equivalently to the original?

8. Use a dual-port RAM to build the FIFO so that simultaneous reads and writes can occur.

The GNOME Microcomputer

Objectives

- Discuss the programmer's model for the GNOME microcomputer.
- Show how to program the GNOME.
- Explain the process of binary multiplication.
- Describe the general organization of the GNOME.
- Show how to build the controller section of the GNOME.
- Show how to build the circuitry of the GNOME datapath.
- Show how the GNOME controller and the datapath work together.

Discussion

You have seen how a RAM is built, but you may be asking, "Why are large memories needed?" After all, the largest memory you have needed so far is only a few bits for storing the state of some of your state machines. Some of you may be replying, "It's obvious why we need large RAMs—they are for storing large amounts of data!" Of course, you are right—RAMs are good for storing large amounts of data.

But just having a memory to store data is rather useless. Imagine having all the data on every purchase made by every person in the United States over the course of a year. Now suppose you want to find the group of people who have purchased Volvos so you can sell them a horn that yodels instead of beeps. Your only recourse is to address each memory location and pick out the data you need (and we are talking trillions of memory locations).

However, you could write down detailed instructions and give them to someone else and tell them to find the information you want. These instructions constitute a program and the paper they are written on is a primitive memory. If you had a way to encode the written instructions as binary numbers, you could store them in a RAM just like the data. Then you could build a state machine to interpret the binary numbers stored in the program RAM (which are called opcodes) and execute the instructions upon the data stored in the data RAM (which are called operands).

Of course, in order to do anything to the data, you will need some type of arithmetic logic unit (ALU) that can do things like add or compare numbers. As you use the ALU to process the data, you will have intermediate results that need further manipulation before the final answer is found. These intermediate results are typically stored in a specialized memory called an accumulator (ACC) that can hold a single number. In addition, there are typically a couple of single-bit flags that are used for storing the result of the carry output of the ALU (C) or whether the result computed by the ALU is zero or not (Z).

Of primary importance when executing the program is to remember what step you are on. The programmer's model for the GNOME microcomputer (shown in Figure 9.1) does this with a program counter (PC). The PC stores the address of the memory location in the program RAM that holds the current instruction being executed. Once this instruction is processed, the address in the PC is incremented and the instruction in the next program memory location is fetched and executed.

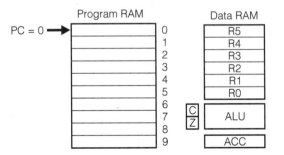

Figure 9.1   A programmer's model of the GNOME microcomputer.

As an example of the execution of a program, the steps involved in adding two 8-bit numbers using a 4-bit ALU are shown in Figure 9.2. Recall that 8-bit numbers can be added by hooking a couple of 4-bit adders together with the carry output of one adder feeding the carry input of the other adder. This is an example of solving a larger problem by increasing the amount of circuitry. With a microcomputer having a fixed-size 4-bit ALU, however, you cannot increase the amount of circuitry to handle 8-bit numbers. Instead, you have to use the same 4-bit ALU to work on the 8-bit numbers by breaking the numbers into pieces and handling one piece of the problem at a time. This is an example of reducing the amount of circuitry needed to solve a problem by spreading the solution process over a longer time. The GNOME program for adding two 8-bit numbers will demonstrate this type of solution.

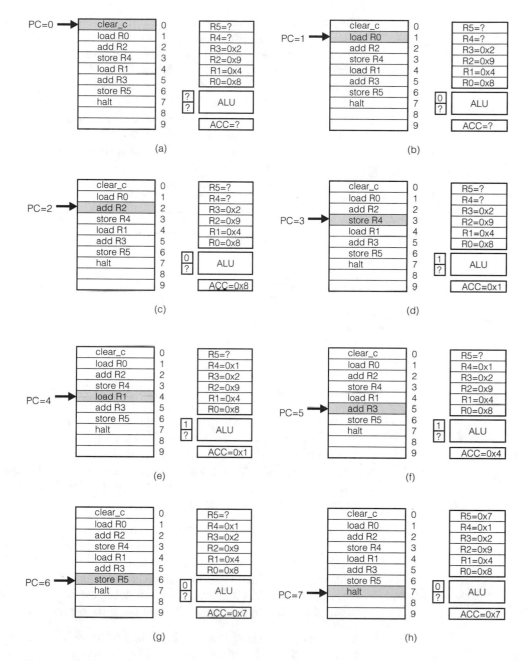

Figure 9.2 Adding two 8-bit numbers using a 4-bit ALU.

The two 8-bit numbers to be added are 0x48 and 0x29 (these numbers are in hexadecimal). All the data memory in the GNOME is 4 bits wide. A 4-bit number is called a nybble. An 8-bit number (or byte) can only be stored by using two memory locations in the data RAM: one for the least-significant nybble (LSN) and one for the most-significant nibble (MSN) (see Figure 9.3). Let's assume that the MSN of 0x48 (which is 4) is stored in register R1 of the data RAM and the LSN is stored in R0. The second operand is stored similarly, with the MSN placed in R3 and the LSN placed in R2. The 8-bit result of adding these two numbers will be stored in R5 (MSN) and R4 (LSN).

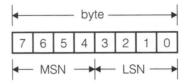

Figure 9.3   The arrangement of bits and nybbles in a byte.

Upon startup, the program RAM is already loaded with the instruction opcodes and the PC is initialized to 0 so it points to the first instruction (Figure 9.2a). (In the figure, the instruction mnemonic is shown instead of the binary opcode in order to make the execution of the program easier to understand.) The operands are already stored in R0, R1, R2, and R3. The RAM for the result, however, is not initialized and contains some random number (represented by ?). The accumulator is also not initialized, and neither are the carry or zero flags. The carry flag acts as both the input carry bit to the ALU and stores the output carry after the ALU adds two 4-bit numbers. For this program, you want to start with C = 0. The clear_c instruction clears the carry bit. Note that in Figure 9.2b the carry bit is now zero and the PC has been incremented so it points to the next instruction.

The next instruction, load R0, is used to load the LSN of the first operand into the accumulator. In effect, the instruction mnemonic means "load the value found in R0 into the ACC." Note that while the address of the RAM location that holds the data (R0) is explicitly mentioned, the place where the data is transferred to (ACC) is only implied. After executing this instruction, the ACC holds the value 0x8 and the PC has been incremented.

The next instruction (Figure 9.2c) is add R2, which you might guess would be an instruction to add the contents of R2 to something. But what? Again, the implied operand is the accumulator. This instruction actually means "add the value found in R2, the value in the ACC, and the bit in the C flag, and store the result in the ACC while loading the C flag with the carry output." In this case, the result (in both hexadecimal and binary representations) will be:

0x0	0	input carry
+0x8	1000	contents of accumulator
+0x9	1001	contents of R2
0x11	10001	5-bit result
0x1	0001	four-bit result in accumulator
0x1	1	1-bit output carry

The addition of 0x8 and 0x9 creates a number that will not fit in the 4-bit accumulator: 0x11. The uppermost bit of this result is the output carry, and this replaces the value that was stored in the C flag (see Figure 9.2d). The carry output will be needed when the MSNs of the operands are added. The accumulator stores the lower four bits (0001 in this case).

The next instruction, store R4, stores the contents of the accumulator in the LSN of the result, R4 (Figure 9.2e). Note that this instruction copies the values; it does not affect the original values held in the accumulator and the carry flag.

Once the PC has reached address 4 (Figure 9.2e), the 8-bit addition is half done. The least-significant nibbles of the operands have been added and the result stored away in R4. The C flag now holds the carry that resulted from this addition, and it will be used in the addition of the most-significant nibbles. This begins by loading the accumulator with the MSN of the first operand from R1. The add R3 instruction (Figure 9.2f) adds the MSN of the second operand (R3) and the carry bit to the ACC as follows:

0x1	1	input carry
+0x4	0100	contents of accumulator
+0x2	0010	contents of R3
0x07	00111	5-bit result
0x7	0111	four-bit result in accumulator
0x0	0	1-bit output carry

The contents of the accumulator are stored in the MSN of the result (R5) after which the program is stopped by the halt instruction (Figures 9.2g and 9.2h, respectively). Converting the contents of the operand and result registers back to decimal, we can see that the addition program came up with the right result:

```
R1,R0 = 0x48 = 4 × 16 + 8 =  72
R3,R2 = 0x29 = 2 × 16 + 9 = +41
R5,R4 = 0x71 = 7 × 16 + 1 = 113
```

The addition program showed some of the operations that the GNOME can do. Table 9.1 lists all the possible instructions for the GNOME. The instructions can be divided into five main categories: (1) load/store instructions; (2) arithmetic instructions; (3) value testing instructions; (4) flag instructions; and (5) branching instructions.

You have already seen the use of some of the load/store, arithmetic, and flag instructions in the addition program. The only new feature is the ability to load or add to the accumulator with a number stored in the opcode instead of having to get it from the RAM. Such instructions are said to have immediate operands. The instruction load #0, for example, loads the accumulator with zero, not with the contents of the RAM location at address zero. The # symbol before the operand indicates the presence of an immediate operand.

The test instruction is used to test bits in the accumulator to see if they are zero. For example, the instruction test #2 will do a bitwise AND of 0010 with the contents of the accumulator. The Z flag will be set if all the bits in the result are zero. If ACC contains 0110 when test #2 is executed, then Z will be cleared to zero because 0110 & 0010 = 0010 and this result is not zero. Note that the value of the accumulator after this operation is still 0110 because the result of the AND operation is not stored—only the Z bit is affected.

The branching instructions are used to alter the sequence in which instructions are executed. In the following code fragment, the third bit of the value in R0 is tested to see if it is zero. If it is, the jump instruction is skipped and execution proceeds from address 0x4. Otherwise, the third bit is not zero so the jump instruction is not skipped and execution proceeds from address 0x15.

```
 0:  load R0    ; ACC ← contents of RAM location 0
 1:  test #4    ; see if third bit of ACC is nonzero
 2:  skip_z     ; skip next instruction if the third bit is zero
 3:  jump #15   ; jump to address 0x15 if third bit is nonzero
 4:  test #2    ; skipped to here. Test the second bit....
     ...
     ...
15:  test #8    ; jumped to here. Test the fourth bit.
     ...
```

Branching instructions allow the program to alter its operations depending on the results of previous operations. The movement of the instruction counter from address to address is similar to the state transitions made by your drink machine controller in response to the actions of the customer.

You might ask, "How was it decided what instructions would be included in the GNOME instruction set?" This is the important issue of instruction set design. The decisions you make when designing an instruction set are affected by many factors (such as the number of logic gates you have available to build the microcomputer), but the most important criteria is what kinds of operations the microcomputer must perform. To determine what these operations are, you typically must write some programs for things you would like the microcomputer to do. By examining what types of instructions you need in these benchmark programs, you can decide what to include or leave out of the instruction set. For example, the addition program demonstrated the need for instructions to clear the carry bit, load and store the accumulator, and add numbers to the accumulator. If you plan for the microcomputer to do a lot of additions, you probably want these instructions in your instruction set.

Many of the instructions of Table 9.1 were selected after I wrote a program to do multiplication. Therefore, the process of doing multiplication with binary numbers will be explained and then a program to do it will be presented.

Before tackling binary multiplication, it might be best to review decimal multiplication. Consider forming the product of 456×123:

$$
\begin{array}{r}
456 \\
\times\ 123 \\
\hline
1368 = 456 \times 3 \times 10^0 \\
9120 = 456 \times 2 \times 10^1 \\
+45600 = 456 \times 1 \times 10^2 \\
\hline
56088
\end{array}
$$

As you can see, the product is formed by summing the subproducts of the multiplicand (456) with each digit of the multiplier (1, then 2, then 3). However, each succeeding digit of the multiplier is also multiplied by a power of 10, and this causes the subproducts to be shifted left by the addition of the zeros in the rightmost digits.

Binary multiplication is exactly the same, only the base (or radix) is changed. For example, multiplying 12 (1100) by 6 (0110) is done as follows:

$$
\begin{array}{r}
1100 \\
\times\ 0110 \\
\hline
0000 = 1100 \times 0 \times 2^0 \\
11000 = 1100 \times 1 \times 2^1 \\
110000 = 1100 \times 1 \times 2^2 \\
+0000000 = 1100 \times 0 \times 2^3 \\
\hline
1001000
\end{array}
$$

Notice that the multiplicand is shifted left and added to the product if the corresponding bit of the multiplier is one. But if the multiplier bit is zero, then a zero is added, which is the same as doing nothing. So there are just a few things that need to be done to perform a binary multiplication:

1. We need to be able to determine if a given bit of the multiplier is zero or one.

2. If a multiplier bit is zero, we need to skip the addition of the multiplicand to the sum.

Table 9.1 GNOME instruction set.

Mnem.	Operations	Description
load Rd	ACC ← Rd	Load the accumulator with the contents of RAM address Rd. d is an address in the range [0..15].
load #d	ACC ← d	Load the accumulator with the value d. d is a value in the range [0..15].
store Rd	Rd ← ACC	Store the value in the accumulator into RAM location Rd. d is an address in the range [0..15].
add Rd	ACC ← ACC + Rd + C; C← carry out	Add the contents of RAM address Rd and the C flag to the accumulator. d is an address in the range [0..15].
add #d	ACC ← ACC + d + C; C ← carry out	Add the value d and the flag C to the accumulator. d is a value in the range [0..15].
xor Rd	ACC ← ACC$Rd	Exclusive-OR the contents of RAM address Rd with the accumulator. d is an address in the range [0..15].
test Rd	Z ← ACC&Rd	Setthe Z flag if the logical-AND of RAM address Rd and the accumulator is zero. Otherwise, clear the Z flag. d is an address in the range [0..15].
clear_c	C ← 0	Clear the C flag to zero.
set_c	C ← 1	Set the C flag to one.
skip_c	PC ← PC + 1 + C	If C = 1, skip the next instruction by incrementing the program counter by two instead of one. If C = 0, execute the next instruction.
skip_z	PC ← PC + 1 + Z	If Z = 1, skip the next instruction by incrementing the program counter by two instead of one. If Z = 0, execute the next instruction.
jump #a	PC ← a	Jump to program address a and execute instructions from that point. a is an address in the range [0..127].

3. If the multiplier bit is one, we need to be able to shift the multiplicand left by the appropriate number of positions and then add it to the sum.

A flowchart for performing a binary multiplication is shown in Figure 9.4. It starts by clearing the product register, which will hold the result of the multiplication when the program is finished. The mask RAM location is set to 0001 and will be used to test each bit of the multiplier to see if it is zero or one.

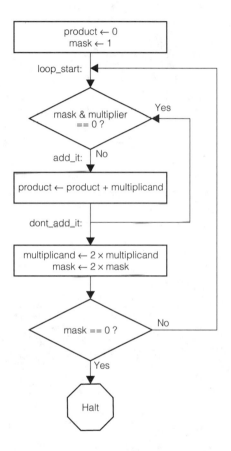

Figure 9.4   A flowchart for binary multiplication of 4-bit numbers.

Next, the mask and the multiplier are logically ANDed together. This will determine if the least-significant bit of the multiplier is zero. If it is not zero, then the multiplicand is added to the product. If it is zero, then the addition is skipped.

Next, the multiplicand has to be multiplied by 2 to prepare it for the next time it is needed. Then the mask is multiplied by 2 so it now contains the value

0010. This value is not equal to zero, so the program execution loops back toward the beginning of the program.

Now the value 0010 is ANDed with the multiplier, thus testing the second bit of the multiplier to see if it is one or zero. If the bit is one, the multiplicand is added to the product again. But the multiplicand has already been multiplied by 2, thus shifting it left by one bit position. So we can see that as the program proceeds, the multiplicand will be shifted left each time through the loop. This shifting ensures that the correct number of zeros will be added onto the right end of the multiplier each time it is added to the product, just as we saw was needed in the example of binary multiplication.

Of course, the looping has to stop sometime, but when? For this program, the multiplier only has 4 bits so the loop should terminate after four iterations. This is handled at the bottom of the program when a test is made to see if the mask is zero. How can the mask ever equal zero if it starts at 1 and gets doubled every time through the loop? Shouldn't it keep progressing in the sequence 1, 2, 4, 8, 16, 32,... ? Since the GNOME handles 4-bit values, when the mask reaches eight (1000) and gets doubled, the leftmost one falls off the end and the result is 0000. This happens after four doublings, which is the same thing as four iterations through the program. So the check for the mask being zero terminates the program at the correct point.

How can the operations shown in Figure 9.4 be performed by the GNOME? Let's start by assigning the various quantities in the flowchart (or variables) to registers:

Register	Variable
R0	multiplier
R1	LSN of multiplicand
R2	MSN of multiplicand
R3	LSN of product
R4	MSN of product
R5	mask

Notice that the product requires two 4-bit registers because multiplying the 4-bit multiplier and the 4-bit multiplicand can give an 8-bit result. Since the multiplicand has to be added to the 8-bit product, the multiplicand is also assigned to two 4-bit registers. The program that operates on the registers and arrives at the product is shown in Listing 9.1.

The various initialization steps for the multiplication program are performed on lines 1–11 in Listing 9.1. The instructions on lines 1–4 load the values to be multiplied into the multiplier and multiplicand registers, respectively. Lines

Listing 9.1   Binary multiplication program.

```
1         load #5            ; multiplier <- 5
2         store R0
3         load #8            ; multiplicand <- 8
4         store R1
5         load #0            ; MSN of multiplicand <- 0
6         store R2
7         load #0            ; clear the product
8         store R3
9         store R4
10        load #1            ; initialize the mask with 0001
11        store R5
12   loop_start:
13        load R0            ; get the multiplier
14        test R5            ; AND it with the mask
15        skip_z             ; skip the next instruction if Z=1
16        jump add_it        ; if Z=0, add multiplicand to product
17        jump dont_add_it   ; if Z=1, skip the addition
18   add_it:
19        clear_c            ; add the multiplicand to the product
20        load R3
21        add R1
22        store R3
23        load R4
24        add R2
25        store R4
26   dont_add_it:
27        clear_c            ; multiply the multiplicand by 2
28        load R1
29        add R1
30        store R1
31        load R2
32        add R2
33        store R2
34        clear_c            ; multiply the mask by 2
35        load R5
36        add R5
37        store R5
38        skip_c             ; skip the next instruction if mask=0
39        jump loop_start    ; otherwise, jump and do another iteration
40   halt:
41        jump halt          ; keep jumping to this same place forever
42                           ;    when the multiplication is done
```

5–6 clear the upper 4-bits of the 8-bit multiplicand register. The registers for holding the product are cleared on lines 7–9. Lines 10–11 initialize the mask to 0001.

The start of the computational loop for the multiplication program begins at the label `loop_start`. (The program is using labels instead of explicit program instruction addresses to improve readability. For the program to work, these labels have to be replaced with the addresses of the instructions they are attached to.) The instructions on lines 13–14 load the accumulator with the multiplier and then test the accumulator to see if the bit selected by the mask is zero or not. The Z flag is set to one if the selected bit in the multiplier is zero.

The conditional addition of the shifted multiplicand to the product is achieved with the instructions on lines 15–17. The `skip_z` instruction causes the program counter to skip the first `jump` instruction (line 16) if the bit tested in the multiplier is zero (which causes the Z flag to be set to 1). This causes the GNOME to execute the second `jump` instruction (line 17) that transfers control to a section of the program that does not add anything to the product.

If the Z bit is zero, then the tested bit of the multiplier is one and the addition must take place. In this case, the `skip_z` instruction does not skip over the `jump` instruction on line 16. Instead, the `jump` instruction is executed and it transfers control to the section of the program that adds the multiplicand to the product (lines 19–25). This is just the program for adding two 8-bit quantities that we discussed previously.

The only thing that is left to do is to shift the mask and the multiplicand left by one bit position. In the flowchart this operation is shown as a multiplication by 2, but there is no multiply instruction for the GNOME. That is no problem because a number can be multiplied by two merely by adding it to itself: $2 \times number = number + number$. The multiplicand is left-shifted by the 8-bit addition of lines 27–33 and the 4-bit mask is left-shifted on lines 34–37.

Note that when the mask has a value of 1000, the addition instruction will cause mask to go to 0000 and the carry flag will be set. This will happen after four iterations through the loop. On line 38, the carry flag is tested and the program either jumps to `main_loop` and does another iteration (line 39, if C = 0) or stops by continually looping to itself (line 41, if C = 1).

The instructions on lines 1–42 are called the assembly code for the program. However, the GNOME can only interpret ones and zeros, not printed instructions on a page. Therefore, the assembly code must be transformed into machine code, which is merely the binary code for each instruction. Table 9.2 lists the encoding for each instruction of the GNOME.

Translating assembly code into machine code is relatively straightforward for the GNOME: Merely look up the instruction in the table, write down the code,

Table 9.2   GNOME instruction set opcodes.

Mnem.	Encoding	Comments
load Rd	0100 d3 d2 d1 d0	d3..d0 are the address bits of a RAM location in the range [0..15].
load #d	0001 d3 d2 d1 d0	d3..d0 are the bits of a value in the range [0..15].
store Rd	0011 d3 d2 d1 d0	d3..d0 are the address bits of a RAM location in the range [0..15].
add Rd	0101 d3 d2 d1 d0	d3..d0 are the address bits of a RAM location in the range [0..15].
add #d	0010 d3 d2 d1 d0	d3..d0 are the bits of a value in the range [0..15].
xor Rd	0110 d3 d2 d1 d0	d3..d0 are the address bits of a RAM location in the range [0..15].
test Rd	0111 d3 d2 d1 d0	d3..d0 are the address bits of a RAM location in the range [0..15].
clear_c	00000000	
set_c	00000001	
skip_c	00000010	
skip_z	00000011	
jump #a	1 a6 a5 a4 a3 a2 a1 a0	a7..a0 are the bits of an address in the range [0..127].

and fill in any immediate data or address bits that are required. Figure 9.5 shows the machine code associated with the instructions for the 8-bit addition program. The first instruction is clear_c, which has an encoding of 00000000. This 8-bit opcode is placed in the program at address zero.

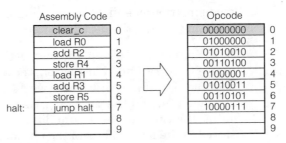

Figure 9.5   GNOME machine code for the 8-bit addition program.

The next instruction, load R0, has 0100 as the most-significant 4 bits of the encoding. The remaining 4 bits encode the address of the data RAM location. In this case that is RAM location zero, so the least-significant 4 bits are 0000 and the total opcode is 01000000.

As a final example, the instruction jump halt performs a program jump to the address labeled halt. The actual address associated with the label halt is seven (0000111 in 7-bit binary). The first bit of a jump instruction opcode is 1. The final 7 bits of the opcode is the address to jump to. So the opcode for the jump halt instruction is 10000111. That's all there is to it.

Once the opcodes are stored in the program memory, the GNOME can read out the bit patterns and execute the program. But what are the individual steps that must occur to execute each instruction? Each instruction has three basic steps:

Fetch: The GNOME must read the current instruction from the program memory and store it into the instruction register.

Decode: The GNOME must figure out which type of instruction is to be executed and get the required operands.

Execute: The GNOME must do the operation on the operands and store the result.

Table 9.3 shows these steps for each of the GNOME instructions. During the fetch step, a new instruction opcode is read from the address in the external RAM that is pointed to by the PC. This opcode is stored in the instruction register (IR). The program counter is also incremented so that it points to the next instruction (since all the GNOME instructions are 1 byte long).

During the decode step, the instructions will require different operations depending upon their addressing mode. For example, an instruction that uses direct addressing, like add Rd, first needs to get its data from memory. This is done by accessing the external memory with the memory address stored in the lower four bits of the opcode. The four-bit value read from memory replaces these address bits in the instruction register. On the other hand, immediate mode instructions like the load immediate data instruction (load #d) don't do anything during the decoding step because they already have their operands.

In the final execute step, the instruction is completed. To continue the description of the direct add instruction, the accumulator is loaded with the sum of the operand data (stored in the lower 4 bits of the IR), the current value in the accumulator, and the carry flag. As another example, the store Rd instruction stores the value in the accumulator into the memory location whose address is stored in the lower 4 bits of the opcode. Finally, the skip_c and skip_z instructions will increment the PC again (which will jump over the next instruction after the skip instruction) if the C or Z flag is set, respectively.

Table 9.3   Steps in executing each GNOME instruction.

Instruction	Fetch	Decode	Execute
load #d	IR ← (PC); PC ← PC+1		ACC ← [IR3..IR0]
load Rd	IR ← (PC); PC ← PC+1	[IR3..IR0] ← ([IR3..IR0])	ACC ← [IR3..IR0]
store Rd	IR ← (PC); PC ← PC+1		([IR3..IR0]) ← ACC
add #d	IR ← (PC); PC ← PC+1		ACC ← ACC + [IR3..IR0] + C
add Rd	IR ← (PC); PC ← PC+1	[IR3..IR0] ← ([IR3..IR0])	ACC ← ACC + [IR3..IR0] + C
xor Rd	IR ← (PC); PC ← PC+1	[IR3..IR0] ← ([IR3..IR0])	ACC ← ACC$[IR3..IR0]
test Rd	IR ← (PC); PC ← PC+1	[IR3..IR0] ← ([IR3..IR0])	Z ← ACC&[IR3..IR0]
clear_c	IR ← (PC); PC ← PC+1		C ← 0
set_c	IR ← (PC); PC ← PC+1		C ← 1
skip_c	IR ← (PC); PC ← PC+1		PC ← PC + C
skip_z	IR ← (PC); PC ← PC+1		PC ← PC + Z
jump #a	IR ← (PC); PC ← PC+1		PC ← [IR6..IR0]

Experimental

GNOME in ABEL

The instruction set for the GNOME was discussed in the last section. It does not do a lot, but it does do enough. Our next task is to build the hardware that will execute the GNOME instruction set.

There are several things that you already know about the hardware needed for the GNOME:

- There needs to be a memory for storing the program instructions. Since the instruction opcodes are 8 bits wide, the program memory should also be 8 bits wide. As to the number of memory locations, the jump instruction contains a 7 bit address, so the program can jump to any location in the range [0..127], inclusive. So the program memory will have 128 bytes. We could build such a memory using the internal RAMs of the XC4005XL, but we would also like to build the GNOME using the XC95108 which has no internal RAMs. Therefore, the program memory will have to be housed in an external RAM.

- A 7-bit program counter (PC) is needed for storing the address of the current instruction.

- An 8-bit instruction register (IR) is needed for storing the current instruction that is being executed.

- There needs to be a RAM for storing data. Since the GNOME operates on 4-bit operands, the RAM should be 4 bits wide. The instruction set can address RAM locations with a 4-bit address, so there should be $2^4 = 16$ RAM locations. Therefore, a 16×4 memory is needed for the data RAM. For the data memory we can use the LSN of 16 locations in the same external RAM that we will also use for the program memory.
- Two flip-flops are needed for storing the carry (C) and zero (Z) flags.
- A 4-bit ALU is needed to operate on two 4-bit operands and produce a 4-bit result.
- A 4-bit accumulator (ACC) is needed to store the result from the ALU.

These pieces of the GNOME are shown in Figure 9.6.

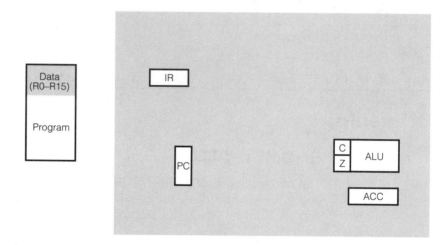

Figure 9.6   Major pieces of the GNOME microcomputer.

Now the communication buses between these pieces have to be determined. The ALU is a relatively simple place to start. Based on the operations needed to perform each instruction as shown in Table 9.3, the ALU only needs to combine the ACC with a 4-bit operand stored in the LSN of the instruction register. Then the result is stored back into the ACC. The buses necessary to do this are shown in Figure 9.7.

We need to get data and opcodes from the external RAM in order to do any operations. The first step for accessing the RAM is to send it an address. This address can either be the address of a data register if a direct mode operand is being processed or the address of the next opcode. Figure 9.8 shows how a multiplexer is used to select the source of the external RAM address. The inputs to the multiplexer are (1) the LSN of the current opcode in the IR, or (2) the 7-bit program counter. One of these inputs is transferred to the address bus of the

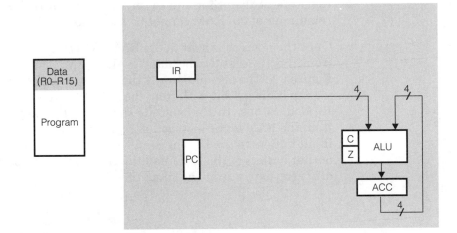

Figure 9.7   Buses for the ALU.

RAM depending upon the value of a selector input for the multiplexer. (The control of this selector input will be discussed later.)

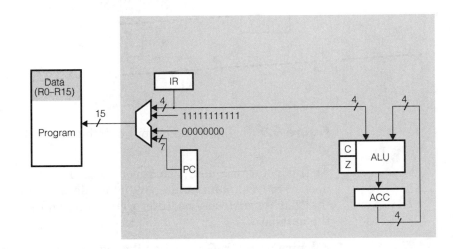

Figure 9.8   Multiplexing and extending addresses for the external RAM.

Remember that the external 32K RAM requires a 15-bit address. Therefore, we have to add 8 extra bits to the 7-bit program counter or 11 bits to the 4-bit operand address in order to get a full RAM address. We can extend the program counter with zeroes and the operand address with ones. This will place the program at one end of the RAM (in the address range [0..127]) and the data

Experimental   329 registers at the other end (with addresses in the range [32752..32767]). The remainder of the RAM cannot be accessed at all.

Once the address is sent to the RAM, GNOME must have a way to accept data from the RAM and send data to the RAM. The buses for this are shown in Figure 9.9. The accumulator is the only source for data written to the RAM, so we can build a 4-bit path from the ACC through the output drivers to the 8-bit RAM data bus. In the opposite direction, the only destination for data read from the RAM is the instruction register, so we place an 8-bit data bus from the input buffers to the IR. The read/write control logic in GNOME will control the output buffers of the RAM and the GNOME data bus so that they do not try to drive the data bus at the same time.

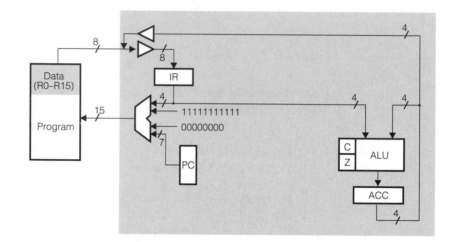

Figure 9.9   Input and output data buses for the external RAM.

At this point, the datapath is complete. The datapath contains the circuitry needed to read, manipulate, and store data. What is lacking is the controller that fetches instructions, interprets them, and then directs the operations of the datapath.

Fetching an instruction is handled by outputing the program counter through the memory address multiplexer and latching the output data from the RAM into the IR. After fetching an instruction, the simple incrementer in Figure 9.10 can advance the PC to the next instruction in memory. The incrementer can also be used when skipping an instruction.

But a program can also jump to a new address instead of proceeding sequentially through memory. Therefore, a multiplexer is added in Figure 9.11 to select whether the PC is loaded with (1) the address of the next instruction in memory, as calculated by the incrementer; or (2) with the address stored in the

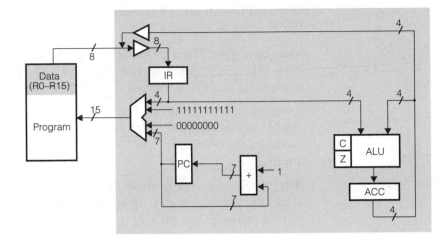

Figure 9.10 Circuitry for incrementing the PC.

lower 7 bits of a jump instruction in the IR. (The control of this selector input will be discussed later.)

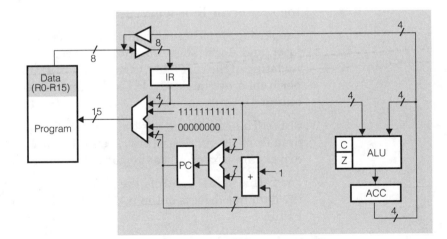

Figure 9.11 Additional circuitry for supporting the jump instruction.

Only one last major component is missing from the GNOME microcomputer: the instruction decoder. The instruction decoder interprets the current instruction opcode along with the carry and zero flags and outputs the control signals which will make the other parts of the GNOME do the required operations. Just what control signals are needed? Figure 9.12 shows the instruction decoder and the 10 control signals which enter the various components of the GNOME. Here is what they do:

sel_data_ram: This signal controls the multiplexer attached to the address bus. When high, the address of a data register is selected. When low, the address in the PC is applied to the RAM.

jump_pc: This signal controls the multiplexer attached to the PC. When high, the new value of the PC comes from the address stored in the current instruction. When low, the PC is updated with the value from the incrementer.

inc_pc: This signal controls the incrementer attached to the PC. When high, the value in the PC is increased by one. When low, the PC retains its current value.

ld_ir: This signal controls the loading of the instruction register. When high, the 8 bits of opcode from the RAM are loaded into the IR. When low, the instruction register retains its contents.

ld_ir_lsn: This signal controls the loading of the LSN of the instruction register with direct address mode operands. When high, the lower 4 bits of the IR are loaded with the lower 4 bits of data from the RAM. When low, the LSN of the IR is unchanged.

[aop2..aop0]: These signals control the operation of the ALU, accumulator, and flags. These three signals control (1) the setting, clearing, and updating of the carry flag, (2) the updating of the zero flag, (3) the operation performed by the ALU, (4) the loading of the accumulator with the ALU output.

write: This signal controls the writing of data to the data RAM. When high, data from the ACC can be written into the RAM. When low, the FPLD I/O drivers are turned off so the data from the RAM can enter.

read: This signal controls the reading of data from the RAM. When high, the data bus drivers of the RAM are enabled so it can force values into the FPLD.

We have an idea of what the controller and datapath sections of the GNOME must do and how they interface to each other and the external RAM. Now we can write the ABEL code for the GNOME (Listing 9.2).

Lines 6–13 declare the interface pins that go between the FPLD and the external RAM and LED digit. There are also clock and reset inputs. On line 15, flip-flops are declared to hold the state of the instruction decoder. Since there are only 3 states (for fetching, decoding, and executing), 2 flip-flops are sufficient. Lines 16–20 define the storage for the instruction register, program

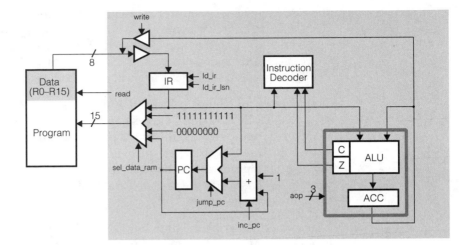

Figure 9.12 The instruction decoder and the control signals.

counter, accumulator, and carry and zero flags. The internal control signals from Figure 9.12 are declared on lines 22–29.

Synonyms for various items are listed on lines 31–48. The bit patterns that correspond to various ALU operations are also listed (lines 38–43), as are the bit patterns for the instruction decoder states (lines 46–48).

Lines 52–63 declare the bit pattern for each opcode as originally defined in Table 9.2. Note the use of the "don't-care" symbol in instructions with immediate-mode or direct-mode operands. Since these operand bits are stored in the opcode and may take on any value, they are not useful to the instruction decoder for figuring out which instruction is being executed. For example, the instruction decoder must decide what to do with a load Rd instruction by looking at only the MSN of the opcode since the LSN will hold the address of the operand.

The interconnection of the GNOME components begins after line 65. Lines 69–74 connect the clock input to all the flip-flops in GNOME. All changes to the state of GNOME occur on the rising edge of the clock. This gives us a nice, synchronous system.

Lines 76–81 connect the reset input to the asynchronous clear inputs of all the flip-flops. Thus, applying a logic 1 to the reset input will put the instruction decoder in the FETCH state (00) and the PC will point to the first location of memory (0000000). This is where GNOME programs always begin.

The interface to the control lines of the external RAM is described on lines 83–91. Line 83 pulls the RAM chip-select line low so the RAM stays enabled at all times. Line 84 lowers the output enable and enables the data bus drivers in the RAM whenever the internal read control line goes high. Line 85 lowers the RAM's write-enable whenever the internal write control line goes high. The write enable is gated by the clock so it only goes low during the last half of each clock cycle. This is done to make sure the RAM's address and data buses stabilize during the first half of the clock cycle before the write occurs.

The address multiplexer for the RAM address is described on lines 87 and 88. When GNOME is trying to access operand values from the data section of the RAM (sel_data_ram=1), the RAM address is driven by the operand address stored in the LSN of the IR with 11111111111 prepended. When GNOME is fetching an opcode from the program section of the RAM (sel_data_ram=0), the PC is gated to the lower 7 bits of the RAM address with 00000000 prepended as the upper 8 bits.

Lines 90–91 describe the data bus GNOME uses to write its accumulator value into the RAM. The data bus drivers from the FPLD to the external RAM are enabled only during the write operation. The value driven into the RAM has the accumulator value in its LSN and 0000 as its MSN.

Lines 93–99 are concerned with the loading of the instruction register. When the ld_ir signal is high, the IR is loaded with the opcode that is on the output pins of the RAM data bus. (The data bus drivers from the FPLD should be disabled at this time so that they do not interfere with the opcode.) The ld_ir_lsn control signal is raised when a direct mode operand is retrieved from RAM. In this case, the LSN of the IR is loaded with the LSN of the data from the RAM and the MSN of the IR retains its value. Finally, if neither the ld_ir or ld_ir_lsn signals are at logic 1, then the instruction register is unchanged.

The circuitry for updating the program counter is described on lines 101–103. When the jump_pc signal is at logic 1, the PC is loaded with the value in the lower 7 bits of the instruction register. If a jump is not being performed but the inc_pc control signal is raised, then the program counter gets loaded with its incremented value. And if neither control signal is high, the PC retains its value.

Lines 106–138 show how the ALU performs various operations. When the instruction decoder requests an addition operation (lines 106–111), the ALU adds the value in the accumulator, the LSN of the instruction register, and the carry flag. The result is returned to the accumulator and the carry flag is updated with the carry output from the addition. Note that we used the addition operator provided in ABEL instead of building our own adder circuitry. To handle the carry flag, we had to do a 5-bit addition instead of 4. The input carry was placed into the least-significant bit following 4 zero bits while the accumulator and operand values were extended to 5 bits by adding a most-significant bit of zero to each. This arrangement places the carry input at the least-significant position while the carry output comes out in the most-significant bit of the result. The lower 4 bits of the result replace the value in the accumulator, while the fifth bit is loaded into the carry flag. The zero flag is just reloaded with its value (line 110).

The exclusive-OR operation (lines 112–116) loads the accumulator with the XOR of the accumulator and the LSN of the instruction register. The carry and zero flags are not changed. The PASS operation does not do much: It just loads the accumulator with the LSN of the instruction register (line 119) and leaves the carry and zero flags alone.

The AND operation (lines 121–126) is used during the execution of a test instruction. The accumulator is bit-wise ANDed with the LSN of the instruction register. Then the 4 product terms are ORed together and the inverse is stored in the zero flag. Thus, the zero flag will contain a logic 1 if all the product terms are zero. Otherwise, it will contain a logic zero. The ALU operation for setting (clearing) the carry flag is described on lines 127–130 (131–134). The carry flag is loaded while the accumulator and zero flag remain unchanged. Finally, if no legal ALU operation is requested by the instruction decoder, the accumulator, carry flag, and zero flag retain their current values (lines 135–138).

The state machine that controls the fetching, decoding, and execution of the instructions is described on lines 140–183. The FETCH state is relatively simple (lines 142–147): The address in the PC is output on the RAM address bus (sel_data_ram=0), the RAM data bus drivers are turned on (read=1), the instruction register is loaded with the opcode output by the RAM (ld_ir=1), and the program counter is incremented (inc_pc=1). Note that these control lines are all driven to their indicated values as soon as the instruction decoder enters the FETCH state. The order of the statements makes no difference. Because the sel_data_ram and read control lines do not control flip-flop elements, their effects are immediately apparent. However, the ld_ir and inc_pc signals affect the values that appear on the inputs to flip-flops but they do not generate clock signals. Therefore, the instruction register and program counter are not updated until the end of the FETCH state when the next rising clock edge occurs. One result of this is that the address in the PC of the current instruction will remain stable for the entire FETCH cycle and will only be incremented after the opcode has been latched into the IR.

The next rising clock edge ends the FETCH state and transitions the instruction decoder to the DECODE state (line 147). The DECODE state is described on lines 149–167. Unlike the FETCH state, the operations carried out in the DECODE state depend upon the opcode in the instruction register. We use the ABEL CASE...ENDCASE; construct to express succinctly the operation of the instruction decoder. After the CASE keyword is a list of Boolean expressions. If a particular expression evaluates to a non-zero logic value, then the statements between its : and the ; delimiters are activated. For example, if the instruction register contains the CLEAR_C opcode, the instruction decoder will not perform any operations other than to move to the EXECUTE state on the next clock cycle. Only the instructions which have to read a direct-mode operand from the external RAM actually do anything in the DECODE state (lines 159–166). In this case, the next instruction decoder state is still EXE-CUTE, but a brace-enclosed set of control signal activations is also listed following the WITH keyword. The RAM address bus is driven by the address stored in the LSN of the instruction register (sel_data_ram=1) and the RAM data bus drivers are enabled (read=1). The LSN of the data from the RAM is latched into the LSN of the IR (ld_ir_lsn=1). At this point, the direct operand from RAM has replaced the register address in the LSN of the instruction register.

The final EXECUTE state of the instruction decoder is described on lines 169–183. Most of the instructions just place the appropriate bit pattern on the alu_op control bus and let the ALU do its thing. However, the SKIP_C and SKIP_Z instructions (lines 173 and 174) make use of the carry and zero flags to affect the flow of the program by incrementing the program counter if the appropriate flag is set. The PC was already incremented at the end of the FETCH phase so that it would point to the next instruction. Thus, another increment causes GNOME to skip the next instruction completely. (This works because all GNOME instructions are 1 byte long.)

The STORE_DIR opcode (line 178) also handles writing data into the RAM. The sel_data_ram control signal is raised so that the RAM is addressed with the address stored in the LSN of the instruction register. The write signal is also raised to enable the data bus drivers in the FPLD and to generate the pulse for the RAM's write-enable control line.

An LED decoder is connected to the accumulator on lines 185–202. This lets us monitor the value in the accumulator as programs are executed. (As an alternative, we could observe the sequence of instruction addresses by connecting it to the LSN of the program counter.)

Now we must compile the GNOME1.ABL file. If you try to compile the ABEL file in Listing 9.2 for the XC95108 CPLD or XC4005XL, you may receive a message from the software saying that the design will not fit. The problem is the complexity of the adder logic for the ALU and the program counter. The fitter tries very hard to pack as much logic into each macrocell as possible. The fitter would rather collapse logic for intermediate signals (i.e., those declared as NODEs) into the CLBs or macrocells that contain the logic driven by these intermediate signals. This is especially true for carry logic in adders. The fitter attempts to do complete carry look-ahead so no macrocells have to be used for intermediate carry bits. Complete carry look-ahead circuitry for 5 or more bits requires complex logic that is hard to fit into single macrocells.

To limit the amount of carry look-ahead, we can place the following statement on line 67: @CARRY 1. This limits the amount of carry look-ahead to 1 adder bit, effectively making ripple-carry adders. Without this statement, the logic synthesis tools will try to create carry look-ahead circuitry for all 7 bits of the PC incrementer and the 5 bits of the ALU adder.

The GNOME1.ABL file should compile successfully for the XC95108 CPLD. The utilization of the XC95108 is :

```
*********************** Resource Summary ***********************
Design     Device          Macrocells   Product Terms Pins
Name       Used            Used         Used          Used
gnom1_95 XC95108-20-PC84 75/108 (69%)  230/540 (42%)  35/69 (50%)
```

You can see that there are 33 macrocells and 310 product terms left free in the CPLD. After limiting the carry chain length, the equivalent statistics when GNOME is mapped to the XC4005XL FPGA are:

```
Design Summary
--------------
    Number of errors:       0
    Number of warnings:     3
    Number of CLBs:                116 out of   196   59%
       CLB Flip Flops:     23
       CLB Latches:         0
       4 input LUTs:      213
       3 input LUTs:       34
    Number of bonded IOBs:          35 out of    65   53%
       IOB Flops:           0
       IOB Latches:         0
Total equivalent gate count for design: 1524
```

The GNOME1.ABL file can be compiled for the XS95 Board using these pin assignment constraints:

```
NET clock        LOC=P46;
NET reset        LOC=P47;
NET csb          LOC=P65;
NET web          LOC=P63;
NET oeb          LOC=P62;
NET data0        LOC=P44;
NET data1        LOC=P43;
NET data2        LOC=P41;
NET data3        LOC=P40;
NET data4        LOC=P39;
NET data5        LOC=P37;
NET data6        LOC=P36;
NET data7        LOC=P35;
NET addr0        LOC=P75;
```

Listing 9.2   ABEL code for the GNOME microprocessor.

```
001-  MODULE GNOME1
002-  TITLE 'GNOME microcontroller -- V1.0'
003-
004-  DECLARATIONS
005-  "----- external signals -----
006-  clock        PIN;                      "GNOME is synchronous
007-  reset        PIN;                      "zeroes the internal state
008-  addr14..addr0 PIN ISTYPE 'COM'; "external RAM address bus
009-  data7..data0 PIN   ISTYPE 'COM'; "external RAM data bus
010-  csb          PIN   ISTYPE 'COM'; "external RAM chip select
011-  web          PIN   ISTYPE 'COM'; "external RAM write control
012-  oeb          PIN   ISTYPE 'COM'; "external RAM read control
013-  s6..s0       PIN   ISTYPE 'COM'; "LED digit drivers
014-  "----- internal storage elements -----
015-  st1..st0     NODE  ISTYPE 'REG'; "GNOME internal state
016-  ir7..ir0     NODE  ISTYPE 'REG'; "instruction register (IR)
017-  pc6..pc0     NODE  ISTYPE 'REG'; "program counter (PC)
018-  acc3..acc0   NODE  ISTYPE 'REG'; "accumulator (ACC)
019-  carry        NODE  ISTYPE 'REG'; "carry flag (C)
020-  zero         NODE  ISTYPE 'REG'; "zero flag (Z)
021-  "----- internal signals -----
022-  read         NODE ISTYPE 'COM'; "1 when reading RAM
023-  write        NODE ISTYPE 'COM'; "1 when writing RAM
024-  sel_data_ram NODE ISTYPE 'COM'; "1 when accessing R0-R15
025-  jump_pc      NODE  ISTYPE 'COM'; "1 when overwriting PC
026-  inc_pc       NODE ISTYPE 'COM'; "1 when incrementing PC
027-  ld_ir        NODE ISTYPE 'COM'; "1 when loading IR
028-  ld_ir_lsn    NODE ISTYPE 'COM'; "1 when loading LSN of IR
029-  aop2..aop0   NODE ISTYPE 'COM'; "ALU operation code
030-  "----- synonyms for various items -----
031-  ir      = [ir7..ir0];
032-  pc      = [pc6..pc0];
033-  address = [addr14..addr0];
034-  data    = [data7..data0];
035-  acc     = [acc3..acc0];
036-  "----- ALU operation codes -----
037-  alu_op  = [aop2..aop0];
038-  PASS      = ^b001; "ALU passes input to output
039-  ADD       = ^b010; "ALU adds inputs
040-  XOR       = ^b011; "ALU exclusive-ORs inputs
041-  AND       = ^b100; "ALU logically ANDs inputs
042-  SET_CARRY = ^b101; "carry flag is set
```

Listing 9.2   ABEL code for the GNOME microprocessor. (Cont'd.)

```
043- CLR_CARRY = ^b110; "carry flag is cleared
044- "----- GNOME control section states -----
045- controller_state = [st1..st0];
046- FETCH   = ^b00;      "GNOME fetching opcode state
047- DECODE  = ^b01;      "GNOME decoding opcode state
048- EXECUTE = ^b11;      "GNOME executing instruction state
049-
050- X = .X.; "synonym for DON'T-CARE
051- "----- GNOME instruction opcodes -----
052- CLEAR_C   = [0,0,0,0,0,0,0,0];
053- SET_C     = [0,0,0,0,0,0,0,1];
054- SKIP_C    = [0,0,0,0,0,0,1,0];
055- SKIP_Z    = [0,0,0,0,0,0,1,1];
056- LOAD_IMM  = [0,0,0,1,X,X,X,X];
057- ADD_IMM   = [0,0,1,0,X,X,X,X];
058- STORE_DIR = [0,0,1,1,X,X,X,X];
059- LOAD_DIR  = [0,1,0,0,X,X,X,X];
060- ADD_DIR   = [0,1,0,1,X,X,X,X];
061- XOR_DIR   = [0,1,1,0,X,X,X,X];
062- TEST_DIR  = [0,1,1,1,X,X,X,X];
063- JUMP      = [1,X,X,X,X,X,X,X];
064-
065- EQUATIONS
066-
067-
068-
069- controller_state.CLK = clock; "synchronously clock
070- pc.CLK               = clock; "  all storage elements
071- ir.CLK               = clock;
072- acc.CLK              = clock;
073- carry.CLK            = clock;
074- zero.CLK             = clock;
075-
076- controller_state.ACLR = reset; "reset all storage
077- pc.ACLR               = reset; "  elements to zero
078- ir.ACLR               = reset; "  on startup
079- acc.ACLR              = reset;
080- carry.ACLR            = reset;
081- zero.ACLR             = reset;
082-
083- csb = 0;                  "always keep RAM selected
084- oeb = !read;              "enable RAM outputs during RAM read
085- web = !(write & !clock); "pulse RAM write in last half of clock
```

Listing 9.2   ABEL code for the GNOME microprocessor. (Cont'd.)

```
086- "address either the data or program sections of the external RAM
087- WHEN sel_data_ram THEN address=[1,1,1,1,1,1,1,1,1,1,1,1,ir3..ir0];
088- ELSE                      address=[0,0,0,0,0,0,0,0,0,pc6..pc0];
089-
090- data.OE = write;              "drive RAM data bus during write ops
091- data = [0,0,0,0,acc3..acc0]; "  with the value in the accumulator
092-
093- WHEN      ld_ir     THEN ir := data.PIN; "IR<=opcode from RAM
094- ELSE WHEN ld_ir_lsn THEN
095-   { "load the lower 4 bits of IR with RAM data...
096-   [ir3..ir0] := [data3..data0].PIN
097-   [ir7..ir4] := [ir7..ir4]; "but keep the upper 4-bits of IR
098-   }
099- ELSE ir := ir;        "IR unchanged
100-
101- WHEN      jump_pc THEN pc := [ir6..ir0]; "PC<=address from opcode
102- ELSE WHEN inc_pc   THEN pc := pc+1;       "increment PC
103- ELSE                   pc := pc;          "PC unchanged
104-
105- "ALU operations
106- WHEN alu_op==ADD THEN
107-   { "add ACC, data from IR, and C and update ACC and C
108-   [carry,acc3..acc0] := [0,acc3..acc0] + [0,ir3..ir0] +
109-                              [0,0,0,0,carry];
110-   zero := zero; "zero flag is unaffected by add operations
111-   }
112- ELSE WHEN alu_op==XOR THEN
113-   { "XOR ACC and data from IR and update the ACC but not the flags
114-   [acc3..acc0] := [acc3..acc0] $ [ir3..ir0];
115-   carry:=carry;  zero:=zero; "flags unaffected by XOR operation
116-   }
117- ELSE WHEN alu_op==PASS THEN
118-   { "pass data from the IR into the ACC but leave flags alone
119-   [acc3..acc0]:=[ir3..ir0];  carry:=carry;  zero:=zero;
120-   }
121- ELSE WHEN alu_op==AND THEN
122-   { "AND ACC and data from IR and update the zero flag
123-   zero := !(acc3&ir3 # acc2&ir2 # acc1&ir1 # acc0&ir0);
124-   [acc3..acc0] := [acc3..acc0]; "don't change the accumulator
125-   carry := carry;                  "carry unaffected by AND operation
126-   }
127- ELSE WHEN alu_op==SET_CARRY THEN
128-   { "set carry flag to 1 but leave ACC and Z unchanged
```

Listing 9.2   ABEL code for the GNOME microprocessor. (Cont'd.)

```
129-    carry:=1;  [acc3..acc0]:=[acc3..acc0];  zero:=zero;
130-    }
131- ELSE WHEN alu_op==CLR_CARRY THEN
132-    { "clear carry flag to 0 but leave ACC and Z unchanged
133-    carry:=0;  [acc3..acc0]:=[acc3..acc0];  zero:=zero;
134-    }
135- ELSE
136-    { "in every other case, leave everything unchanged
137-    [acc3..acc0]:=[acc3..acc0];  carry:=carry;  zero:=zero;
138-    }
139-
140- STATE_DIAGRAM controller_state   "controls instruction execution
141-
142-    STATE FETCH: "get the opcode from external RAM
143-        sel_data_ram = 0; "select PROGRAM section of RAM
144-        read = 1;           "enable output from RAM
145-        ld_ir = 1;          "load RAM data into IR
146-        inc_pc = 1;         "incremented the PC
147-        goto DECODE;        "now move on to the next state
148-
149-    STATE DECODE: "decode instruction and get operand if necessary
150-      CASE          "  then move on to the next state
151-        (ir==CLEAR_C)  : EXECUTE; "instructions with implied or
152-        (ir==SET_C)    : EXECUTE; "  immediate operands don't
153-        (ir==SKIP_C)   : EXECUTE; "  need to do anything during
154-        (ir==SKIP_Z)   : EXECUTE; "  this state
155-        (ir==LOAD_IMM) : EXECUTE;
156-        (ir==ADD_IMM)  : EXECUTE;
157-        (ir==JUMP)     : EXECUTE;
158-        (ir==STORE_DIR): EXECUTE;
159-        ((ir==LOAD_DIR) # (ir==ADD_DIR) #  "need to get operands
160-         (ir==XOR_DIR) # (ir==TEST_DIR))    "  for these opcodes
161-                       : EXECUTE WITH
162-           { "read direct operands from the RAM's data section
163-           sel_data_ram = 1; "drive RAM with address in LSN of IR
164-           read = 1;          "enable output from RAM
165-           ld_ir_lsn = 1;     "load 4-bits from RAM into LSN of IR
166-           }
167-      endcase;
168-
169-    STATE EXECUTE: "execute operations on the operands
170-      CASE           " (mostly just give the ALU the right opcode)
```

Listing 9.2   ABEL code for the GNOME microprocessor. (Cont'd.)

```
171-        (ir==CLEAR_C)  : FETCH WITH alu_op   = CLR_CARRY;
172-        (ir==SET_C)    : FETCH WITH alu_op   = SET_CARRY;
173-        (ir==SKIP_C)   : FETCH WITH WHEN carry THEN inc_pc = 1;
174-        (ir==SKIP_Z)   : FETCH WITH WHEN zero  THEN inc_pc = 1;
175-        (ir==LOAD_IMM) : FETCH WITH alu_op   = PASS;
176-        (ir==ADD_IMM)  : FETCH WITH alu_op   = ADD;
177-        (ir==JUMP)     : FETCH WITH jump_pc = 1;
178-        (ir==STORE_DIR): FETCH WITH {sel_data_ram=1; write=1;}
179-        (ir==LOAD_DIR) : FETCH WITH alu_op   = PASS;
180-        (ir==ADD_DIR)  : FETCH WITH alu_op   = ADD;
181-        (ir==XOR_DIR)  : FETCH WITH alu_op   = XOR;
182-        (ir==TEST_DIR) : FETCH WITH alu_op   = AND;
183-     endcase;
184-
185- TRUTH_TABLE  "LED displays the value in the accumulator
186- (acc -> [s6, s5, s4, s3, s2, s1, s0])
187-     0 -> [1,  1,  1,  0,  1,  1,  1 ];
188-     1 -> [0,  0,  1,  0,  0,  1,  0 ];
189-     2 -> [1,  0,  1,  1,  1,  0,  1 ];
190-     3 -> [1,  0,  1,  1,  0,  1,  1 ];
191-     4 -> [0,  1,  1,  1,  0,  1,  0 ];
192-     5 -> [1,  1,  0,  1,  0,  1,  1 ];
193-     6 -> [1,  1,  0,  1,  1,  1,  1 ];
194-     7 -> [1,  0,  1,  0,  0,  1,  0 ];
195-     8 -> [1,  1,  1,  1,  1,  1,  1 ];
196-     9 -> [1,  1,  1,  1,  0,  1,  1 ];
197-    10-> [1,  1,  1,  1,  1,  1,  0 ];
198-    11-> [0,  1,  0,  1,  1,  1,  1 ];
199-    12-> [1,  1,  0,  0,  1,  0,  1 ];
200-    13-> [0,  0,  1,  1,  1,  1,  1 ];
201-    14-> [1,  1,  0,  1,  1,  0,  1 ];
202-    15-> [1,  1,  0,  1,  1,  0,  0 ];
203-
204- END GNOME1
```

```
             NET addr1        LOC=P79;
             NET addr2        LOC=P82;
             NET addr3        LOC=P84;
             NET addr4        LOC=P1;
             NET addr5        LOC=P3;
             NET addr6        LOC=P83;
             NET addr7        LOC=P2;
             NET addr8        LOC=P58;
```

```
NET addr9          LOC=P56;
NET addr10         LOC=P54;
NET addr11         LOC=P55;
NET addr12         LOC=P53;
NET addr13         LOC=P57;
NET addr14         LOC=P61;
NET s0             LOC=P21;
NET s1             LOC=P23;
NET s2             LOC=P19;
NET s3             LOC=P17;
NET s4             LOC=P18;
NET s5             LOC=P14;
NET s6             LOC=P15;
```

And here are the pin assignments to use when targeting GNOME to the XS40 Board:

```
NET clock          LOC= P44;
NET reset          LOC= P45;
NET addr0          LOC= P3;
NET addr1          LOC= P4;
NET addr2          LOC= P5;
NET addr3          LOC= P78;
NET addr4          LOC= P79;
NET addr5          LOC= P82;
NET addr6          LOC= P83;
NET addr7          LOC= P84;
NET addr8          LOC= P59;
NET addr9          LOC= P57;
NET addr10         LOC= P51;
NET addr11         LOC= P56;
NET addr12         LOC= P50;
NET addr13         LOC= P58;
NET addr14         LOC= P60;
NET data0          LOC= P41;
NET data1          LOC= P40;
NET data2          LOC= P39;
NET data3          LOC= P38;
NET data4          LOC= P35;
NET data5          LOC= P81;
NET data6          LOC= P80;
NET data7          LOC= P10;
NET csb            LOC= P65;
NET oeb            LOC= P61;
NET web            LOC= P62;
NET s0             LOC= P25;
```

```
NET s1              LOC= P26;
NET s2              LOC= P24;
NET s3              LOC= P20;
NET s4              LOC= P23;
NET s5              LOC= P18;
NET s6              LOC= P19;
```

Once the design is compiled, you can download the GNOME1 microprocessor to the XS95 or XS40 Board. But what about a program for it to execute? That has to be loaded into the external RAM so GNOME can fetch instructions. Let's use the following addition program to test GNOME. The address and opcode for each instruction are listed in the two columns on the left.

```
0000 18        load   #8      ; initialize the data registers
0001 30        store  R0      ;   with the numbers to be added
0002 14        load   #4      ; we will do 0x29 + 0x48 = 0x71
0003 31        store  R1
0004 19        load   #9
0005 32        store  R2
0006 12        load   #2
0007 33        store  R3
0008 00        clear_c        ; start carry flag with zero
0009 40        load   R0      ; add the LSNs of each number and
000A 52        add    R2      ;   also set carry for next addition
000B 34        store  R4      ; store LSN of sum
000C 41        load   R1
000D 53        add    R3      ; add MSNs of both numbers + carry
000E 35        store  R5      ; store MSN of sum
            loop:
000F 44        load   R4      ; display LSN of sum
0010 45        load   R5      ; display MSN of sum
0011 8F        jump   #loop   ; display sum repetitively
```

The opcodes for the addition program can be stored in a file called ADD.HEX. The file contains a sequence of lines, each line beginning with a dash (-), the number of opcodes it contains, the beginning memory address at which to start placing the opcodes, and the list of opcodes. (All the numbers are in hexadecimal.) Here are the contents of the ADD.HEX file:

```
- 12 0000 18 30 14 31 19 32 12 33 00 40 52 34 41 53 35 44 45 8F
```

The first (and only) line of ADD.HEX says: "There are 18 opcodes in this line (0x12=18). Begin loading them into memory at address 0x0000=0. Here is the list of opcodes."

Now you can download the program and the GNOME microprocessor into the XS95 Board using the command:

```
C:\XCPROJ\GNOM1_95>  XSLOAD ADD.HEX GNOM1_95.SVF
```

Or use this command for the XS40 Board:

```
C:\XCPROJ\GNOM1_40>  XSLOAD ADD.HEX GNOM1_40.BIT
```

Then you can use the following command sequences to reset GNOME to the beginning of the program and single-step it through the instructions:

Action	XSPORT Command Sequence
Reset	XSPORT 10 XSPORT 00
Instruction single-step	XSPORT 00 XSPORT 01 XSPORT 00 XSPORT 01 XSPORT 00 XSPORT 01

The reset action first raises the reset input so that the GNOME program counter gets loaded with 0. (This is good because that is the address where our program starts in the external RAM.) Then the reset is lowered to logic 0 so the program counter can increment through the opcodes.

The command sequence that makes GNOME execute a single instruction just pulses the clock input 3 times. This sequences the GNOME instruction decoder through the FETCH, DECODE, and EXECUTE phases of each instruction. (To make things easier, you might place this set of 6 commands into a batch file. Then you can single-step GNOME using a single command.)

The leddigit LED decoder module is connected to the accumulator. You should see intermediate results displayed on the LED digit as the program executes. Eventually, GNOME will enter the loop at the end of the program and will repetitively display the 2-nybble sum: 1..7...1...7....

GNOME in VHDL

The VHDL version of the GNOME microcprocessor is presented in Listing 9.3. I have kept the same same major circuit blocks, buses, and control lines shown in Figure 9.12.

Lines 3–6 provide access to the IEEE and xse libraries. The USE statement on line 5 lets us use the concise arithmetic operations from the std_logic_unsigned package of the IEEE library.

Lines 11 and 12 declare the main clock input and the asynchronous reset input. Lines 13–18 declare the interface pins that go between the FPLD and the external RAM and LED digit.

The architecture description begins by declaring the low-skew clock buffers and buffered clock signals that are used when GNOME is targeted to the XS40 Board (lines 26–28). These can be removed for the XS95 Board. The flip-flops for storing the state of the instruction execution process are declared on line 31. There are only 3 possible states (which are defined as CONSTANTs on lines 34–36), so a 2-element flip-flop array is sufficient.

Lines 39–43 define the storage for the program counter, accumulator, instruction register, and carry and zero flags. Each of these quantities is declared with both a current value and a next value. The current value will be used to control the operations of GNOME during a given state, and the next value will become the current value as GNOME transitions between states.

The internal control signals from Figure 9.12 are declared on lines 46–54. This includes a 5-element signal array, sum, that will hold the output from the GNOME ALU adder. Line 54 declares the 3-bit array which conducts operation codes from the instruction decoder to the ALU. The 3-bit patterns for the various ALU operations are defined as CONSTANTs on lines 57–62. This is followed on lines 65–76 with definitions of the bit patterns for all the opcodes as originally defined in Table 9.2. The operand bits have not been included in the bit strings. Since the immediate-mode and direct-mode operand bits in the opcode may take on any value, they are not useful to the instruction decoder for figuring out which instruction is being executed. For example, the instruction decoder must decide what to do with a load Rd instruction by looking at the upper 4 bits of the opcode since the lower 4 bits will hold the address of the operand.

The interface to the control lines of the external RAM is described on lines 81–83. Line 81 pulls the RAM chip-select line low so the RAM stays enabled at all times. Line 82 lowers the output enable and enables the data bus drivers in the RAM whenever the internal read control line goes high. Line 83 lowers the RAM's write-enable whenever the internal write control line goes high. The write enable is gated by the clock so it only goes low during the last half of each clock cycle. This is done to make sure the RAM's address and data buses stabilize during the first half of the clock cycle before the write occurs.

The address multiplexer for the RAM address is described on lines 86 and 87. When GNOME is trying to access operand values from the data section of the RAM (sel_data_ram=1), the RAM address is driven by the operand address stored in the LSN of the IR with 11111111111 prepended. Otherwise when GNOME is fetching an opcode from the program section of the RAM (sel_data_ram=0), the current value of the PC is gated to the lower 7 bits of the RAM address with 00000000 prepended as the upper 8 bits.

Line 91 describes the data bus that GNOME uses to write its accumulator value into the RAM. The data bus drivers from the FPLD to the external RAM are enabled only during the write operation (write=1). The value driven into the RAM has the accumulator value in its LSN and 0000 as its MSN. At all other times, the FPLD data bus drivers are placed in a high impedance state.

Lines 94–98 are concerned with the loading of the instruction register. When the ld_ir signal is high, the IR is loaded with the opcode that is on the output pins of the RAM data bus. (The data bus drivers from the FPLD should be disabled at this time so that they do not interfere with the opcode.) The ld_ir_lsn control signal is raised when a direct mode operand is retrieved from RAM. In this case, the LSN of the IR is loaded with the LSN of the data from the RAM and the MSN of the IR retains its value. Finally, if neither the ld_ir or ld_ir_lsn signals are at logic 1, then the instruction register is just reloaded with its current contents.

The circuitry for updating the program counter is described on lines 101–105. When the jump_pc signal is at logic 1, the PC is loaded with the value in the lower 7 bits of the instruction register. If a jump is not being performed but the inc_pc control signal is raised, then the program counter gets loaded with its incremented value (note the use of the + operation provided by the std_logic_unsigned package). And if neither control signal is high, the PC retains its value.

Lines 108–159 show how the ALU performs various operations. This section uses the VHDL CASE…END CASE; construct. The value in the alu_op signal array is compared with the constant following the WHEN keyword on lines 117, 124, 129, 134, 143, and 148. If a match is found, the statements following the => delimiter are activated. If no match is found, activation falls through to the statements following the default match on line 153.

When the instruction decoder requests an addition operation (lines 117–123), the ALU adds the value in the accumulator, the LSN of the instruction register, and the carry flag. The result is returned to the accumulator and the carry flag is updated with the carry output from the addition. Again we used the addition operator provided in the std_logic_unsigned package instead of building our own adder circuitry. To handle the carry flag, we had to do a 5-bit addition instead of 4. The current carry flag was placed into the least-significant bit following 4 zero bits while the accumulator and operand values were extended to 5 bits by adding a most-significant bit of zero to each. This arrangement places the carry input at the least-significant position while the carry output comes out in the most-significant bit of the sum result. The lower 4 bits of the result will replace the value in the accumulator, while the fifth bit will be loaded into the carry flag. The zero flag is not changed by the addition operation (line 123).

The exclusive-OR operation (lines 124–128) loads the accumulator with the XOR of the accumulator and the LSN of the instruction register. The carry and zero flags are not changed. The PASS operation on lines 129–133 does not do much: It just loads the accumulator with the LSN of the instruction register (line 131) and leaves the carry and zero flags alone.

The AND operation (lines 134–142) is used during the execution of a test instruction. The accumulator is bit-wise ANDed with the LSN of the instruction register. Then the 4 product terms are ORed together and the inverse is stored in the zero flag. Thus, the zero flag will contain a logic 1 if all the product terms are zero. Otherwise, it will contain a logic zero. The accumulator and carry flag are unaffected.

The ALU operation for setting (clearing) the carry flag is described on lines 143–147 (148–152). The carry flag is loaded while the accumulator and zero flag remain unchanged. Finally, if no legal ALU operation is requested by the instruction decoder, the accumulator, carry flag, and zero flag retain their current values (lines 153–157).

The state machine that controls the fetching, decoding, and execution of the instructions is described by the process on lines 163–248. Once again, a CASE...END CASE; construct is used where the value in the curr_st array selects one of three groups of signal assignments. All the control signals are assigned default values on lines 166–173 to prevent the synthesis of implied latches if none of the cases match the current state. The FETCH state is relatively simple (lines 176–181): The address in the PC is output on the RAM address bus, the RAM data bus drivers are turned on, the instruction register is loaded with the opcode output by the RAM, and the program counter is incremented. Note that these control lines are all driven to their indicated values as soon as the instruction decoder enters the FETCH state. The order of the statements makes no difference. Because the sel_data_ram and read control lines do not control flip-flop elements, their effects are immediately apparent. However, the ld_ir and inc_pc signals affect the values that appear on the inputs to flip-flops but they do not generate clock signals. Therefore, the instruction register and program counter are not updated until the end of the FETCH state when the next rising clock edge occurs. One result of this is that the address in the PC of the current instruction will remain stable for the entire FETCH cycle and will only be incremented after the opcode has been latched into the IR.

The next rising clock edge ends the FETCH state and transitions the instruction decoder to the DECODE state (lines 182–206). Unlike the FETCH state, the operations carried out in the DECODE state depend upon the opcode in the instruction register. We use a set of IF...END IF; constructs to express the decoding operation. Since the instruction opcode bit patterns are distinct, there is no chance that more than one block of statements will be activated. Most of the GNOME instructions simply pass through the DECODE to the EXECUTE state. Only the instructions which have to read a direct-mode operand from the external RAM actually do anything in the DECODE state. In these cases the next instruction decoder state is still EXECUTE. The RAM address bus is driven by the address stored in the LSN of the instruction register and the RAM data bus drivers are enabled. The LSN of the data from the RAM is latched into the LSN of the IR. At this point, the direct operand from RAM has replaced the register address in the LSN of the instruction register.

The final EXECUTE state of the instruction decoder is described on lines 207–245. Most of the instructions just place the appropriate bit pattern on the alu_op control bus and let the ALU do its thing. However, the SKIP_C (lines 214–216) and SKIP_Z (lines 217–219) instructions make use of the carry and zero flags to affect the flow of the program by incrementing the program counter if the appropriate flag is set. The PC was already incremented at the end of the FETCH phase so that it would point to the next instruction. Thus, another increment causes GNOME to skip the next instruction completely. (This works because all GNOME instructions are 1 byte long.)

The STORE_DIR opcode (lines 229–232) also handles writing data into the RAM. The sel_data_ram control signal is raised so that the data RAM is addressed with the address stored in the LSN of the instruction register. The write signal is also raised to enable the data bus drivers in the FPLD and to generate the pulse for the RAM's write-enable control line.

Lines 251–252 connect the clock input to the low-skew buffer. The buffered clock drives the state changes in the process on lines 255–274. All changes to the state of GNOME occur on the rising edge of the clock (see line 266). This gives us a well behaved, synchronous system. Lines 259–264 are activated whenever the reset input is held high. Thus, applying a logic 1 to the reset input will put the instruction decoder in the FETCH state and the PC will point to the first location of memory (0000000). This is where GNOME programs always begin.

An LED decoder is connected to the current value of the accumulator on line 277. This lets us monitor the value in the accumulator as programs are executed. (As an alternative, we could observe the sequence of instruction addresses by connecting it to the LSN of the program counter.)

The VHDL in Listing 9.3 is targeted at an XC4005XL FPGA, but it can also be used with an XC95108 CPLD by removing the clock buffers on the clock input. The constraint files used for the ABEL version can also be used with the VHDL version, but all the bus indices have to be surrounded with "<" and ">" (e.g., addr4 should be changed to addr<4> in the VHDL constraint file). Once it is compiled, the VHDL version of GNOME will run the same programs and respond to the same XSPORT command sequences as the ABEL version does.

Observing the device utilizations, the GNOME microprocessor now consumes 60% of an XC95108 CPLD as shown below:

Listing 9.3   VHDL code for the GNOME microprocessor.

```
001- -- GNOME microcomputer
002-
003- LIBRARY IEEE,xse;
004- USE IEEE.std_logic_1164.ALL;
005- USE IEEE.std_logic_unsigned.ALL;
006- USE xse.led.ALL;
007-
008- ENTITY gnome IS
009- PORT
010- (
011-   clock: IN STD_LOGIC;   -- clock (naturally)
012-   reset: IN STD_LOGIC;   -- reset control input
013-   addr: OUT STD_LOGIC_VECTOR (14 DOWNTO 0);-- address to RAM
014-   data: INOUT STD_LOGIC_VECTOR (7 DOWNTO 0);-- data bus to RAM
015-   csb: OUT STD_LOGIC;    -- active-low chip-select for RAM
016-   web: OUT STD_LOGIC;    -- active-low write-enable for RAM
017-   oeb: OUT STD_LOGIC;    -- active-low output-enable for RAM
018-   s: OUT STD_LOGIC_VECTOR (6 DOWNTO 0)-- drivers for 7-seg LED
019- );
020- END gnome;
021-
022-
023- ARCHITECTURE gnome_arch OF gnome IS
024-
025- -- buffered, low-skew clock signal
026- COMPONENT IBUF PORT(I: IN STD_LOGIC; O: OUT STD_LOGIC); END COMPONENT;
027- COMPONENT BUFG PORT(I: IN STD_LOGIC; O: OUT STD_LOGIC); END COMPONENT;
028- SIGNAL buf_clock, bufg_clock: STD_LOGIC;
029-
030- -- current and next GNOME state-machine state
031- SIGNAL curr_st, next_st: STD_LOGIC_VECTOR (1 DOWNTO 0);
032-
033- -- possible GNOME state-machine states and their definitions
034- CONSTANT FETCH:  STD_LOGIC_VECTOR (1 DOWNTO 0) := "00"; -- fetch instr.
035- CONSTANT DECODE: STD_LOGIC_VECTOR (1 DOWNTO 0) := "01"; -- decode instr.
036- CONSTANT EXECUTE: STD_LOGIC_VECTOR (1 DOWNTO 0) := "11"; -- execute instr.
037-
038- -- current and next PC, ACC, IR, carry, and zero flag
039- SIGNAL curr_pc, next_pc: STD_LOGIC_VECTOR (6 DOWNTO 0);
040- SIGNAL curr_acc, next_acc: STD_LOGIC_VECTOR (3 DOWNTO 0);
041- SIGNAL curr_ir, next_ir: STD_LOGIC_VECTOR (7 DOWNTO 0);
042- SIGNAL curr_carry, next_carry: STD_LOGIC;
043- SIGNAL curr_zero, next_zero: STD_LOGIC;
```

Listing 9.3   VHDL code for the GNOME microprocessor. (Cont'd.)

```
044-
045- -- control signals
046- SIGNAL read: STD_LOGIC;              -- 1 when reading RAM
047- SIGNAL write: STD_LOGIC;             -- 1 when writing RAM
048- SIGNAL sel_data_ram: STD_LOGIC;      -- 1 when accessing RO-R15
049- SIGNAL jump_pc: STD_LOGIC;           -- 1 when overwriting PC
050- SIGNAL inc_pc: STD_LOGIC;            -- 1 when incrementing PC
051- SIGNAL ld_ir: STD_LOGIC;             -- 1 when loading IR
052- SIGNAL ld_ir_lsn: STD_LOGIC;         -- 1 when loading LSN of IR
053- SIGNAL sum: STD_LOGIC_VECTOR (4 DOWNTO 0);  -- for adder output
054- SIGNAL alu_op: STD_LOGIC_VECTOR (2 DOWNTO 0); -- ALU operation code
055-
056- -- possible ALU opcodes
057- CONSTANT PASS_OP: STD_LOGIC_VECTOR (2 DOWNTO 0) := "001"; -- in to out
058- CONSTANT ADD_OP: STD_LOGIC_VECTOR (2 DOWNTO 0) := "010"; -- add inputs
059- CONSTANT XOR_OP: STD_LOGIC_VECTOR (2 DOWNTO 0) := "011"; -- XOR inputs
060- CONSTANT AND_OP: STD_LOGIC_VECTOR (2 DOWNTO 0) := "100"; -- test input=0
061- CONSTANT SET_CARRY_OP: STD_LOGIC_VECTOR (2 DOWNTO 0) := "101"; -- set C
062- CONSTANT CLR_CARRY_OP: STD_LOGIC_VECTOR (2 DOWNTO 0) := "110"; -- clear C
063-
064- -- possible instruction opcodes
065- CONSTANT CLEAR_C:  STD_LOGIC_VECTOR (7 DOWNTO 0) := "00000000";
066- CONSTANT SET_C:    STD_LOGIC_VECTOR (7 DOWNTO 0) := "00000001";
067- CONSTANT SKIP_C:   STD_LOGIC_VECTOR (7 DOWNTO 0) := "00000010";
068- CONSTANT SKIP_Z:   STD_LOGIC_VECTOR (7 DOWNTO 0) := "00000011";
069- CONSTANT LOAD_IMM: STD_LOGIC_VECTOR (3 DOWNTO 0) := "0001";
070- CONSTANT ADD_IMM:  STD_LOGIC_VECTOR (3 DOWNTO 0) := "0010";
071- CONSTANT STORE_DIR:STD_LOGIC_VECTOR (3 DOWNTO 0) := "0011";
072- CONSTANT LOAD_DIR: STD_LOGIC_VECTOR (3 DOWNTO 0) := "0100";
073- CONSTANT ADD_DIR:  STD_LOGIC_VECTOR (3 DOWNTO 0) := "0101";
074- CONSTANT XOR_DIR:  STD_LOGIC_VECTOR (3 DOWNTO 0) := "0110";
075- CONSTANT TEST_DIR: STD_LOGIC_VECTOR (3 DOWNTO 0) := "0111";
076- CONSTANT JUMP:     STD_LOGIC                     := '1';
077-
078- BEGIN
079-
080- -- external RAM control signals
081- csb <= '0';-- always keep the RAM selected
082- oeb <= NOT(read);-- enable RAM drivers during RAM read operations
083- web <= NOT(write AND NOT(bufg_clock)); -- pulse write in last 1/2 of clock
084-
085- -- address either the data or program sections of the external RAM
```

Listing 9.3   VHDL code for the GNOME microprocessor. (Cont'd.)

```
086- addr <= "11111111111" & curr_ir(3 DOWNTO 0) WHEN sel_data_ram='1' ELSE
087-    "00000000" & curr_pc;
088-
089- -- drive the accumulator into the RAM during write operations
090- -- but disable the drivers into high-impedance state at all other times
091- data <= "0000" & curr_acc WHEN write='1' ELSE "ZZZZZZZZ";
092-
093- -- load the instruction register with a new opcode
094- next_ir <= data WHEN ld_ir='1' ELSE
095-    -- or load only the lower 4 bits of the IR with data
096-    curr_ir(7 DOWNTO 4) & data(3 DOWNTO 0) WHEN ld_ir_lsn='1' ELSE
097-    -- or else don't change the IR
098-    curr_ir;
099-
100- -- load the PC with an address to jump to
101- next_pc <= curr_ir(6 DOWNTO 0) WHEN jump_pc='1' ELSE
102-    -- or increment the PC
103-    curr_pc+1 WHEN inc_pc='1' ELSE
104-    -- or else don't change the PC
105-    curr_pc;
106-
107- -- this process describes the operations of the ALU
108- PROCESS (alu_op,curr_zero,curr_carry,curr_acc,curr_ir,sum)
109- BEGIN
110-     -- set the defaults for these signals to avoid synthesis of latches
111- sum <= "00000";
112- next_acc <= "0000";
113- next_carry <= '0';
114- next_zero <= '0';
115-
116- CASE alu_op IS
117-    WHEN ADD_OP =>
118-             -- add acc with the lower 4 bits of the IR and the carry
119-             sum <= ('0' & curr_acc) + ('0' & curr_ir(3 DOWNTO 0))
120-                          + ("0000" & curr_carry);
121-             next_acc <= sum(3 DOWNTO 0); -- ACC gets low 4 bits of sum
122-             next_carry <= sum(4); -- carry is most sig bit of the sum
123-             next_zero <= curr_zero; -- zero flag is not changed
124-    WHEN XOR_OP =>
125-             -- XOR the accumulator with the lower 4 bits of the IR
126-             next_acc <= curr_acc XOR curr_ir(3 DOWNTO 0);
127-             next_carry <= curr_carry;-- carry flag is not changed
128-             next_zero <= curr_zero;-- zero flag is not changed
```

Listing 9.3   VHDL code for the GNOME microprocessor. (Cont'd.)

```
129-    WHEN PASS_OP =>
130-            -- pass lower 4 bits of IR into ACC
131-            next_acc <= curr_ir(3 DOWNTO 0);
132-            next_carry <= curr_carry;-- carry flag is not changed
133-            next_zero <= curr_zero;-- zero flag is not changed
134-    WHEN AND_OP =>
135-            -- test the ACC for zeroes in unmasked bit positions
136-            next_acc <= curr_acc;-- ACC is not changed
137-            next_carry <= curr_carry;-- carry is not changed
138-            -- zero flag is set if ACC has zeroes where IR has ones
139-            next_zero <= NOT(  (curr_acc(3) AND curr_ir(3))
140-                           OR (curr_acc(2) AND curr_ir(2))
141-                           OR (curr_acc(1) AND curr_ir(1))
142-                           OR (curr_acc(0) AND curr_ir(0)));
143-    WHEN SET_CARRY_OP =>
144-            -- set the carry bit
145-            next_acc <= curr_acc;-- ACC is not changed
146-            next_carry <= '1';-- set carry bit
147-            next_zero <= curr_zero;-- zero flag is not changed
148-    WHEN CLR_CARRY_OP =>
149-            -- clear the carry bit
150-            next_acc <= curr_acc;-- ACC is not changed
151-            next_carry <= '0';-- clear carry bit
152-            next_zero <= curr_zero;-- zero flag is not changed
153-    WHEN OTHERS =>
154-            -- don't do anything for undefined ALU opcodes
155-            next_acc <= curr_acc;
156-            next_carry <= curr_carry;
157-            next_zero <= curr_zero;
158- END CASE;
159- END PROCESS;
160-
161- -- this process describes the transitions of the GNOME state machine
162- -- and sets the control signals that are activated in each state
163- PROCESS(curr_st,curr_carry,curr_zero,curr_ir)
164- BEGIN
165-    -- set the defaults for these signals to avoid synthesis of latches
166- sel_data_ram <= '0';
167- read <= '0';
168- write <= '0';
169- ld_ir <= '0';
170- ld_ir_lsn <= '0';
171- inc_pc <= '0';
```

Listing 9.3   VHDL code for the GNOME microprocessor. (Cont'd.)

```
172- jump_pc <= '0';
173- alu_op <= "000";
174- next_st <= FETCH;
175- CASE curr_st IS
176-        WHEN FETCH => -- fetch an instruction from external RAM
177-            sel_data_ram <= '0'; -- select the instruction RAM
178-            read <= '1';   -- read from the RAM
179-            ld_ir <= '1';  -- load the instr. reg with the new opcode
180-            inc_pc <= '1'; -- increment the PC to the next instruction
181-            next_st <= DECODE; -- then decode the new opcode
182-        WHEN DECODE =>
183-            -- decode the instruction. Actually, this state is used to
184-            -- read a direct-address operand from the data section of
185-            -- external RAM and store it in the lower 4 bits of the IR.
186-            IF curr_ir(7 DOWNTO 4)=LOAD_DIR THEN
187-                    sel_data_ram <= '1'; -- select data RAM
188-                    read <= '1'; -- enable read of RAM
189-                    ld_ir_lsn <= '1'; -- load lower 4 bits of IR
190-            END IF;
191-            IF curr_ir(7 DOWNTO 4)=ADD_DIR THEN
192-                    sel_data_ram <= '1';
193-                    read <= '1';
194-                    ld_ir_lsn <= '1';
195-            END IF;
196-            IF curr_ir(7 DOWNTO 4)=XOR_DIR THEN
197-                    sel_data_ram <= '1';
198-                    read <= '1';
199-                    ld_ir_lsn <= '1';
200-            END IF;
201-            IF curr_ir(7 DOWNTO 4)=TEST_DIR THEN
202-                    sel_data_ram <= '1';
203-                    read <= '1';
204-                    ld_ir_lsn <= '1';
205-            END IF;
206-            next_st <= EXECUTE;-- then execute the instruction
207-        WHEN EXECUTE => -- execute the instruction.
208-            IF curr_ir=CLEAR_C THEN
209-                    alu_op <= CLR_CARRY_OP;-- clear the carry flag
210-            END IF;
211-            IF curr_ir=SET_C THEN
212-                    alu_op <= SET_CARRY_OP;-- set the carry flag
213-            END IF;
214-            IF curr_ir=SKIP_C THEN  -- skip the next instruction
```

Listing 9.3   VHDL code for the GNOME microprocessor. (Cont'd.)

```
215-                                inc_pc <= curr_carry;  -- if the carry flag is set
216-            END IF;
217-            IF curr_ir=SKIP_Z THEN   -- skip the next instruction
218-                                inc_pc <= curr_zero;   -- if the zero flag is set
219-            END IF;
220-            IF curr_ir(7 DOWNTO 4)=LOAD_IMM THEN -- load ACC with imm.
221-                                alu_op <= PASS_OP;-- data from low 4 bits of IR
222-            END IF;
223-            IF curr_ir(7 DOWNTO 4)=ADD_IMM THEN -- add low 4 bits of
224-                                alu_op <= ADD_OP;    -- IR to the ACC
225-            END IF;
226-            IF curr_ir(7)=JUMP THEN-- jump to address in the
227-                                jump_pc <= '1';-- lower 7 bits of the IR
228-            END IF;
229-            IF curr_ir(7 DOWNTO 4)=STORE_DIR THEN -- write ACC to RAM
230-                                sel_data_ram <= '1';
231-                                write <= '1';
232-            END IF;
233-            IF curr_ir(7 DOWNTO 4)=LOAD_DIR THEN  -- load ACC with the
234-                                alu_op <= PASS_OP;-- data read from RAM
235-            END IF;
236-            IF curr_ir(7 DOWNTO 4)=ADD_DIR THEN  -- add the RAM data
237-                                alu_op <= ADD_OP;    -- to the ACC
238-            END IF;
239-            IF curr_ir(7 DOWNTO 4)=XOR_DIR THEN  -- XOR the RAM data
240-                                alu_op <= XOR_OP;    -- with the ACC
241-            END IF;
242-            IF curr_ir(7 DOWNTO 4)=TEST_DIR THEN -- mask the ACC with
243-                                alu_op <= AND_OP;-- data read from RAM and set
244-            END IF;         -- the zero flag if all bits =0
245-            next_st <= FETCH;-- execution complete, so go fetch
246-    WHEN OTHERS =>
247- END CASE;
248- END PROCESS;
249-
250- -- generate the low-skew clock signal
251- buf1: IBUF PORT MAP(I=>clook, O=>buf_clock);
252- buf2: BUFG PORT MAP(I=>buf_clock, O=>bufg_clock);
253-
254- -- this process updates all these signals on the rising clock edge
255- PROCESS (bufg_clock,reset)
256- BEGIN
257- -- asynchronously reset the state of the GNOME microcomputer
```

Listing 9.3   VHDL code for the GNOME microprocessor. (Cont'd.)

```
258- IF reset='1' THEN
259-    curr_st <= FETCH;       -- start by fetching instructions
260-    curr_pc <= "0000000";   -- start at beginning of instr. RAM
261-    curr_ir <= "00000000";
262-    curr_acc <= "0000";     -- clear accumulator
263-    curr_carry <= '0';      -- clear carry flag
264-    curr_zero <= '0';       -- clear zero flag
265- -- otherwise, update state on the rising clock edge
266- ELSIF (bufg_clock'event AND bufg_clock='1') THEN
267-    curr_st <= next_st;
268-    curr_pc <= next_pc;
269-    curr_ir <= next_ir;
270-    curr_acc <= next_acc;
271-    curr_carry <= next_carry;
272-    curr_zero <= next_zero;
273- END IF;
274- END PROCESS;
275-
276- -- connect ACC to an LED decoder so we can monitor what is happening
277- u0: leddcd PORT MAP (d=>curr_acc,s=>s);
278-
279- END gnome_arch;
280-
```

```
*********************** Resource Summary ***********************
Design     Device         Macrocells   Product Terms  Pins
Name       Used           Used         Used           Used
gnom1_95   XC95108-20-PC84 65/108 (60%) 328/540 (60%) 35/69 (50%)
```

and it only uses about one-quarter of an XC4005XL FPGA:

```
Design Summary
--------------
     Number of errors:      0
     Number of warnings:    6
     Number of CLBs:              48 out of   196   24%
        CLB Flip Flops:      23
        CLB Latches:         0
        4 input LUTs:        90
        3 input LUTs:        11 (5 used as route-throughs)
     Number of bonded IOBs:       35 out of    65   53%
        IOB Flops:           0
        IOB Latches:         0
     Number of BUFGLSs:            2 out of     8   25%
```

```
Total equivalent gate count for design: 813
Additional JTAG gate count for IOBs:   1680
```

Why does the VHDL version of GNOME consume fewer gates than the ABEL version? The typical answers are are: (1) The language constructs used to describe the operations of the circuitry are lead to more efficient circuitry in VHDL than ABEL; and (2) The VHDL synthesis algorithms are more sophisticated than those in ABEL. In this case, the organization and method of description for the GNOME microprocessor has been kept very similar between the VHDL and ABEL versions. Therefore, most of the difference in device utilization may lie with the synthesis algorithms themselves.

Projects

1. Hand assemble the following instructions:

```
start: load #0
start1: add R5
        skip_c
        jump #start
        add R0
        skip_z
        jump #start
        jump #start1
```

Load the opcodes into the program memory. Download the code to your XS40 or XS95 Board and verify its operation.

2. You have seen the multiplication flowchart and the associated instructions. Compute the number of instructions that must be executed to multiply 1111 by 1111. Compute the number of instructions that must be executed to multiply 0000 by 0000.

3. Create a multiplication program that executes fewer instructions than the program listed in this chapter. Write the complete sequence of instructions and then generate the machine code that meets the following requirements:

   a. The program must accept two 4-bit numbers from which it will compute their 8-bit product and store it in RAM.

   b. You can use only 16 or fewer RAM locations in your program: R0, R1, ..., Re, Rf.

   c. The RAM locations will contain random numbers at the start of the program. You must initialize whatever RAM locations you are using to the correct values. This includes the two numbers you will be multiplying. The multiplicand for this problem is 5 and the multiplier is 14.

4. Hand assemble your multiplication program. Download the code and the GNOME bitstream to your XS95 or XS40 board and verify its operation.

The DWARF
Microcomputer

Objectives

- Extend the GNOME datapath to 8 bits.
- Add indirect and indexed addressing to the GNOME microprocessor.
- Add subroutine call and return instructions to the GNOME microprocessor.
- Add input and output instructions to the GNOME microprocessor.
- Add interrupt handling to the GNOME microprocessor.

Discussion

In this chapter, we will extend the capabilities of the GNOME microprocessor to create the DWARF microprocessor. The capabilities we will add are as follows:

1. The ability to handle 8-bit data

2. A register set

3. Indexed and indirect addressing

4. Input and output ports

5. Subroutine call and return instructions

6. Interrupt handling.

8-bit Datapath

The datapath width can easily be increased to handle 8-bit data by adding more bits to the ALU and the accumulator. However, this means that immediate-mode data can no longer be stored directly in the instruction opcode because it would consume all 8 bits. Therefore, immediate-mode data will have to be stored in the memory location just after the instruction opcode. Once we make that modification, we might as well increase the data memory addresses from 4 to 8 bits. Then we can increase the data memory to 256 bytes. With these changes, many of the instructions will consume 2 bytes and require two fetches to read from the external RAM. These complications will have to be accounted for in the DWARF instruction decoder.

Register Set

The GNOME operated on its accumulator using operands fetched from external RAM. A microprocessor can execute more swiftly if it can get operands from a set of internal registers rather than bringing them in from outside the chip. These registers can store intermediate values and provide new addressing modes (see the next subsection). A set of eight registers provides adequate internal storage and the register addresses can be stored in the opcode using only three of the eight bits.

Indexed and Indirect Addressing

The addition instruction for the GNOME can add any data RAM location to the accumulator. However, the address of the data RAM location must be stored in the lower 4 bits of the add opcode. So if you want to sum the data stored in data RAM locations R0–R7, you have to write the code as follows:

```
clear_c
load R0 ; ACC <- R0
add   R1 ; ACC <- R0 + R1
add   R2 ; ACC <- R0 + R1 + R2
add   R3 ; ACC <- R0 + R1 + R2 + R3
add   R4 ; ACC <- R0 + R1 + … + R4
add   R5 ; ACC <- R0 + R1 + … + R5
add   R6 ; ACC <- R0 + R1 + … + R6
add   R7 ; ACC <- R0 + R1 + … + R7
```

This becomes very tedious if you have to sum many values. Instead, imagine that a microprocessor has an index register, X, that stores the address of a data RAM location. X can be used to supply the address of the operand that will be added to the accumulator. Then X can be incremented so that it points to the next data RAM location. This is the indexed addressing mode.

Indirect addressing is similar to indexed addressing except the index register is stored in external RAM. In this mode of addressing, the instruction contains the address of a memory location that, in turn, contains the address of the memory location having the operand. So an instruction with an indirect-mode operand has to fetch the address from RAM and then use this address to fetch the data value from RAM. Then the address stored in RAM can be incremented. So it is like indexed addressing, but it is slower because the microprocessor has to keep accessing external RAM instead of an internal register.

Input and Output Ports

A microprocessor needs the ability to gather input data so it can sense changes in the environment. It must also have outputs that allow it to affect the external world. For example, it may need to read the value from a digital thermometer and output control signals to a furnace.

A set of eight I/O lines can easily interface to the internal datapath of a microprocessor. In effect, the I/O port operates just like the data bus interface to the external RAM. The microprocessor just needs appropriate control signals so the digital values from the I/O port do not collide with data read in from the RAM. The I/O port could be addressed just as if it were one of the internal registers. This would allow the microprocessor to interact with the I/O port using all the same resources that internal registers can access.

Subroutine Call and Return Instructions

A program that performs a particular operation at several points in the program will be smaller if the common operation is placed in a subroutine. An example of a likely candidate for a subroutine is the code for doing 8-bit addition. Then whenever 8-bit addition is needed, the program can jump to the subroutine and perform the operation. After the subroutine is completed, the program returns to where it was previously and continues. This saves the programmer from having to replicate the code wherever the addition operation is needed.

The sequence of calling and returning from a subroutine is shown in Figure 10.1. The jsr add instruction causes the PC to be loaded with the address of the addition subroutine (step A). Thus, the next instruction to be executed is at address add. After all the instructions in the addition subroutine have been performed (step B), the ret instruction causes the program to return to the instruction right after the jsr instruction (step C). So the program can branch off and do the addition and then resume processing right after the subroutine call.

For subroutines to work, the program counter must be stored somewhere so the microprocessor can remember which address to return to when the subroutine is finished. A simple way to do this is to build a special SAVE register to hold the PC value when a subroutine is called. The SAVE register is loaded with the incremented value of the program counter whenever a jsr instruction is executed. (Thus, SAVE holds the address of the instruction that comes immediately after the jsr instruction.) At the same time, the subroutine address (stored as an operand in the instruction) is loaded into the program counter. After the subroutine is done, the ret instruction activates the control signals that cause the multiplexer to load the PC with the value stored in the SAVE register. Then the program begins executing the instructions that follow the jsr instruction.

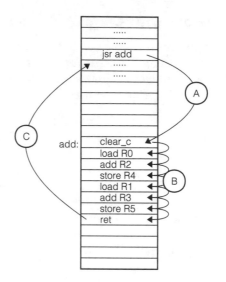

Figure 10.1 Calling and returning from a subroutine.

The circuitry just discussed will work only if a single subroutine is called. But what if the subroutine calls another subroutine? This is called subroutine nesting. In this case, the SAVE register will be overwritten with an address from the subroutine code and the return address to the main program will be lost. Therefore, the microprocessor will not remember which address to return to after all the subroutines have finished. To support nesting of N subroutines, N multiple registers, (SAVE$_1$, SAVE$_2$, ..., SAVE$_N$) must be built. Whenever a jsr instruction is executed, the following transfers of addresses occur:

$$\text{SAVE}_N \Leftarrow \text{SAVE}_{N-1}$$
. . . .
$$\text{SAVE}_2 \Leftarrow \text{SAVE}_1$$
$$\text{SAVE}_1 \Leftarrow \text{PC}$$

and the reverse operation occurs whenever a ret instruction is executed:

$$\text{PC} \Leftarrow \text{SAVE}_1$$
$$\text{SAVE}_1 \Leftarrow \text{SAVE}_2$$
. . . .
$$\text{SAVE}_{N-1} \Leftarrow \text{SAVE}_N$$

The multiple SAVE registers form what is called a subroutine stack. Addresses are pushed on the stack whenever a jsr instruction is executed. These addresses are popped off the stack and into the PC whenever a ret instruction is executed. Microprocessors often store their stack in external RAM and use a dedicated register (called a stack pointer) to keep track of where the top of the stack is in memory.

Interrupt Handling

Subroutine calls and returns occur under the direction of the program. There are situations, however, when external conditions demand immediate attention and the flow of control must be abruptly changed. These interrupts are not events that the microprocessor can anticipate and plan for. Instead, it must save enough information about its current state, jump to a section of code that handles the interrupt (called an interrupt service routine or ISR), and then restore its state and continue its original flow of instructions.

The state information that the microprocessor must store after an interrupt is received is, at minimum, the value of the program counter. With this information, the microprocessor can return to where it was after it finishes processing the interrupt. But the microprocessor must also save the values in registers and flags as well if these might be altered by the instructions that service the interrupt. The values of the PC, registers, and flags at any given time are typically called the context. A common technique for saving the context is to push the values onto the stack using the same techniques employed during subroutine calls. Then the context can be restored by popping it off the stack at the end of the ISR.

Experimental

We have touched on the basic ideas, so now it is time to implement them in the DWARF microprocessor. This design is oriented toward the XC4005XL FPGA because it has more left-over resources than the XC95108 CPLD, so it will be an easier fit. The XC4005XL also has internal tristate buffers and built-in RAMs that we will use in the DWARF design. This is not to say that the DWARF architecture cannot be implemented in an XC9500 CPLD, but we will not do it here.

Table 10.1 shows the DWARF instruction set. There are twice as many instructions as found in the GNOME. Here are the main differences from the GNOME instruction set:

1. The load and store instructions have been extended with register, indexed, and indirect addressing modes.

2. An in instruction was added that loads the value on an input port into one of the eight internal registers. The out instruction sends the value in the accumulator to an output port.

3. The xor, add, and test instructions have been restricted to using the internal registers as operands.

4. The skip_c and skip_z instructions were replaced with the jc and jz instructions, which can jump to any location in the PROGRAM region if the carry or zero flag is set, respectively.

5. The jump-to-subroutine (jsr) instruction was added that saves the address of the next instruction on the stack and then transfers to a new address in the PROGRAM region. The ret instruction was added that removes the saved address from the stack and restores it to the PC.

6. The return-from-interrupt (reti) instruction was added that restores the accumulator and program counter values from the stack and then resumes instruction execution at the address in the PC. This instruction reverses the operations performed when an interrupt service routine is initiated (i.e., the PC and ACC are saved on the stack).

Table 10.1 DWARF instruction set.

Mnem.	Operations	Descriptions
load Rd	ACC ← Rd	Load the accumulator with the contents of register Rd. d is an address in the range [0..7].
load (Rd)	ACC ← RAM[Rd]	Load the accumulator with the contents of the DATA RAM location whose address is found in register Rd.
load #d	ACC ← d	Load the accumulator with the value d. d is in the range [0..255].
load d	ACC ← RAM[d]	Load the accumulator with the contents of the DATA RAM location whose address is d. d is in the range [0..255].
load (d)	ACC ← RAM[RAM[d]]	Load the accumulator with the contents of the DATA RAM location whose address is found in DATA RAM location d. d is in the range [0..255].
store Rd	Rd ← ACC	Store the accumulator into register Rd. d is an address in the range [0..7].
store (Rd)	RAM[Rd] ← ACC	Store the accumulator into DATA RAM at the address found in register Rd. d is an address in the range [0..7].
store d	RAM[d] ← ACC	Store the accumulator into DATA RAM at the address d. d is in the range [0..255].
store (d)	RAM[RAM[Rd]] ← ACC	Store the accumulator into the DATA RAM location whose address is found in DATA RAM at the address found in register Rd. d is in the range [0..7].
in Rd	Rd ← IN_PORT	Load register Rd with the value on the input port.
out	OUT_PORT ← ACC	Output the accumulator value on the output port.
xor Rd	ACC ← ACC$Rd	Exclusive-OR the register Rd with the accumulator. d is an address in the range [0..7].
add Rd	ACC ← ACC + Rd + C; C ← carry out	Add the contents of register Rd and the C flag to the accumulator. d is an address in the range [0..7].

Table 10.1   DWARF instruction set. (Cont'd.)

Mnem.	Operations	Descriptions
test Rd	Z ← ACC&Rd	AND the contents of register Rd with the accumulator, but do not store the result in the accumulator. Instead, set the Z flag if the result is 0; otherwise clear the Z flag. d is an address in the range [0..7].
clear_c	C ← 0	Clear the C flag to zero.
set_c	C ← 1	Set the C flag to one.
jc #a	IF C = 1 → PC ← a ELSE PC ← PC + 2	If C = 1, branch to address a. If C = 0, continue to the next instruction. a is in the range [0..255].
jz #a	IF Z = 1 → PC ← a ELSE PC ← PC + 2	If Z = 1, branch to address a. If Z = 0, continue to the next instruction. a is in the range [0..255].
jump #a	PC ← a	Unconditionally branch to address a. a is in the range [0..255].
jsr #a	RAM[SP] ← PC + 2 SP ← SP + 1 PC ← a	Save the address of the next instruction on the stack and increment the stack pointer. Then jump to address a. a is in the range [0..255].
ret	SP ← SP – 1 PC ← RAM[SP]	Decrement the stack pointer and load the PC with the address found on the top of the stack.
reti	SP ← SP – 1 ACC ← RAM[SP] SP ← SP – 1 PC ← RAM[SP]	Decrement the stack pointer and load the ACC with the value found on the top of the stack. Then decrement the SP and load the PC with the address at which to resume execution of the noninterrupt program.

The encodings of these instructions are shown in Table 10.2. The last three bits of the opcode are used to store the address of the internal register that is used by the instruction. (Some instructions do not operate on the registers, so these bits are not used.) The remaining five bits are used to distinguish one instruction from another. Note that the use of 8-bit data and addresses makes it necessary to store some of the instructions in two bytes instead of one.

The basic architecture for the DWARF is shown in Figure 10.2. You are already familiar with most of these basic components from the GNOME design such as the instruction decoder, program counter (PC), instruction register (IR), arithmetic logic unit (ALU), accumulator (ACC), and carry and zero flags (C and Z). The new components are as follows:

Stack pointer (SP): The stack pointer contains the address of a location in data memory where the PC and the ACC can be stored during subroutine calls and interrupts.

Table 10.2 DWARF instruction set encoding.

Mnem.	# Bytes	Encoding
load Rd	1	00100 d2 d1 d0
load (Rd)	1	00101 d2 d1 d0
load #d	2	01000000 d7 d6 d5 d4 d3 d2 d1 d0
load d	2	00110000 d7 d6 d5 d4 d3 d2 d1 d0
load (d)	2	00111000 d7 d6 d5 d4 d3 d2 d1 d0
store Rd	1	00000 d2 d1 d0
store (Rd)	1	00001 d2 d1 d0
store d	2	00010000 d7 d6 d5 d4 d3 d2 d1 d0
store (d)	2	00011000 d7 d6 d5 d4 d3 d2 d1 d0
in Rd	1	01100 d2 d1 d0
out	1	01101000
xor Rd	1	10000 d2 d1 d0
add Rd	1	10001 d2 d1 d0
test Rd	1	10010 d2 d1 d0
clear_c	1	10100000
set_c	1	10101000
jc #a	2	11000000 a7 a6 a5 a4 a3 a2 a1 a0
jz #a	2	11001000 a7 a6 a5 a4 a3 a2 a1 a0
jump #a	2	11010000 a7 a6 a5 a4 a3 a2 a1 a0
jsr #a	2	11011000 a7 a6 a5 a4 a3 a2 a1 a0
ret	1	11100000
reti	1	11101000

Register set (R0–R7): DWARF has 16 internal registers for holding data values. Eight of these are available for normal use, and the other eight are used during interrupts. The register set is built from a dual-port memory with two 8-bit outputs (A_OUT and B_OUT) and a single 8-bit input (A_IN). This lets the register set output two values and overwrite one of them in a single clock cycle.

I/O port (IOP): DWARF has eight output pins and eight input pins dedicated to general-purpose input/output.

Address selector: DWARF divides the external memory into three regions: PROGRAM (256 bytes), DATA (256 bytes), and STACK (256 bytes). The address selector enables one of these regions and sends it an

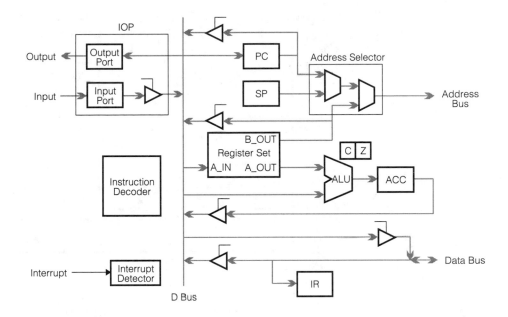

Figure 10.2 DWARF architecture.

address from one of three sources: The PROGRAM region is addressed by the PC; the STACK region is addressed by the SP; and the DATA region is addressed by the value held in one of the registers.

Interrupt detector: A rising edge on the interrupt pin (INT) triggers the interrupt circuitry, which alerts DWARF of the need to alter its sequence of instructions.

D bus: This bus ties most of the other components together. It can be driven by the PC, the B output of the register set, the ACC, the input side of the IOP, and the external data bus that connects to the external RAM. In turn, the value on the D-bus can be loaded into the PC, the A_IN input of the register set, the ACC, the output side of the IOP, and the external RAM.

Now we can look at the details of each of these components, starting with the datapath built from the register set and ALU (Figure 10.3, left side). The dual-port register set (DWRF_REG) takes two 3-bit addresses (A_ADDR and B_ADDR), but these are both connected to the lower 3 bits of the instruction register ([IR2..IR0]). The A_ADDR and B_ADDR inputs control which register will output its contents on the A_OUT and B_OUT outputs, respectively, of the register set. The A_ADDR input also selects which register will be written with the value entering the A_IN input from the D bus. The new value will be written on the next rising edge of CLOCK if the LD_REG input is at logic one (the register set is built from a synchronous memory).

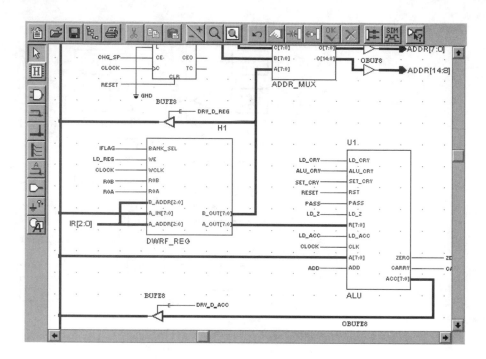

Figure 10.3 DWARF's register set and ALU.

When executing instructions, DWARF often needs a "scratch-pad" register where it can store intermediate values and addresses. R0 is selected for this purpose. But the instruction decoder needs a means of addressing this register when it needs to. For this reason, a logic one applied to the R0A or R0B input forces the address of R0 onto the A_ADDR or B_ADDR input, respectively, regardless of the register address stored in the instruction register.

The register set actually contains 16 registers, so the BANK_SEL input selects one of two banks of eight registers. The IFLAG signal is high during the execution of an interrupt service routine, and this is used to switch the register banks. This gives the interrupt service routine its own set of registers to use and prevents it from altering the values in the registers used during DWARF's normal operations.

The value on the B_OUT output of the register set can be driven onto the D bus by raising the DRV_D_REG signal. This enables the drivers of the internal tristate buffer (BUFE8) so the register value can pass through. This operation is needed when a value from the register set is being transferred to another section of DWARF's circuitry or into external RAM.

The details of the DWARF register set are shown in Figure 10.4. The heart of the register set is a dual-port synchronous RAM (RAM16X8D) from the standard

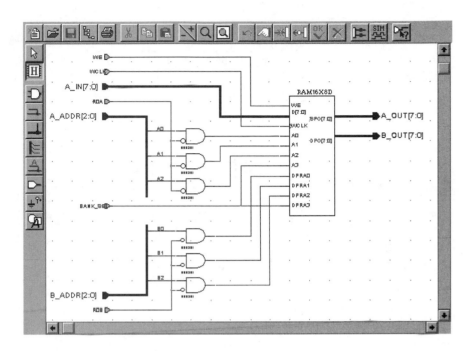

Figure 10.4   Detailed circuitry for DWARF's register set.

part list. The BANK_SEL input drives the most-significant bit of both register addresses. Two sets of AND gates are provided to force the register addresses to zero (the address of R0) under the control of the R0A and R0B inputs.

The ALU (Figure 10.3, right side) contains the circuitry for adding and XORing values and for storing the accumulator and carry and status flags. The 8-bit operands enter the ALU through the [R7..R0] inputs from the register set's A_OUT output and through the [A7..A0] inputs from the D bus. The ADD and PASS inputs control the arithmetic operation performed on these operands as follows:

PASS	ADD	ALU Operation
0	0	ACC ← ACC$REGA
0	1	ACC ← ACC + REGA + CARRY
1	X	ACC ← D Bus

The details of the arithmetic circuitry are shown in Figure 10.5. The PASS and ADD inputs control a pair of byte-wide multiplexers (identifiers H2 and H5). The multiplexers select either the [A7..A0] input, the output of an 8-bit adder (ADD8), or the output of a module that does a bitwise XOR on two 8-bit values (XOR_8X8). The accumulator (module ACCUM0) is loaded with the output

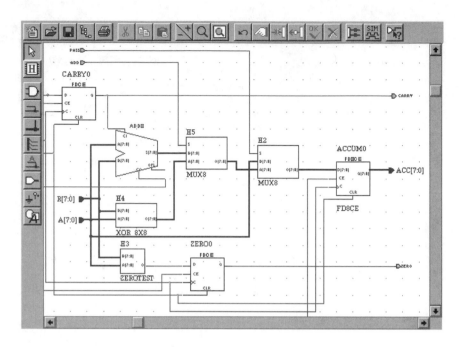

Figure 10.5   Details of DWARF's arithmetic circuitry.

from the H2 multiplexer when LD_ACC is a logic one and there is a rising edge on the CLOCK input.

The zero and carry flags are also updated when there is a rising edge on CLOCK and LD_Z or LD_CRY are at logic one, respectively. The zero flag is always loaded with the NOR of all the bits that result from a bitwise AND of the 8-bit values from the D bus and the register set (Figure 10.6 shows the details of the ZEROTEST module).

The carry flag (Figure 10.7) is loaded with the carry output from the 8-bit adder (CO of the ADD8 module) if the ALU_CRY input is high. This permits the propagation of carry bits in multibyte arithmetic operations. But if ALU_CRY=0, the carry flag is loaded with the value on the SET_CRY input, thus allowing the carry flag to be forced to one or zero.

A logic one on the RESET input clears the accumulator and status flags to zero.

The carry and zero flags exit the ALU module through the CARRY and ZERO outputs. The accumulator value appears on the [ACC7..ACC0] outputs. Setting DRV_D_ACC=1 lets the accumulator drive the D bus (Figure 10.3). This provides a path for storing the accumulator into external memory, the register set, or for feeding the accumulator back into the ALU as an operand.

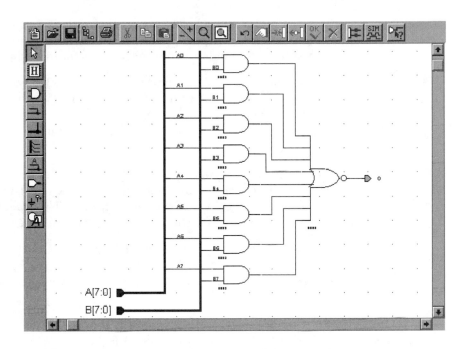

Figure 10.6 Detailed circuitry for DWARF's ZEROTEST module.

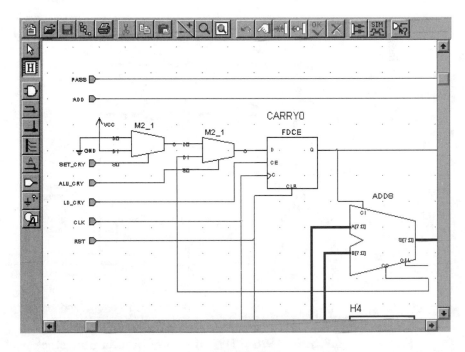

Figure 10.7 Detailed circuitry for DWARF's carry flag.

DWARF's memory address generation circuitry is shown in Figure 10.8. The program counter (top left) shows that the PC (with reference identifier of PC0) is just an 8-bit, loadable counter from the standard part list (CB8CLE).

A logic one on the LD_PC input will load the PC with a new value from the D bus on the next rising edge of CLOCK. This operation is used for jumps within the program. On the other hand, a logic one on the INC_PC input will cause the PC to increment its value. This operation is needed during the normal sequential execution of instructions.

The value in the program counter can be driven onto the D bus by raising the DRV_D_PC signal. This enables the drivers of the internal tristate buffer (BUFE8) so the PC value can pass through. This operation is needed when the value in the PC is being stored on the stack in external RAM.

The PC is cleared by raising the RESET or CLR_PC signals. This operation is needed when DWARF first starts so that program execution can begin at address 0 in the PROGRAM region. The program counter is also cleared at the beginning of an ISR.

Looking next at the stack pointer (Figure 10.8, bottom left) shows that the SP (with reference identifier of SP0) is just an 8-bit, loadable, up/down counter from the standard part list in the schematic editor (CC8CLED).

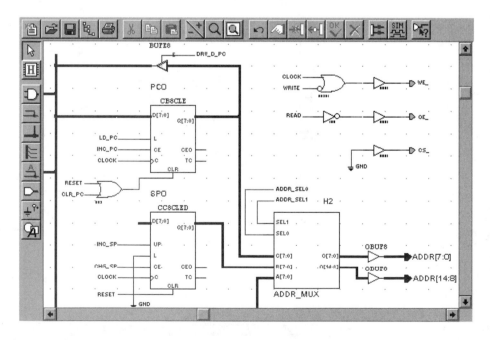

Figure 10.8 DWARF's program counter, stack pointer, address selector, and RAM control circuitry.

A logic one on the CHG_SP input will cause the stack pointer to change its value on a rising edge of the CLOCK signal. If a logic one is applied on the INC_SP input, the change will be an increment. Otherwise, the value in the SP will be decremented. The increment (decrement) operation is needed when values are being pushed onto (popped off) the stack.

The SP is cleared by raising the RESET signal. This operation is needed when DWARF starts up so the stack begins growing upward from address 0 in the STACK region. There is no need to randomly address the contents of the stack, so the D[7:0] input to the counter is left unconnected and the load signal (L) is tied to a logic zero.

The B output of the register set, SP, and PC are the A, B, and C inputs to the address multiplexer module (ADDR_MUX in Figure 10.8, bottom right). The address multiplexer selects one of its inputs as the source of the lower 8 bits of the RAM address and sets the upper 7 bits so that the DATA, STACK, and PROGRAM memory regions are separated. ADDR_MUX outputs the following values given the settings of the SEL0 and SEL1 inputs:

SEL1	SEL0	[ADDR14..ADDR8]	[ADDR7..ADDR0]	Region
0	0	1111111	[REGB7..REGB0]	DATA
0	1	1111110	[SP7..SP0]	STACK
1	0	0000001	[PC7..PC0]	unused
1	1	0000000	[PC7..PC0]	PROGRAM

The DATA region of the external RAM is accessed when SEL0=SEL1=0. The B output of the register set supplies the lower 8 bits of the address while the upper 7 bits are set to 0x7F. Thus, the DATA region lies at the top of the RAM between hexadecimal addresses 0x7F00 and 0x7FFF. In a similar manner, the STACK region (SEL0=1, SEL1=0) occupies the address range 0x7E00–0x7EFF with the stack pointer supplying the lower eight address bits. The PROGRAM region (SEL0 = SEL1 = 1) occupies the lowest 256 bytes of RAM (0x0000–0x00FF) with the PC, naturally, acting as the source for the lower 8 bits of address in this region.

The detailed circuitry for the address multiplexer is shown in Figure 10.9. It consists mainly of two 8-bit multiplexers that work in series to steer one of the three byte-wide A, B, or C inputs to the [O7..O0] output under the direction of the SEL0 and SEL1 inputs. SEL0 and SEL1 also go to a set of inverters that generate the correct bit pattern for each memory region on the upper 7 bits of the memory address.

Miscellaneous circuitry for driving the control inputs of the external RAM is shown in the upper right portion of Figure 10.8. The RAM is permanently enabled by tying the chip select (CS_) to logic zero (GND). The output enable of

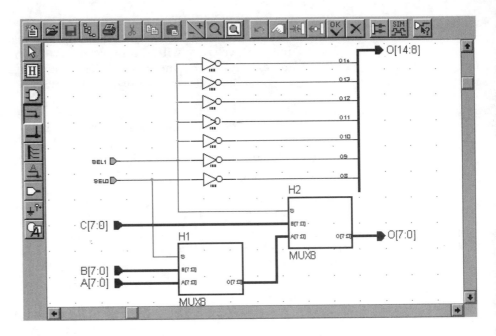

Figure 10.9 Detailed circuitry for DWARF's address multiplexer.

the RAM (OE_) is pulled low during read operations so that the data bus drivers of the RAM are enabled. The circuitry for driving the write-enable input of the RAM (WE_) generates a logic-zero pulse during the second half of a write cycle when CLOCK=0 and WRITE=1. This timing gives DWARF sufficient time to set up the address and data values to the RAM before the write pulse arrives.

The I/O port circuitry is presented in Figure 10.10. The byte-wide output port is built from eight flip-flops located in the IOBs along the periphery of the FPGA (module OFDEX8). Thus, these flip-flops do not consume resources in the FPGA's CLB array. When LD_PORT=1 and a rising edge occurs on the CLOCK signal, the [OUT_PORT7..OUT_PORT0] pins will retain the value that was on the D bus at that time. The value is always actively driven onto the pins because the tristate buffers at the outputs of the flip-flops have their enable signals tied to logic one (VCC).

The input side of the I/O circuitry consists of eight input signals whose values are loaded into flip-flops in the IOBs (IFD8) of the FPGA on every rising edge of the clock. The value clocked into the flip-flops is driven onto the D bus when a logic one is applied to the DRV_D_PORT signal. Why use flip-flops if their value is always replaced during every clock cycle? Because the flip-flops synchronize the input signal transitions to the same clock that drives the rest of DWARF. This reduces the chance that one of DWARF's internal flip-flops will

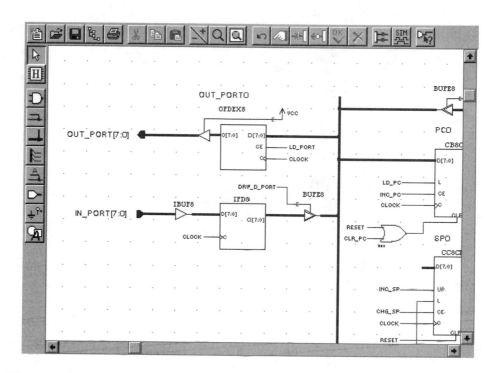

Figure 10.10 DWARF's I/O port circuitry.

enter a metastable state due to a change on one of the inputs just as a clock edge occurs.

Figure 10.11 shows the circuitry for detecting and recording the state of interrupts. A rising edge on the interrupt input (INT) loads a logic one into the first D flip-flop. The logic one propagates through the INTRPT_DFF0 flip-flop and appears on the INTRPT output on the next rising edge of CLOCK. This synchronizes the interrupt to DWARF's clock and reduces the chances of entering a metastable state.

In addition to a flip-flop to record the occurence of an interrupt, DWARF also needs a flip-flop to indicate when it is executing an interrupt service routine. The IFLAG0 flip-flop serves this purpose. IFLAG0 is set to a logic one state when SET_IFLAG=1. The IFLAG output from the flip-flop is used to switch to the alternate set of eight registers in the DWRF_REG module. Also, both IFLAG and INTRPT are inputs to the instruction decoder. If INTRPT=1 and IFLAG=0, the instruction decoder will alter the flow of instructions by initiating the interrupt service routine. But if INTRPT=1 and IFLAG=1, DWARF will ignore the pending interrupt because it is already processing an interrupt.

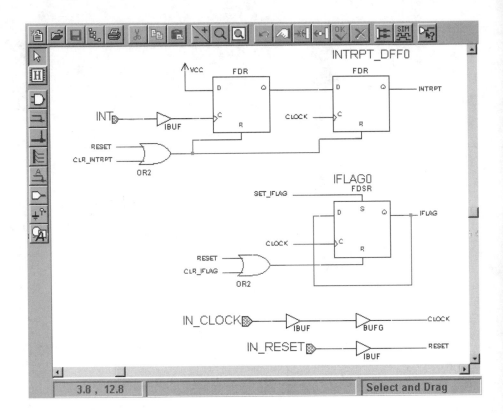

Figure 10.11 DWARF's interrupt-handling circuitry.

The interrupt flip-flops are cleared when the CLR_INTRPT and CLR_IFLAG are high. The CLR_INTRPT signal is asserted to clear the INTRPT flag when the interrupt routine is entered. This minimizes the amount of time that passes before the circuitry can detect another interrupt. The CLR_IFLAG signal, however, is asserted as the processor is leaving the ISR. The RESET signal also clears the INTRPT and IFLAG flip-flops so that DWARF does not erroneously begin processing an interrupt when it first starts up.

Figure 10.11 also shows how the clock and reset signals enter DWARF. Both the IN_RESET and IN_CLOCK inputs enter through normal IBUFs. But the IN_CLOCK input then passes through a primary global buffer (symbol BUFGP). The use of such a buffer allows the clock signal to travel over a global net that does not suffer from the delays incurred by signals that travel through the routing switches. Thus, the CLOCK signal has very low skew and clock edges will reach all the flip-flops at nearly the same instant.

The bidirectional data bus interface to the external RAM is shown in Figure 10.12. Data values on the D bus are driven out of DWARF onto the

[DATA7..DATA0] pins when WRITE=1. This enables the tristate output buffers (OBUFE8). When the tristate gates are disabled, it is possible for an external device (the RAM in this case) to drive a value onto the DATA bus. The value can reach the internal circuitry of DWARF by placing a logic one on the DRV_D_DATA signal. That enables the internal tristate buffers (BUFE8) so the value on the DATA bus will appear on the internal D bus.

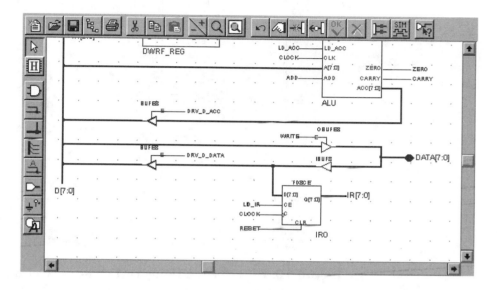

Figure 10.12 DWARF's instruction register and data bus interface.

The instruction register (module IR0) is connected directly to the DATA bus before it reaches the internal tristate buffers. Therefore, the instruction register is loaded with whatever value is on the [DATA7..DATA0] pins (this will be an opcode fetched from the RAM). The loading occurs on the rising edge of CLOCK when LD_IR=1. The instruction register is cleared when RESET is asserted.

The final module is the DWARF instruction decoder (Figure 10.13). It receives the carry and zero flags, interrupt status signals, and the 8 bits of the instruction register as inputs. The instruction decoder outputs 29 signals that control the circuitry we have already discussed.

The instruction decoder module was created by importing a netlist synthesized from the ABEL code in Listing 10.1 or the VHDL code in Listing 10.2. The listing is quite long because DWARF has twice as many instructions as GNOME. We will go over a few of the instruction sequences (with line numbers for the for the VHDL version of the code indicated in **bold**). The remainder of the code should be decipherable after that.

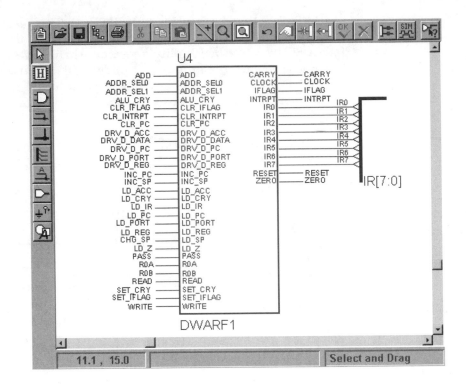

Figure 10.13 The DWARF instruction decoder module.

The instruction decoder starts with the list of inputs (lines 6–12, **lines 10–16**), state storage flip-flops (line 14, **line 51**), and control outputs (lines 16–43, **lines 17–44**) for the instruction decoder. These are followed by various definitions including the encodings for the DWARF's opcodes (lines 45–79, **lines 54–89**).

There are seven states that the instruction decoder state machine can be in (lines 49–51, **lines 54–60**). Two of these states (R0 and R1) are occupied only when DWARF is initializing after being reset. Four of the states (T0, T1, T2, and T3) are entered as instructions are executed. The interrupt pin is also checked during state T0 and the state machine transfers to state I1 to set up for executing an interrupt service routine if an interrupt is pending.

The description of how the instruction decoder transitions between states and sets values on the control lines for the rest of the DWARF circuitry follows (lines 86–296, **lines 93–312**). When a reset occurs, DWARF enters state R0 by clearing the state flip-flops to 000 which is the code for state R0 (line 84, **lines 95–96**). After a reset, the program counter is cleared to zero. In state R0 (lines 88–90, **lines 134–136**), the PC is incremented once and the state machine moves to state R1. In state R1 (lines 92–94, **lines 137–139**), the PC is incremented once more (so now it holds an address of 2) and the state machine transitions to state T0 (lines 96–109, **lines 140–152**). In T0, an instruction is fetched and program execution begins. This means that all DWARF programs start at address 2. Why go to all this trouble when the programs could start at address zero?

The answer to this question has to do with the need to process interrupts. In T0, a check is made (line 97, **line 141**) to see if an interrupt has occured (intrpt=1) and that an interrupt service routine is not already being executed (iflag=0). If these conditions are true, the pending interrupt is cleared (line 99, **line 142**), iflag is set to indicate that an ISR is now being executed (line 100, **line 143**), the address of the current instruction is written to the external RAM at the address stored in the stack pointer (line 101, **line 144**), and the stack pointer is incremented at the end of the current clock cycle (line 102, **line 145**). Then the instruction decoder moves to state I1. Here, it stores the accumulator onto the stack (line 112, **line 154**), increments the stack pointer again (line 113, **line 155**), and clears the program counter to zero (line 114, **line 156**). Then the state machine jumps to state T0 and begins to process instructions starting at address 0. Thus, all interrupt service routines start at address 0 which is why the normal DWARF programs cannot begin there. Note that DWARF will not respond to any further interrupts after it enters an ISR because it has set its iflag to logic one (recall the conditional test on line 97 or **line 141**). This saves DWARF from a condition where it continually sets up to process interrupts but never actually gets to run the service routine.

Now that we have examined how DWARF enters an ISR, it makes sense to see how it gets out of one. This is done using the reti instruction (return-from-interrupt). The execution of this instruction begins in state T0 with the fetch of the instruction from the PROGRAM region of the RAM (line 106, **line 148**). The opcode from the RAM enters DWARF and is loaded into the instruction register (line 107, **line 149**). The program counter is incremented at the end of this clock cycle (line 108, **line 150**), but this does not matter much in this case since the purpose of the reti instruction is to return the flow of execution to the address stored on the stack.

The instruction decoder next moves to state T1. The actions performed in T1 depend upon the opcode stored in IR. Since the opcode is the one for reti, the actions described on lines 137–140 or **lines 174–176** are enabled. The only action is to decrement the stack pointer. This points SP at the memory location where the accumulator value was saved when the ISR was initiated. However, restoring the accumulator value must wait until state T2 is entered (lines 246–251, **lines 264–268**) because the stack pointer is not updated until the end of the T1 state. On line 248 or **line 265**, the STACK region of the RAM is addressed with the value in the SP. The value output by the RAM is driven onto the D bus, passed through the ALU, and loaded into the accumulator (line

249, **line 266**). The stack pointer is decremented again at the end of T2 (line 250, **line 267**) and the instruction decoder enters state T3.

Upon entering state T3 (lines 280–285, **lines 293–297**), the SP is pointing to the location where the program counter was saved upon entering the ISR. The program counter is restored in the same way as the accumulator (lines 282–283, **lines 294–295**). The iflag is cleared (line 284, **line 296**) to indicate an ISR is no longer being executed. This will allow new interrupts to be processed. Then control passes to state T0 where a new instruction is fetched. Only the instruction will be fetched from the address stored in the PC. This instruction is the one that was going to be processed just as the interrupt was detected. Thus, normal program execution resumes right where it left off as if nothing had happened.

As a final example of instruction execution, we will examine the indirect addressing mode version of the load instruction. Starting in state T1 after the opcode has already been fetched, the instruction decoder must read in the address that follows the opcode (line 187, **line 215**). This address is written to register R0 by forcing the port A address to zero (R0A=1) and raising the LD_REG control signal (line 188, **line 216**). The program counter is incremented so that the PC points to the next instruction after this one (line 189, **line 217**). The instruction decoder moves on to state T2.

In state T2, the instruction decoder addresses the DATA region of the RAM with the contents of R0 (line 259, **line 274**). The value output by the RAM is passed onto the internal D bus where it is loaded into R0 (line 260, **line 275**). Thus, the contents of R0 are replaced with the data held at the address R0 was originally storing. In the case of indirect addressing, R0 is now holding another address. The instruction decoder now moves to state T3.

In state T3, the instruction decoder once again addresses the DATA region of RAM with the contents of R0 (line 288, **line 299**). Now the value output by the RAM, however, is looked upon as data bound for the accumulator. So the RAM data is gated through to the D bus, passed through the ALU, and loaded into the accumulator (line 289, **line 300**). This completes the execution of the load instruction with an indirect operand. The instruction decoder returns to state T0 and fetches another opcode.

You can interpret the actions performed for the other instructions in much the same way as we did here. Just look for a chosen opcode under the case statement in each of the T1, T2, and T3 states. By examining the values asserted onto the control signals, you can determine which multiplexers and tristate buffers are enabled and where the data flows.

Below are the pin assignments for DWARF when it is targeted to the XS40 Board. Three of the outputs from the PC parallel port are connected to DWARF's clock, interrupt, and reset inputs. The lower seven bits of DWARF's

Listing 10.1   The ABEL version of the DWARF instruction decoder description.

```
001- MODULE DWARF1
002- TITLE 'DWARF controller -- V1.0'
003-
004- DECLARATIONS
005- "----- inputs -----
006- clock         PIN;
007- reset         PIN; "zeroes the internal state
008- ir7..ir0      PIN; "instruction register (IR)
009- carry         PIN; "carry flag (C)
010- zero          PIN; "zero flag (Z)
011- intrpt        PIN; "interrupt
012- iflag         PIN; "set when processing interrupt-handling code
013- "----- internal storage elements -----
014- st2..st0      NODE ISTYPE 'REG'; "DWARF internal state
015- "----- outputs -----
016- read          PIN  ISTYPE 'COM'; "1 when reading RAM
017- write         PIN  ISTYPE 'COM'; "1 when writing RAM
018- addr_sel1..addr_sel0 PIN  ISTYPE 'COM'; "2=PROG;1=DATA;0=STACK
019- inc_pc        PIN  ISTYPE 'COM'; "1 when incrementing PC
020- ld_pc         PIN  ISTYPE 'COM'; "1 when changing PC
021- clr_pc        PIN  ISTYPE 'COM'; "1 when clearing PC to zero
022- inc_sp        PIN  ISTYPE 'COM'; "0=--sp; 1=sp++
023- ld_sp         PIN  ISTYPE 'COM'; "1 when changing stack pointer
024- ld_ir         PIN  ISTYPE 'COM'; "1 when loading IR
025- ld_port       PIN  ISTYPE 'COM'; "1 when loading I/O port
026- drv_d_data    PIN  ISTYPE 'COM'; "1 when D bus driven by data bus
027- drv_d_acc     PIN  ISTYPE 'COM'; "1 when D bus driven by accumulator
028- drv_d_pc      PIN  ISTYPE 'COM'; "1 when D bus driven by PC
029- drv_d_port    PIN  ISTYPE 'COM'; "1 when D bus is driven by I/O port
030- drv_d_reg     PIN  ISTYPE 'COM'; "1 when D bus driven by register
031- ld_reg        PIN  ISTYPE 'COM'; "1 when loading a register
032- R0A           PIN  ISTYPE 'COM'; "1 forces A reg address to 0
033- R0B           PIN  ISTYPE 'COM'; "1 forces B reg address to 0
034- set_cry       PIN  ISTYPE 'COM'; "0=reset carry; 1=set carry
035- alu_cry       PIN  ISTYPE 'COM'; "1 when ALU operation updates carry
036- ld_cry        PIN  ISTYPE 'COM'; "1 when changing carry flag
037- ld_z          PIN  ISTYPE 'COM'; "1 when changing zero flag
038- add           PIN  ISTYPE 'COM'; "0=XOR; 1=ADD
039- pass          PIN  ISTYPE 'COM'; "0=adder output; 1=operand passes thru
040- ld_acc        PIN  ISTYPE 'COM'; "1 when changing accumulator
041- set_iflag     PIN  ISTYPE 'COM'; "1 when setting interrupt flag
042- clr_iflag     PIN  ISTYPE 'COM'; "1 when clearing interrupt flag
043- clr_intrpt    PIN  ISTYPE 'COM'; "1 when clearing interrupt flip-flop
```

Listing 10.1   The ABEL version of the DWARF instruction decoder description. (Cont'd.)

```
044- "----- synonyms for various items -----
045- ir = [ir7..ir0];
046- addr_sel = [addr_sel1..addr_sel0];
047- "----- DWARF control section states -----
048- controller_state = [st2..st0];
049- R0,R1       = ^b000,^b001;              "reset states
050- T0,T1,T2,T3 = ^b011,^b010,^b110,^b100;  "normal instruction states
051- I1          = ^b111;                    "interrupt handling state
052- "----- memory regions -----
053- PROGRAM = ^b11;    "memory region for program opcodes & operands
054- DATA    = ^b00;    "memory region for data values
055- STACK   = ^b01;    "memory region for stack
056- X = .X.; "synonym for DON'T-CARE
057- "----- DWARF instruction opcodes -----
058- STORE_REG = [0,0,0,0,0,X,X,X];
059- STORE_INX = [0,0,0,0,1,X,X,X];
060- STORE_DIR = [0,0,0,1,0,X,X,X];
061- STORE_IND = [0,0,0,1,1,X,X,X];
062- LOAD_REG  = [0,0,1,0,0,X,X,X];
063- LOAD_INX  = [0,0,1,0,1,X,X,X];
064- LOAD_DIR  = [0,0,1,1,0,X,X,X];
065- LOAD_IND  = [0,0,1,1,1,X,X,X];
066- LOAD_IMM  = [0,1,0,0,0,X,X,X];
067- IN_REG    = [0,1,1,0,0,X,X,X];
068- OUT       = [0,1,1,0,1,X,X,X];
069- XOR       = [1,0,0,0,0,X,X,X];
070- ADD       = [1,0,0,0,1,X,X,X];
071- TEST      = [1,0,0,1,0,X,X,X];
072- CLEAR_C   = [1,0,1,0,0,X,X,X];
073- SET_C     = [1,0,1,0,1,X,X,X];
074- JC        = [1,1,0,0,0,X,X,X];
075- JZ        = [1,1,0,0,1,X,X,X];
076- JUMP      = [1,1,0,1,0,X,X,X];
077- JSR       = [1,1,0,1,1,X,X,X];
078- RET       = [1,1,1,0,0,X,X,X];
079- RETI      = [1,1,1,0,1,X,X,X];
080-
081- EQUATIONS
082-
083- controller_state.CLK = clock; "synchronously clock state machine
084- controller_state.ACLR = reset; "reset state machine
085-
```

```
086-    STATE_DIAGRAM controller_state   "controls instruction execution
087-
088-       STATE R0: "handles reset by incrementing PC to 0x02
089-          ld_pc=1; inc_pc=1; "increments PC to 0x01
090-          goto R1;
091-
092-       STATE R1: "finishes incrementing PC to 0x02
093-          ld_pc=1; inc_pc=1; "increments PC to 0x02
094-          goto T0;                "now t executing instructions
095-
096-       STATE T0: "handles interrupts or normal instruction flow
097-          IF (intrpt & !iflag) THEN I1 WITH
098-            { "handle interrupt -- store PC on stack
099-            clr_intrpt=1;       "clear current interrupt
100-            set_iflag=1;        "and indicate that it is being processed
101-            drv_d_pc=1; addr_sel=STACK; write=1; "store PC on stack
102-            ld_sp=1; inc_sp=1; "increment the stack pointer
103-            }
104-          ELSE T1 WITH
105-            { "handle normal instruction flow -- fetch opcode
106-            addr_sel=PROGRAM; read=1; "read from program memory
107-            drv_d_data=1; ld_ir=1;     "load opcode into instruction register
108-            ld_pc=1; inc_pc=1;         "increment program counter
109-            }
110-
111-       STATE I1: "handles interrupts
112-          drv_d_acc=1; addr_sel=STACK; write=1; "store ACC on stack
113-          ld_sp=1; inc_sp=1; "increment the stack pointer
114-          clr_pc=1;          "reset PC to start of interrupt-handling code
115-          goto T0;           "start processing interrupt code instructions
116-
117-       STATE T1:
118-          CASE
119-            (ir==IN_REG) : T0 WITH
120-              {
121-              drv_d_port=1; ld_reg=1; "register gets value on input port
122-              }
123-            (ir==OUT) : T0 WITH
124-              {
125-              drv_d_acc=1; ld_port=1; "output register gets value from ACC
128-              { "get subroutine address and increment PC to next instr.
129-              addr_sel=PROGRAM; read=1; "read subr. address from RAM
130-              drv_d_data=1; ld_reg=1; R0A=1; "load address into R0
```

```
131-              ld_pc=1; inc_pc=1; "increment PC to next instruction
132-              }
133-         (ir==RET) : T2 WITH
134-            { "decrement stack pointer to point at return address
135-            ld_sp=1; inc_sp=0; "decrement stack pointer
136-            }
137-         (ir==RETI) : T2 WITH
138-            { "decrement the stack pointer to point at the stored ACC
139-            ld_sp=1; inc_sp=0; "decrement stack pointer
140-            }
141-         (ir==JUMP) : TO WITH
142-            {
143-            addr_sel=PROGRAM; read=1; "read jump address from RAM
144-            drv_d_data=1; ld_pc=1;     "load it into program counter
145-            }
146-         ((ir==JC) & carry) : TO WITH
147-            { "if carry flag is set, load PC with the jump address
148-            addr_sel=PROGRAM; read=1; "read jump address from RAM
149-            drv_d_data=1; ld_pc=1;     "load it into program counter
150-            }
151-         ((ir==JC) & !carry) : TO WITH
152-            { "otherwise, step over the jump address to the next instr.
153-            ld_pc=1; inc_pc=1;
154-            }
155-         ((ir==JZ) & zero) : TO WITH
156-            { "if zero flag is set, load PC with the jump address
157-            addr_sel=PROGRAM; read=1; "read jump address from RAM
158-            drv_d_data=1; ld_pc=1;     "load it into program counter
159-            }
160-         ((ir==JZ) & !zero) : TO WITH
161-            { "otherwise, step over the jump address to the next instr.
162-            ld_pc=1; inc_pc=1;
163-            }
164-         (ir==LOAD_REG) : TO WITH
165-            { "load ACC with data from a register
166-            pass=1; ld_acc=1; "load ACC with data from register
167-            }
168-         (ir==LOAD_INX) : TO WITH
169-            { "load ACC with data whose memory address is in register
170-            addr_sel=DATA; read=1; "address memory with register value
171-            drv_d_data=1; pass=1; ld_acc=1; "store RAM data into ACC
172-            }
```

```
173-          (ir==LOAD_IMM) : TO WITH
174-            { "load ACC with data from program memory
175-            addr_sel=PROGRAM; read=1; drv_d_data=1; "read data from RAM
176-            pass=1; ld_acc=1;   "pass it thru the ALU and into the ACC
177-            ld_pc=1; inc_pc=1; "increment the PC to the next instr.
178-            }
179-          (ir==LOAD_DIR) : T2 WITH
180-            { "load R0 with the address of the data
181-            addr_sel=PROGRAM; read=1; drv_d_data=1; "get address from RAM
182-            ROA=1; ld_reg=1;    "store address in R0
183-            ld_pc=1; inc_pc=1; "increment PC to the next instruction
184-            }
185-          (ir==LOAD_IND) : T2 WITH
186-            { "load R0 with the address of the address of the data
187-            addr_sel=PROGRAM; read=1; drv_d_data=1; "get address from RAM
188-            ROA=1; ld_reg=1;    "store address in R0
189-            ld_pc=1; inc_pc=1; "increment PC to the next instruction
190-            }
191-          (ir==STORE_REG) : TO WITH
192-            { "store ACC into a register
193-            drv_d_acc=1; ld_reg=1;
194-            }
195-          (ir==STORE_INX) : TO WITH
196-            { "store ACC into memory whose address is in a register
197-            addr_sel=DATA; "address RAM with the contents of a register
198-            drv_d_acc=1; write=1; "output ACC value on data bus
199-            }
200-          (ir==STORE_DIR) : T2 WITH
201-            { "get address at which to store accumulator
202-            addr_sel=PROGRAM; read=1; drv_d_data=1; "get address from RAM
203-            ROA=1; ld_reg=1;    "store address in R0
204-            ld_pc=1; inc_pc=1; "increment PC to the next instruction
205-            }
206-          (ir==STORE_IND) : T2 WITH
207-            { "get address containing address at which to store ACC
208-            addr_sel=PROGRAM; read=1; drv_d_data=1; "get address from RAM
209-            ROA=1; ld_reg=1;    "store address in R0
210-            ld_pc=1; inc_pc=1; "increment PC to the next instruction
211-            }
212-          (ir==ADD) : TO WITH
213-            { "ACC <- ACC + Register
214-            pass=0; add=1; ld_acc=1;
```

```
215-          ld_cry=1; alu_cry=1; "update carry flag with ALU carry out
216-          }
217-      (ir==XOR) : TO WITH
218-        { "ACC <- ACC $ Register
219-        pass=0; add=0; ld_acc=1;
220-        }
221-      (ir==TEST) : TO WITH
222-        { "bitwise-AND test of ACC against a register
223-        ld_z = 1; "zero flag is set if the result is 0
224-        }
225-      (ir==SET_C) : TO WITH
226-        { "set carry flag
227-        ld_cry=1; set_cry=1;
228-        }
229-      (ir==CLEAR_C) : TO WITH
230-        { "clear carry flag
231-        ld_cry=1; set_cry=0;
232-        }
233-    endcase;
234-  STATE T2:
235-    CASE
236-      (ir==JSR) : T3 WITH
237-        { "store program counter on the stack and increment SP
238-        drv_d_pc=1; addr_sel=STACK; write=1; "store PC onto stack
239-        ld_sp=1; inc_sp=1;                    "increment stack pointer
240-        }
241-      (ir==RET) : TO WITH
242-        { "load return address into program counter
243-        addr_sel=STACK; read=1; "read return address from stack
244-        drv_d_data=1; ld_pc=1;  "load address into PC
245-        }
246-      (ir==RETI) : T3 WITH
247-        { "restore ACC from the stack and decrement SP again
248-        addr_sel=STACK; read=1;          "read ACC value from the stack
249-        drv_d_data=1; pass=1; ld_acc=1; "load it into ACC
250-        ld_sp=1; inc_sp=0;               "decrement stack pointer
251-        }
252-      (ir==LOAD_DIR) : TO WITH
253-        { "load ACC with the data whose address is in R0
254-        R0B=1; addr_sel=DATA; read=1; "read value at RAM address in R0
255-        drv_d_data=1; pass=1; ld_acc=1; "load value into ACC
256-        }
257-      (ir==LOAD_IND) : T3 WITH
```

```
258-          { "load RO with the address (pointed-to by RO) of the data
259-          ROB=1; addr_sel=DATA; read=1;  "get address from RAM
260-          drv_d_data=1; ROA=1; ld_reg=1; "store address in RO
261-          }
262-        (ir==STORE_DIR) : TO WITH
263-          { "store ACC in RAM at address stored in RO
264-          ROB=1; addr_sel=DATA; write=1; "RAM address is value in RO
265-          drv_d_acc=1; write=1; "output ACC value into RAM
266-          }
267-        (ir==STORE_IND) : T3 WITH
268-          { "get address at which to store ACC
269-          ROB=1; addr_sel=DATA; read=1;  "RAM address is stored in RO
270-          drv_d_data=1; ROA=1; ld_reg=1; "store new address in RO
271-          }
272-      endcase;
273-    STATE T3:
274-      CASE
275-        (ir==JSR) : TO WITH
276-          { "load subroutine address into program counter
277-          drv_d_reg=1; ROB=1; "read subroutine address from RO
278-          ld_pc=1;            "and load it into the PC
279-          }
280-        (ir==RETI) : TO WITH
281-          { "PC gets return address and interrupt flag is cleared
282-          addr_sel=STACK; read=1; "read return address from stack
283-          drv_d_data=1; ld_pc=1;  "load address into PC
284-          clr_iflag=1;       "indicate interrupt processing is done
285-          }
286-        (ir==LOAD_IND) : TO WITH
287-          { "finally, load ACC with the data whose address is in RO
288-          ROB=1; addr_sel=DATA; read=1; "read value at RAM address in RO
289-          drv_d_data=1; pass=1; ld_acc=1; "load value into ACC
290-          }
291-        (ir==STORE_IND) : TO WITH
292-          { "store ACC in RAM at address stored in RO
293-          ROB=1; addr_sel=DATA; write=1; "RAM address is value in RO
294-          drv_d_acc=1; write=1; "output ACC value into RAM
295-          }
296-      endcase;
297-
298- END DWARF1
299-
300-
```

Listing 10.2 The VHDL version of the DWARF instruction decoder description.

```
001- -- DWARF controller -- V1.0
002-
003- LIBRARY IEEE;
004- USE IEEE.std_logic_1164.ALL;
005- USE IEEE.std_logic_unsigned.ALL;
006-
007- ENTITY dwarf_vhdl IS
008-     PORT
009-     (
010-       clock: IN STD_LOGIC;
011-       reset: IN STD_LOGIC; -- zeroes the internal state
012-       ir: IN STD_LOGIC_VECTOR (7 DOWNTO 0); -- instruction register (IR)
013-       carry: IN STD_LOGIC; -- carry flag (C)
014-       zero: IN STD_LOGIC; -- zero flag (Z)
015-       intrpt: IN STD_LOGIC; -- interrupt
016-       iflag: IN STD_LOGIC; -- set when processing interrupt routine
017-       read: OUT STD_LOGIC; -- 1 when reading RAM
018-       write: OUT STD_LOGIC; -- 1 when writing RAM
019-       addr_sel: OUT STD_LOGIC_VECTOR (1 DOWNTO 0); -- 2=PROG;1=DATA;0=STACK
020-       inc_pc: OUT STD_LOGIC; -- 1 when incrementing PC
021-       ld_pc: OUT STD_LOGIC; -- 1 when changing PC
022-       clr_pc: OUT STD_LOGIC; -- 1 when clearing PC to zero
023-       inc_sp: OUT STD_LOGIC; -- 0=--sp; 1=sp++
024-       ld_sp: OUT STD_LOGIC; -- 1 when changing stack pointer
025-       ld_ir: OUT STD_LOGIC; -- 1 when loading IR
026-       ld_port: OUT STD_LOGIC; -- 1 when loading I/O port
027-       drv_d_data: OUT STD_LOGIC; -- 1 when D bus driven by data bus
028-       drv_d_acc: OUT STD_LOGIC; -- 1 when D bus driven by accumulator
029-       drv_d_pc: OUT STD_LOGIC; -- 1 when D bus driven by PC
030-       drv_d_port: OUT STD_LOGIC; -- 1 when D bus is driven by I/O port
031-       drv_d_reg: OUT STD_LOGIC; -- 1 when D bus driven by register
032-       ld_reg: OUT STD_LOGIC; -- 1 when loading a register
033-       ROA: OUT STD_LOGIC; -- 1 forces A reg address to 0
034-       ROB: OUT STD_LOGIC; -- 1 forces B reg address to 0
035-       set_cry: OUT STD_LOGIC; -- 0=reset carry; 1=set carry
036-       alu_cry: OUT STD_LOGIC; -- 1 when ALU operation updates carry
037-       ld_cry: OUT STD_LOGIC; -- 1 when changing carry flag
038-       ld_z: OUT STD_LOGIC; -- 1 when changing zero flag
039-       add: OUT STD_LOGIC; -- 0=XOR; 1=ADD
040-       pass: OUT STD_LOGIC; -- 0=adder output; 1=operand passes thru
041-       ld_acc: OUT STD_LOGIC; -- 1 when changing accumulator
042-       set_iflag: OUT STD_LOGIC; -- 1 when setting interrupt flag
```

Listing 10.2 The VHDL version of the DWARF instruction decoder description. (Cont'd.)

```
043-      clr_iflag: OUT STD_LOGIC; -- 1 when clearing interrupt flag
044-      clr_intrpt: OUT STD_LOGIC -- 1 when clearing interrupt flip-flop
045-   );
046- END dwarf_vhdl;
047-
048-
049- ARCHITECTURE dwarf_arch OF dwarf_vhdl IS
050- -- DWARF internal state
051- SIGNAL curr_st, next_st: STD_LOGIC_VECTOR (2 DOWNTO 0);
052-
053- -- DWARF control section states -----
054- CONSTANT R0: STD_LOGIC_VECTOR (2 DOWNTO 0) := "000";-- reset state 0
055- CONSTANT R1: STD_LOGIC_VECTOR (2 DOWNTO 0) := "001";-- reset state 1
056- CONSTANT T0: STD_LOGIC_VECTOR (2 DOWNTO 0) := "011";-- fetch state
057- CONSTANT T1: STD_LOGIC_VECTOR (2 DOWNTO 0) := "010";--
058- CONSTANT T2: STD_LOGIC_VECTOR (2 DOWNTO 0) := "110";--
059- CONSTANT T3: STD_LOGIC_VECTOR (2 DOWNTO 0) := "100";--
060- CONSTANT I1: STD_LOGIC_VECTOR (2 DOWNTO 0) := "111";-- interrupt state
061-
062- -- memory regions -----
063- CONSTANT PROGRAM: STD_LOGIC_VECTOR (1 DOWNTO 0) := "11"; -- program region
064- CONSTANT DATA:    STD_LOGIC_VECTOR (1 DOWNTO 0) := "00"; -- data region
065- CONSTANT STACK:   STD_LOGIC_VECTOR (1 DOWNTO 0) := "01"; -- stack region
066-
067- -- DWARF instruction opcodes -----
068- CONSTANT STORE_REG: STD_LOGIC_VECTOR (4 DOWNTO 0) := "00000";
069- CONSTANT STORE_INX: STD_LOGIC_VECTOR (4 DOWNTO 0) := "00001";
070- CONSTANT STORE_DIR: STD_LOGIC_VECTOR (4 DOWNTO 0) := "00010";
071- CONSTANT STORE_IND: STD_LOGIC_VECTOR (4 DOWNTO 0) := "00011";
072- CONSTANT LOAD_REG: STD_LOGIC_VECTOR (4 DOWNTO 0)  := "00100";
073- CONSTANT LOAD_INX: STD_LOGIC_VECTOR (4 DOWNTO 0)  := "00101";
074- CONSTANT LOAD_DIR: STD_LOGIC_VECTOR (4 DOWNTO 0)  := "00110";
075- CONSTANT LOAD_IND: STD_LOGIC_VECTOR (4 DOWNTO 0)  := "00111";
076- CONSTANT LOAD_IMM: STD_LOGIC_VECTOR (4 DOWNTO 0)  := "01000";
077- CONSTANT IN_REG: STD_LOGIC_VECTOR (4 DOWNTO 0)    := "01100";
078- CONSTANT OUT_OP: STD_LOGIC_VECTOR (4 DOWNTO 0)    := "01101";
079- CONSTANT XOR_OP: STD_LOGIC_VECTOR (4 DOWNTO 0)    := "10000";
080- CONSTANT ADD_OP: STD_LOGIC_VECTOR (4 DOWNTO 0)    := "10001";
081- CONSTANT TEST: STD_LOGIC_VECTOR (4 DOWNTO 0)      := "10010";
082- CONSTANT CLEAR_C: STD_LOGIC_VECTOR (4 DOWNTO 0)   := "10100";
083- CONSTANT SET_C: STD_LOGIC_VECTOR (4 DOWNTO 0)     := "10101";
084- CONSTANT JC: STD_LOGIC_VECTOR (4 DOWNTO 0)        := "11000";
085- CONSTANT JZ: STD_LOGIC_VECTOR (4 DOWNTO 0)        := "11001";
```

Listing 10.2 The VHDL version of the DWARF instruction decoder description. (Cont'd.)

```
086- CONSTANT JUMP: STD_LOGIC_VECTOR (4 DOWNTO 0)        := "11010";
087- CONSTANT JSR: STD_LOGIC_VECTOR (4 DOWNTO 0)         := "11011";
088- CONSTANT RET: STD_LOGIC_VECTOR (4 DOWNTO 0)         := "11100";
089- CONSTANT RETI: STD_LOGIC_VECTOR (4 DOWNTO 0)        := "11101";
090-
091- BEGIN
092-
093- PROCESS(clock,curr_st,reset)
094- BEGIN
095-    IF reset='1' THEN            -- async reset to reset state
096-         curr_st <= RO;
097-    ELSIF (clock'event AND clock='1') THEN
098-         curr_st <= next_st; -- go to next controller state on rising edge
099-    END IF;
100- END PROCESS;
101-
102- PROCESS(curr_st,ir,carry,zero,iflag,intrpt)
103- BEGIN
104-    -- set defaults of all control outputs to prevent latch synthesis
105-    read<='0';
106-    write <= '0';
107-    addr_sel <= PROGRAM;
108-    inc_pc <= '0';
109-    ld_pc <= '0';
110-    clr_pc <= '0';
111-    inc_sp <= '0';
112-    ld_sp <= '0';
113-    ld_ir <= '0';
114-    ld_port <= '0';
115-    drv_d_data <= '0';
116-    drv_d_acc <= '0';
117-    drv_d_pc <= '0';
118-    drv_d_port <= '0';
119-    drv_d_reg <= '0';
120-    ld_reg <= '0';
121-    ROA <= '0';
122-    ROB <= '0';
123-    set_cry <= '0';
124-    alu_cry <= '0';
125-    ld_cry <= '0';
126-    ld_z <= '0';
127-    add <= '0';
```

```
128-    pass <= '0';
129-    ld_acc <= '0';
130-    set_iflag <= '0';
131-    clr_iflag <= '0';
132-    clr_intrpt <= '0'; next_st <= R0;
133-    CASE curr_st IS
134-    WHEN R0 =>    -- handles reset by incrementing PC to 0x02
135-        ld_pc <= '1'; inc_pc <= '1';-- increments PC to 0x01
136-        next_st <= R1;
137-    WHEN R1 =>    -- finishes incrementing PC to 0x02
138-        ld_pc <= '1'; inc_pc <= '1';-- increments PC to 0x02
139-        next_st <= T0;-- start fetching and executing instructions
140-    WHEN T0 =>     -- handles interrupts or normal instruction flow
141-        IF (intrpt='1' AND iflag='0') THEN-- store context
142-            clr_intrpt <= '1';-- clear current interrupt
143-            set_iflag <= '1';-- and flag that it is being processed
144-            drv_d_pc<='1'; addr_sel<=STACK; write<='1';-- PC on stack
145-            ld_sp<='1'; inc_sp<='1';-- increment the stack pointer
146-            next_st <= I1;-- go to interrupt handling state
147-        ELSE -- handle normal instruction flow: fetch opcode
148-            addr_sel<=PROGRAM; read<='1'; -- read from program memory
149-            drv_d_data<='1'; ld_ir<='1';  -- load opcode into IR
150-            ld_pc<='1'; inc_pc<='1';-- increment program counter
151-            next_st <= T1;-- execute instruction
152-        END IF;
153-    WHEN I1 =>    -- handles interrupts
154-        drv_d_acc<='1'; addr_sel<=STACK; write<='1';  -- ACC on stack
155-        ld_sp<='1'; inc_sp<='1';-- increment the stack pointer
156-        clr_pc<='1';-- reset PC to start of interrupt-handling code
157-        next_st <= T0; -- start processing interrupt code instructions
158-    WHEN T1 =>    -- instruction processing
159-        CASE ir(7 DOWNTO 3) IS
160-            WHEN IN_REG =>
161-                drv_d_port<='1'; ld_reg<='1';-- reg gets value on input port
162-                next_st <= T0;-- done with this instruction; get another
163-            WHEN OUT_OP =>
164-                drv_d_acc<='1'; ld_port<='1';-- output reg gets val from ACC
165-                next_st <= T0;-- done with this instruction; get another
166-            WHEN JSR =>-- get subrout addr and increment PC to next instr.
167-                addr_sel<=PROGRAM; read<='1';-- read subr. address from RAM
168-                drv_d_data<='1'; ld_reg<='1'; R0A<='1';-- load addr into R0
169-                ld_pc<='1'; inc_pc<='1';-- increment PC past address
170-                next_st <= T2;-- instruction not done; go on
```

```
171-    WHEN RET =>         -- decrement stack pointer to point at return addr
172-        ld_sp<='1'; inc_sp<='0';-- decrement stack pointer
173-        next_st <= T2;    -- not done with this instruction; keep going
174-    WHEN RETI =>        -- dec the stack ptr to point at the stored ACC
175-        ld_sp<='1'; inc_sp<='0';-- decrement stack pointer
176-        next_st <= T2;    -- not done with this instruction; keep going
177-    WHEN JUMP =>
178-        addr_sel<=PROGRAM; read<='1';-- read jump address from RAM
179-        drv_d_data<='1'; ld_pc<='1';-- load it into program counter
180-        next_st <= T0;    -- done with this instruction; get another
181-    WHEN JC =>
182-        IF carry='1' THEN -- if carry then load PC with jump addr
183-            addr_sel<=PROGRAM; read<='1';-- read jump address from RAM
184-            drv_d_data<='1'; ld_pc<='1';-- load it into program cnter
185-        ELSE     -- else step over the jump address to the next instr.
186-            ld_pc<='1'; inc_pc<='1';-- increment PC past address
187-        END IF;
188-        next_st <= T0;    -- done with this instruction; get another
189-    WHEN JZ =>
190-        IF zero='1' THEN  -- if zero then load PC with the jump address
191-            addr_sel<=PROGRAM; read<='1';-- read jump address from RAM
192-            drv_d_data<='1'; ld_pc<='1';-- load it into program cnter
193-        ELSE     -- else step over the jump address to the next instr.
194-            ld_pc<='1'; inc_pc<='1';-- increment PC past address
195-        END IF;
196-        next_st <= T0;    -- done with this instruction; get another
197-    WHEN LOAD_REG =>        -- load ACC with data from a register
198-        pass<='1'; ld_acc<='1';
199-        next_st <= T0;    -- done with this instruction; get another
200-    WHEN LOAD_INX =>  -- load ACC with data whose memory addr is in reg
201-        addr_sel<=DATA; read<='1';-- address memory with reg value
202-        drv_d_data<='1'; pass<='1'; ld_acc<='1'; -- RAM data into ACC
203-        next_st <= T0;    -- done with this instruction; get another
204-    WHEN LOAD_IMM =>        -- load ACC with data from program memory
205-        addr_sel<=PROGRAM; read<='1'; drv_d_data<='1';-- read RAM data
206-        pass<='1'; ld_acc<='1'; -- pass it thru the ALU and into ACC
207-        ld_pc<='1'; inc_pc<='1';-- inc the PC to the next instr.
208-        next_st <= T0;    -- done with this instruction; get another
209-    WHEN LOAD_DIR =>        -- load R0 with the address of the data
210-        addr_sel<=PROGRAM; read<='1'; drv_d_data<='1'; -- get RAM addr
211-        R0A<='1'; ld_reg<='1';-- store address in R0
212-        ld_pc<='1'; inc_pc<='1';-- increment PC to the next instr.
```

```
213-                next_st <= T2;-- not done with this instruction; keep going
214-            WHEN LOAD_IND =>-- load R0 with the addr of the addr of the data
215-                addr_sel<=PROGRAM; read<='1'; drv_d_data<='1';-- get addr
216-                R0A<='1'; ld_reg<='1';-- store address in R0
217-                ld_pc<='1'; inc_pc<='1';-- increment PC to the next instr.
218-                next_st <= T2;-- not done with this instruction; keep going
219-            WHEN STORE_REG =>-- store ACC into a register
220-                drv_d_acc<='1'; ld_reg<='1';
221-                next_st <= T0;-- done with this instruction; get another
222-            WHEN STORE_INX =>-- store ACC into memory whose addr is in reg
223-                addr_sel<=DATA;-- address RAM with the contents of a register
224-                drv_d_acc<='1'; write<='1';-- output ACC value on data bus
225-                next_st <= T0;-- done with this instruction; get another
226-            WHEN STORE_DIR =>-- get address at which to store accumulator
227-                addr_sel<=PROGRAM; read<='1'; drv_d_data<='1';-- get RAM addr
228-                R0A<='1'; ld_reg<='1';-- store address in R0
229-                ld_pc<='1'; inc_pc<='1';-- increment PC to the next instr.
230-                next_st <= T2;-- not done with this instruction; keep going
231-            WHEN STORE_IND =>-- get addr of addr at which to store ACC
232-                addr_sel<=PROGRAM; read<='1'; drv_d_data<='1';-- get address
233-                R0A<='1'; ld_reg<='1';-- store address in R0
234-                ld_pc<='1'; inc_pc<='1';--increment PC to the next instr.
235-                next_st <= T2;-- not done with this instruction; keep going
236-            WHEN ADD_OP =>-- ACC <- ACC + Register
237-                pass<='0'; add<='1'; ld_acc<='1';
238-                ld_cry<='1'; alu_cry<='1';-- update carry with ALU carry out
239-                next_st <= T0;-- done with this instruction; get another
240-            WHEN XOR_OP =>-- ACC <- ACC $ Register
241-                pass<='0'; add<='0'; ld_acc<='1';
242-                next_st <= T0;-- done with this instruction; get another
243-            WHEN TEST =>-- bitwise-AND test of ACC against a register
244-                ld_z <= '1';-- zero flag is set if the result is 0
245-                next_st <= T0;-- done with this instruction; get another
246-            WHEN SET_C =>-- set carry flag
247-                ld_cry<='1'; set_cry<='1';
248-                next_st <= T0;-- done with this instruction; get another
249-            WHEN CLEAR_C =>-- clear carry flag
250-                ld_cry<='1'; set_cry<='0';
251-                next_st <= T0;-- done with this instruction; get another
252-            WHEN OTHERS =>
253-          END CASE;
254-      WHEN T2 =>-- instruction processing
```

```
255-      CASE ir(7 DOWNTO 3) IS
256-        WHEN JSR =>-- store program counter on stack and increment SP
257-            drv_d_pc<='1'; addr_sel<=STACK; write<='1';--store PC onto stk
258-            ld_sp<='1'; inc_sp<='1';--increment stack pointer
259-            next_st <= T3;-- not done with this instruction; keep going
260-        WHEN RET =>-- load return address into program counter
261-            addr_sel<=STACK; read<='1';--read return address from stack
262-            drv_d_data<='1'; ld_pc<='1';--load address into PC
263-            next_st <= T0;-- done with this instruction; get another
264-        WHEN RETI =>-- restore ACC from the stack and decrement SP again
265-            addr_sel<=STACK; read<='1';--read ACC value from the stack
266-            drv_d_data<='1'; pass<='1'; ld_acc<='1';--load it into ACC
267-            ld_sp<='1'; inc_sp<='0';--decrement stack pointer
268-            next_st <= T3;-- not done with this instruction; keep going
269-        WHEN LOAD_DIR =>-- load ACC with the data whose address is in R0
270-            ROB<='1'; addr_sel<=DATA; read<='1';--read value at RAM[R0]
271-            drv_d_data<='1'; pass<='1'; ld_acc<='1'; --load value into ACC
272-            next_st <= T0;-- done with this instruction; get another
273-        WHEN LOAD_IND =>-- load R0 with addr (poitd-to by R0) of the data
274-            ROB<='1'; addr_sel<=DATA; read<='1';--get address from RAM
275-            drv_d_data<='1'; ROA<='1'; ld_reg<='1';--store address in R0
276-            next_st <= T3;-- not done with this instruction; keep going
277-        WHEN STORE_DIR =>-- store ACC in RAM at address stored in R0
278-            ROB<='1'; addr_sel<=DATA; write<='1';--RAM addr is in R0
279-            drv_d_acc<='1'; write<='1';--output ACC value into RAM
280-            next_st <= T0;-- done with this instruction; get another
281-        WHEN STORE_IND =>-- get address at which to store ACC
282-            ROB<='1'; addr_sel<=DATA; read<='1';--RAM addr is stored in R0
283-            drv_d_data<='1'; ROA<='1'; ld_reg<='1';--store new addr in R0
284-            next_st <= T3;-- not done with this instruction; keep going
285-        WHEN OTHERS =>
286-      END CASE;
287-    WHEN T3 =>-- finish instruction processing
288-      CASE ir(7 DOWNTO 3) IS
289-        WHEN JSR =>-- load subroutine address into program counter
290-            drv_d_reg<='1'; ROB<='1';--read subroutine address from R0
291-            ld_pc<='1';     --and load it into the PC
292-            next_st <= T0;-- done with this instruction; get another
293-        WHEN RETI =>-- PC gets return addr and interrupt flag is cleared
294-            addr_sel<=STACK; read<='1';--read return address from stack
295-            drv_d_data<='1'; ld_pc<='1';--load address into PC
296-            clr_iflag<='1';--indicate interrupt processing is done
297-            next_st <= T0;-- instruction done; get another
```

Listing 10.2   The VHDL version of the DWARF instruction decoder description. (Cont'd.)

```
298-          WHEN LOAD_IND =>-- load ACC with the data whose addr is in R0
299-            ROB<='1'; addr_sel<=DATA; read<='1';--read value at RAM[R0]
300-            drv_d_data<='1'; pass<='1'; ld_acc<='1'; --load value into ACC
301-            next_st <= T0;-- done with this instruction; get another
302-          WHEN STORE_IND =>-- store ACC in RAM at address stored in R0
303-            ROB<='1'; addr_sel<=DATA; write<='1'; --RAM addr is in R0
304-            drv_d_acc<='1'; write<='1';--output ACC value into RAM
305-            next_st <= T0;-- done with this instruction; get another
306-          WHEN OTHERS =>
307-        END CASE;
308-    WHEN OTHERS =>
309-    END CASE;
310- END PROCESS;
311-
312- END dwarf_arch;
313-
```

output port are connected to the seven-segment LED display so you have a limited means of displaying status information.

```
NET IN_CLOCK      LOC=P44;    # B0 argument of XSPORT
NET INT           LOC=P45;    # B1 argument of XSPORT
NET IN_RESET      LOC=P46;    # B2 argument of XSPORT
NET ADDR<0>       LOC=P3;
NET ADDR<1>       LOC=P4;
NET ADDR<2>       LOC=P5;
NET ADDR<3>       LOC=P78;
NET ADDR<4>       LOC=P79;
NET ADDR<5>       LOC=P82;
NET ADDR<6>       LOC=P83;
NET ADDR<7>       LOC=P84;
NET ADDR<8>       LOC=P59;
NET ADDR<9>       LOC=P57;
NET ADDR<10>      LOC=P51;
NET ADDR<11>      LOC=P56;
NET ADDR<12>      LOC=P50;
NET ADDR<13>      LOC=P58;
NET ADDR<14>      LOC=P60;
NET DATA<0>       LOC=P41;
NET DATA<1>       LOC=P40;
NET DATA<2>       LOC=P39;
NET DATA<3>       LOC=P38;
NET DATA<4>       LOC=P35;
```

```
NET DATA<5>        LOC=P81;
NET DATA<6>        LOC=P80;
NET DATA<7>        LOC=P10;
NET CS_            LOC=P65;
NET OE_            LOC=P61;
NET WE_            LOC=P62;
NET OUT_PORT<0>    LOC=P25;        # S0 LED segment
NET OUT_PORT<1>    LOC=P26;        # S1 LED segment
NET OUT_PORT<2>    LOC=P24;        # S2 LED segment
NET OUT_PORT<3>    LOC=P20;        # S3 LED segment
NET OUT_PORT<4>    LOC=P23;        # S4 LED segment
NET OUT_PORT<5>    LOC=P18;        # S5 LED segment
NET OUT_PORT<6>    LOC=P19;        # S6 LED segment
NET OUT_PORT<7>    LOC=P6;
NET IN_PORT<0>     LOC=P7;
NET IN_PORT<1>     LOC=P8;
NET IN_PORT<2>     LOC=P9;
NET IN_PORT<3>     LOC=P77;
NET IN_PORT<4>     LOC=P70;
NET IN_PORT<5>     LOC=P66;
NET IN_PORT<6>     LOC=P67;
NET IN_PORT<7>     LOC=P69;
```

Now the question is: With all the new additions and features, will it fit in the FPGA? Here is the utilization of an XC4005XL FPGA when loaded with the version of DWARF containing the ABEL version of the controller:

```
Design Summary
--------------
    Number of errors:        0
    Number of warnings:      14
    Number of CLBs:              188 out of    196    95%
        CLB Flip Flops:      40
        CLB Latches:          0
        4 input LUTs:       323 (7 used as route-throughs)
        3 input LUTs:        62 (4 used as route-throughs)
        Dual Port RAMs:       8
    Number of bonded IOBs:        45 out of     65    69%
        IOB Flops:           16
        IOB Latches:          0
    Number of TBUFs:              40 out of    448     8%
    Number of BUFGLSs:             1 out of      8    12%
    Number of RPM macros:         2
Total equivalent gate count for design: 3784
```

And here is the utilization of an XC4005XL FPGA when loaded with the version of DWARF containing the VHDL version of the controller:

```
Design Summary
--------------
   Number of errors:        0
   Number of warnings:     13
   Number of CLBs:             109 out of   196    55%
      CLB Flip Flops:       40
      CLB Latches:           0
      4 input LUTs:        174 (11 used as route-throughs)
      3 input LUTs:         54 (2 used as route-throughs)
      Dual Port RAMs:        8
   Number of bonded IOBs:       45 out of    65    69%
      IOB Flops:            16
      IOB Latches:           0
   Number of TBUFs:             40 out of   448     8%
   Number of BUFGLSs:            1 out of     8    12%
   Number of RPM macros:         2
Total equivalent gate count for design: 2854
```

While both versions fit in the XC4005XL FPGA, the VHDL version uses slightly more than half of the FPGA while the ABEL version nearly fills the FPGA.

Now we need a test program for DWARF. A simple program that adds a list of numbers is shown next. Notice that the program starts at address 2, since this is the RAM address where DWARF begins fetching instructions after a reset.

```
0002 40 00     load  #0      ; R1 <- address of start of number
0004 01        store R1      ;       list in DATA region
0005 40 10     load  #10     ; R2 <- number of items in the list
0007 02        store R2
0008 40 00     load  #0      ; R3 <- 0 (clear the sum register)
000A 03        store R3
            loop:
000B 29        load  (R1)    ; ACC <- the next item on the list
000C A0        clear_c       ; erase previous carry result
000D 8B        add   R3      ; add sum register to ACC
000E 03        store R3      ; store result back into sum register
000F 40 01     load  #1      ; now, increment R1 so it points to
0011 A0        clear_c       ;    the next item on the list
0012 89        add   R1
0013 01        store R1
0014 40 FF     load  #FF     ; and decrement R2 since we've just
0016 A0        clear_c       ;    added one more number
```

```
0017 8A          add   R2
0018 02          store R2
0019 40 FF       load  #FF      ; now test R2 to see if it is zero
001B 92          test  R2       ;    (i.e., we've finished the list)
001C C8 20       jz    show     ; exit the loop if item counter is 0
001E D0 0B       jump  loop     ; otherwise, keep processing the list
            show:
0020 23          load  R3       ; ACC <- sum stored in register R3
0021 68          out            ; display sum bits on the LED display
            halt:
0022 D0 22       jump  halt     ; just loop here forever
```

The instruction opcodes are stored in file ADDLIST.HEX as follows:

```
- OE 0002 40 00 01 40 10 02 40 00 03 29 A0 8B 03 40
- 10 0010 01 A0 89 01 40 FF A0 8A 02 40 FF 92 C8 20 D0 0B
- 04 0020 23 68 D0 22
- 10 7F00 01 23 45 67 89 AB CD EF FE DC BA 98 76 54 32 10
```

The first three lines of ADDLIST.HEX just list the instruction opcodes and operands for the program. These are loaded into DWARF's PROGRAM region starting at hexadecimal address 0x0002. The 16 values on the last line, however, get loaded into the DATA region of DWARF's memory, which starts at hexadecimal address 0x7F00. This will initialize the data list that the program will operate on.

You can download the DWARF microprocessor and the ADDLIST.HEX program to the XS40 Board just like we did for the GNOME microprocessor:

```
C:\XCPROJ\DWARF_40> XSLOAD ADDLIST.HEX DWARF_40.BIT
```

Then you can use the following command sequences to reset DWARF to the beginning of the program and interrupt it. (We did not load an ISR into DWARF's memory, so an interrupt will not be handled correctly.) You can also single-step DWARF *through an instruction cycle*. Since different instructions require differing numbers of clock cycles, it is not possible to give a fixed command sequence for single-stepping through an entire instruction. Therefore, we have to settle for stepping through instruction cycles and you will have to manually keep track of where DWARF is in terms of instruction execution.

Action	XSPORT Command Sequence
Reset	XSPORT 100 XSPORT 000
Generate Interrupt	XSPORT 010 XSPORT 000
Instruction cycle single-step	XSPORT 001 XSPORT 000

You will not see any intermediate results displayed on the LED digit as the DWARF program executes. This only happens when the program reaches the end of the addition loop and sends the sum to the output port that is connected to the seven-segment LED. The sum should be 0x7F8, which will be truncated to 0xF8 or 11111000. The lower 7 bits of this result are 1111000, so you should see the upper four segments of the display light up while the lower three remain dark.

If you change the clock input constraint as follows:

```
NET IN_CLOCK LOC=P13;    # 12 MHz oscillator on the XS40 Board
```

then you can clock DWARF at 12 MHz.

Projects

1. Add increment and decrement instructions to DWARF. Should these instructions operate on the accumulator or the registers?

2. Change the load instructions so they load the registers instead of the accumulator.

3. Develop a way to save the carry and zero flags upon entering an interrupt service routine.

4. Write some programs for some common operations like multiplication, division, and string comparison. Determine what instructions could be added to DWARF to make these operations easier and add them to DWARF.

5. Build a simple assembler that will turn a set of DWARF instruction mnemonics into opcodes.

6. Can you construct a 16-bit datapath version of the DWARF? What are the implications on the architecture? On the external memory system?

Final Word

Well, we've come a long way. We learned about simple logic gates, progressed on to flip-flops and state machines, and finished up with the construction a pair of actual, working microcomputers.

Hopefully, you've done the experiments and projects as you went through the book. Some knowledge can only be transmitted by the act of doing. If you have worked through the experiments, then three months from now you should still be able to design and build an edge-triggered flip-flop or a complete microcomputer. If you haven't done the experiments, then three months from now you'll be lucky if you can even remember where you left this book.

This book got you acquainted with CPLDs and FPGAs, but there are a lot of topics I skimmed over and even more I left out completely (like power estimation, pipelining, ...). In my defense, I told you at the start I wouldn't cover everything. But I hope I've given you enough so you can read through the data sheet of a programmable device and understand 90% of it. Even better, you should be able to understand and extend the designs in the application notes provided by the device manufacturers. And with the Web in place, you can publish and share designs with others (after all, a design is just data files). The most important thing is to keep designing and building. Experience is a great teacher, and nothing gives you as many experiences per day as FPLDs.

That's all!

Building the XS40 and XS95 Lite Boards

Objectives

- To describe how to build both XC95108 CPLD and XC4005XL FPGA-based prototyping breadboards that support the experiments in the rest of this book.
- To show how to install and use the XSLOAD and XSPORT utility programs which download and test circuits for the XS40 and XS95 Boards.

Discussion

This appendix describes how to build both a CPLD and an FPGA-based prototyping board. These prototyping boards replicate enough of the functionality of the commercially-available XS40 and XS95 Boards so that you can do all the design examples in this book. For the purpose of discussion, let's call these the XS40 Lite and XS95 Lite Boards.

The most important component of the prototyping board is either an XC95108 CPLD or XC4005XL FPGA. The FPLD is loaded with bitstream files created by the XILINX F1 software tools. The bitstreams are downloaded through the PC parallel port into the Lite board. The parallel port is also used to apply signals to the prototyping board for testing a downloaded logic circuit. A seven-segment LED attached to the FPLD provides a visual indication of how the circuit is acting. A 32 KByte static RAM is also attached to the FPLD to provide external data storage. Then there are miscellaneous components to provide clocks and regulated power to the other components.

Experimental

The following sections will describe how to construct the XS40 and XS95 Lite boards. Compared to buying an XS40 or XS95 Board, it probably won't save you any money to build your own prototyping board unless you already have most of the parts lying around. And if you already have most of the parts on hand, then I'm assuming you have some experience in hooking chips together, using a solderless breadboard, and that you can tell which pin is #1 on a chip. So the instructions that follow will be brief.

Building the XS40 Lite Board

The schematic for the XS40 Lite Board is shown in Figure A.1. You can divide it into the following sections, each of which will be described below:

- Power supply
- XC4005XL FPGA
- Parallel port interface
- LED display
- Memory
- Clock circuit

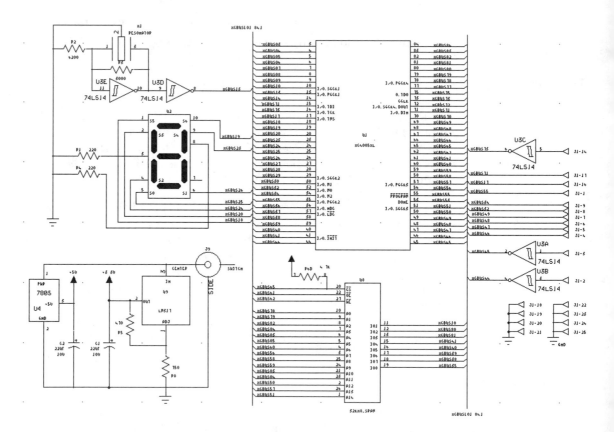

Figure A.1   XS40 Lite Board schematic.

Table A.1   XS40 Lite Board parts list.

B1	ACE 336 SOLDERLESS BREADBOARD BY 3M
C1,C2	22uF TANTALUM CAPACITOR
J1	FEMALE DB25 CONNECTOR
J3	84-PIN PLCC/DIP ADAPTER
J9	2.1MM POWER PLUG
R1,R4	1/8W, 220Ω RESISTOR
R2	1/8W, 6.2KΩ RESISTOR
R3	1/8W, 3.0KΩ RESISTOR
R5	1/8W, 470Ω RESISTOR
R6	1/8W, 4.7KΩ RESISTOR
R8	1/8W, 750Ω RESISTOR
U1	XC4005XL FPGA
U2	MAN74 7-SEGMENT LED
U4	7805 +5V LINEAR REGULATOR
U3	74LS14 HEX SCHMITT-TRIGGER INVERTER
U8	32K × 8 STATIC RAM
U9	LM317 PROGRAMMABLE LINEAR REGULATOR
X1	12 MHz RESONATOR
X2	9V DC, 300 MA, +CENTER, WALL XFORMER

Power Supply

The XS40 Lite draws its power from a wall-mounted, center-positive, 300 mA, 9V DC transformer. As shown in Figure A.2, an LM317 programmable regulator converts the 9V to the 3.3V needed by the XC4005XL FPGA. (Actually, the voltage divider formed by the 470Ω and 750Ω resistors sets the output voltage to 3.24V, which is within the tolerance of the FPGA.) The 5V required by the rest of the XS40 Lite circuitry is supplied by the 7805 linear regulator. Tantalum capacitors on the outputs of both regulators remove most of the voltage ripples.

XC4005XL FPGA

The XC4005XL FPGA is the heart of the prototyping board. It is contained in an 84-pin PLCC package. PLCCs don't plug directly into solderless breadboards, so you have to use a PLCC prototyping adapter with a socket for the XC4005XL. The adapter provides 84 pins that plug into the breadboard.

Once the XC4005XL is plugged into the breadboard through the adapter, you will have access to all the pins of the FPGA. The first thing you should do is connect the VCC and GND pins of the FPGA to the 3.3V supply and the system ground. *Make sure you do not have power applied while you do this!* The VCC and GND pins are listed in Table A.2.

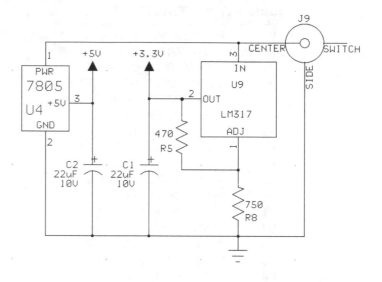

Figure A.2   XS40 Lite Board power supply schematic.

Table A.2   Power (+3.3V) and ground pins for the XC4005XL FPGA.

VCC Pins	GND Pins
2	1
11	12
22	21
33	31
42	43
54	52
63	64
74	76

Next, the TDI input (pin 15) is connected to the DIN input (pin 71), and the TCK input (pin 16) is connected to the CCLK input (pin 73). This connects the data and clock inputs of both the JTAG and configuration circuitry together.

All the other pins of the FPGA are collected together into the 84-line XCBUS. You don't have to physically construct this bus right now—it just serves as a notational convenience that makes it easy to see which FPGA pins are connected to pins on the other chips.

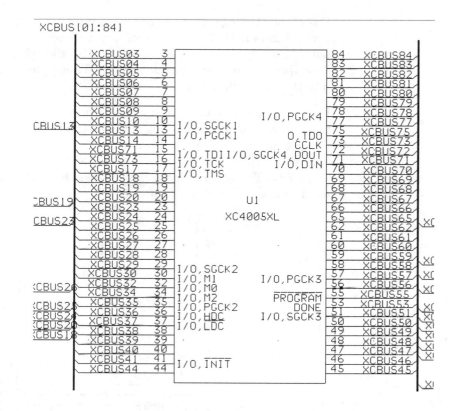

XCBUS[01:84]

Figure A.3   XS40 Lite Board XC4005XL FPGA bus connections.

Parallel Port Interface

The XS40 Lite interface to the PC parallel port is shown in Figure A.4. A female DB25 connector (J1) accepts the 25-wire cable from the parallel port.

Pins 1, 14, and 17 of J1 are used to download bitstreams into the XC4005XL FPGA. A low-going pulse on pin 1 of J1 is transmitted to the /PROGRAM input of the XC4005XL (pin 55). This clears the FPGA and prepares it to accept a new circuit configuration. The bitstream data enters through pin 17 of J1 and appears on the DIN input of the FPGA (pin 71). The data is loaded into the FPGA whenever a rising edge appears on the CCLK input of the XC4005XL (pin 73). CCLK is driven by pin 14 of J1 through a 74LS14 Schmitt-trigger inverter. The 74LS14 provides hysteresis which cleans up the clock edges that are degraded by the trip through the parallel port cable. The XSLOAD program accounts for the signal inversion introduced by the 74LS14.

(When making these connections, don't forget to connect pins 7 and 14 of the 74LS14 chip to ground and 5V, respectively.)

Pin 16 of J1 is connected to the TMS input (pin 17) of the XC4005XL's JTAG port. Pins 14 and 17 of J1 also connect to the TCK and TDI inputs of the JTAG port, since TCK is connected to CCLK and TDI is connected to DIN (as described in the previous subsection). These pins are used to download data into the 32 KByte RAM.

Pins 2-9 of J1 are used to apply test signals to the FPGA once a new circuit has been downloaded. The signals applied through pins 2 and 3 of J1 are suitable for use as clock signals for sequential circuitry since they pass through Schmitt-trigger inverters. (The XSPORT program accounts for these inverters.) The other signals can serve as level inputs, but they should not be used as clocks.

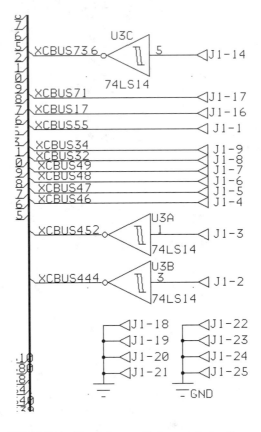

Figure A.4   XS40 Lite Board parallel port interface schematic.

The correspondence between the pins of J1, the pins of the XC4005XL FPGA, and the arguments of the XSPORT program are shown in Table A.3.

Table A.3 Correspondence between the pins of J1 and the FPGA and the arguments of XSPORT.

J1 Pin	XC4005XL Pin	XSPORT Arg.
2	44	B0
3	45	B1
4	46	B2
5	47	B3
6	48	B4
7	49	B5
8	32	B6
9	34	B7

LED Display

The connections between the LED digit and the XC4005XL FPGA are depicted in Figure A.5. Two 220Ω resistors are placed in the ground returns of the LED digit to limit the amount of current sourced from the FPGA outputs. The connections between the FPGA pins and the LED digit segments are listed in Table A.4.

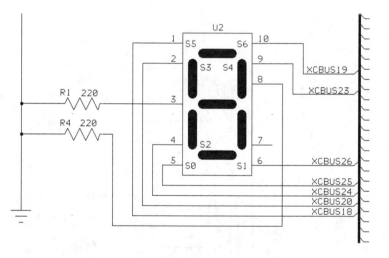

Figure A.5   XS40 Lite Board LED display schematic.

Table A.4　Connections between the FPGA and the LED digit segments.

LED Segment	XC4005XL Pin
S0	25
S1	26
S2	24
S3	20
S4	23
S5	18
S6	19

Memory

The connections between the XC4005XL FPGA and the 32 KByte static RAM are shown in Figure A.6. Not shown are the connections to ground and 5V on pins 14 and 28 of the RAM package, respectively.

The chip-select is pulled high through a 4.7KΩ resistor to disable the RAM when the FPGA is being configured. This lets us load values into the RAM and then reconfigure the FPGA without corrupting the RAM data while the FPGA pins are tristated.

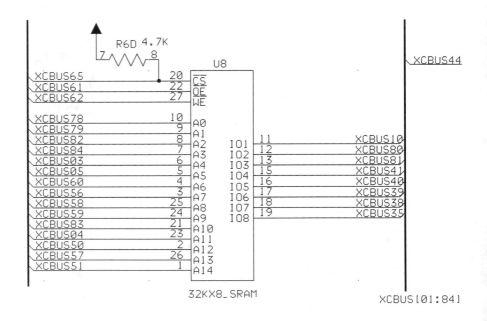

Figure A.6　　XS40 Lite Board 32 KByte RAM schematic.

Table A.5 lists the assignment of RAM address, data, and control lines to FPGA pins as viewed by the portion of the XSLOAD program which downloads data to the RAM. Note that the order of the data and address pins in the list differs from the ordering of the pins on the RAM part in the schematic. However, it makes no difference to the actual RAM chip which address or data pin is the

Table A.5 The RAM pin functions assigned to the pins of the XC4005XL FPGA.

Pin Function	XC4005XL Pin
A0	3
A1	4
A2	5
A3	78
A4	79
A5	82
A6	83
A7	84
A8	59
A9	57
A10	51
A11	56
A12	50
A13	58
A14	60
D0	41
D1	40
D2	39
D3	38
D4	35
D5	81
D6	80
D7	10
/CS	65
/OE	61
/WE	62

least-significant or most-significant as long as every device that accesses the RAM uses the same ordering. The ordering shown in Table A.5 matches the one used in the XS40 Board which was chosen based upon layout considerations for the printed wiring board.

Clock Circuit

The oscillator circuit shown in Figure A.7 generates a 12 MHz clock signal for the XC4005XL FPGA. The 12 MHz resonator connected between the input and output of the first inverter causes it to oscillate. (The two resistors bias the inverter to make sure the oscillation starts up.) The second inverter cleans-up the output of the first inverter and then applies the signal to one of the primary global clock inputs of the FPGA (PGCK1 on pin 13).

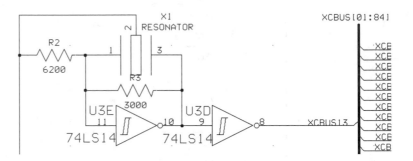

Figure A.7   XS40 Lite Board clock circuit schematic.

Building the XS95 Lite Board

The schematic for the XS95 Lite Board is shown in Figure A.8. You can divide it into the following sections, each of which will be described below:

- Power supply

- XC95108 CPLD

- Parallel port interface

- LED display

- Memory

- Clock circuit

Table A.6   XS95 Lite Board parts list.

```
zB1      ACE 336 SOLDERLESS BREADBOARD BY 3M
C1,C2    22uF TANTALUM CAPACITOR
J1       FEMALE DB25 CONNECTOR
J3       84-PIN PLCC/DIP ADAPTER
J9       2.1MM POWER PLUG
R1,R4    1/8W, 220Ω RESISTOR
R2       1/8W, 6.2KΩ RESISTOR
R3       1/8W, 3.0KΩ RESISTOR
R5       1/8W, 470Ω RESISTOR
R6       1/8W, 10KΩ RESISTOR
R7       1/8W, 2.2KΩ RESISTOR
R8       1/8W, 750Ω RESISTOR
U1       XC95108 CPLD
U2       MAN74 7-SEGMENT LED
U6       7805 +5V LINEAR REGULATOR
U3       74LS14 HEX SCHMITT-TRIGGER INVERTER
U8       32K × 8 STATIC RAM
U9       LM317 PROGRAMMABLE LINEAR REGULATOR
X1       12 MHz RESONATOR
X2       9V DC, 300 MA, +CENTER, WALL XFORMER
```

Power Supply

The XS95 Lite draws its power from a wall-mounted, center-positive, 300 mA, 9V DC transformer. As shown in Figure A.9, a 7805 linear regulator converts the 9V to the 5V needed by the XC95108 CPLD and the rest of the circuitry. A tantalum capacitor on the output of the regulator removes most of the voltage ripple.

XC95108 CPLD

The XC95108 CPLD is the heart of the prototyping board. It is contained in an 84-pin PLCC package. PLCCs don't plug into solderless breadboards, so you have to use a PLCC prototyping adapter with a socket for the XC95108. The adapter provides 84 pins that plug into the breadboard.

Once the XC95108 is plugged into the breadboard through the adapter, you will have access to all the pins of the CPLD. The first thing you should do is connect the VCC and GND pins of the CPLD to the 5V supply and the system ground. ***Make sure you do not have power applied while you do this!*** The VCC and GND pins are listed in Table A.7.

All the pins of the CPLD are collected together into the 84-line XCBUS. You don't have to physically construct this bus right now—it just serves as a notational convenience that makes it easy to see which CPLD pins are connected to pins on the other chips.

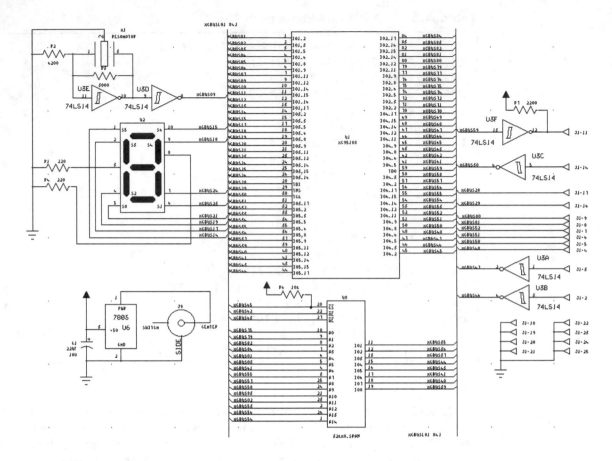

Figure A.8   XS95 Lite Board schematic.

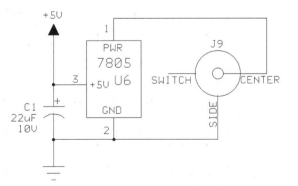

Figure A.9   XS95 Lite Board power supply schematic.

Building the XS40 and XS95 Lite Boards   Appendix A

Table A.7 Power and ground pins for the XC95108 CPLD.

VCC Pins	GND Pins
22	8
38	16
64	27
73	42
78	49
	60

XCBUS[01:84]

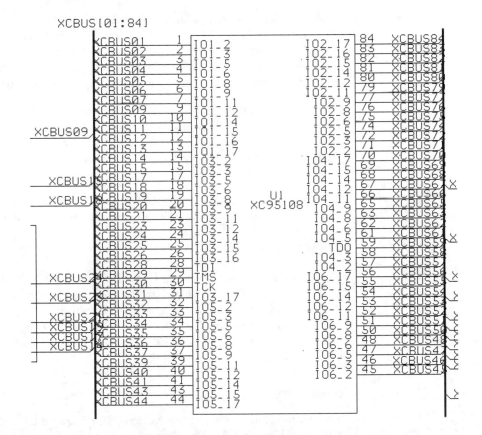

Figure A.10   XS95 Lite Board XC95108 CPLD bus connections.

Parallel Port Interface

The XS95 Lite interface to the PC parallel port is shown in Figure A.11. A female DB25 connector (J1) accepts the 25-wire cable from the parallel port.

Pins 14, 16, and 17 of J1 are used to download bitstreams into the XC95108 CPLD. The XS95 Lite uses the JTAG port to configure the CPLD. Bitstream data enters through pins 16 and 17 of J1 and appears on the TMS and TDI inputs (pins 29 and 28) of the XC95108, respectively. The data is loaded into the CPLD whenever a rising edge appears on the TCK input of the XC95108 (pin 30). TCK is driven by pin 14 of J1 through a 74LS14 Schmitt-trigger inverter. The 74LS14 provides hysteresis which cleans up the clock edges that are degraded by the trip through the parallel port cable. The XSLOAD program accounts for the signal inversion introduced by the 74LS14.

(When making these connections, don't forget to connect pins 7 and 14 of the 74LS14 chip to ground and 5V, respectively.)

There is also a return path from the XC95108 back to the PC parallel port. The TDO output of the CPLD's JTAG port (pin 59) drives through another 74LS14 inverter to pin 11 of J1. A pull-up resistor is attached to the output of the inverter to provide extra current for driving the cable to the PC parallel port. The PC can monitor the status of the XC95108 through pin 11 of the parallel port.

Pins 2-9 of J1 are used to apply test signals to the CPLD once a new circuit has been downloaded. The signals applied through pins 2 and 3 of J1 are suitable for use as clock signals for sequential circuitry since they pass through Schmitt-trigger inverters. The other signals can serve as level inputs, but they should not be used as clocks.

The correspondence between the pins of J1, the pins of the XC95108 CPLD, and the arguments of the XSPORT program are shown in Table A.8.

LED Display

The connections between the LED digit and the XC95108 CPLD are depicted in Figure A.12. Two 220Ω resistors are placed in the ground returns of the LED digit to limit the amount of current sourced from the CPLD outputs. The connections between the CPLD pins and the LED digit segments are listed in Table A.9. (Note that the XS95 Lite, unlike the XS40 Lite, provides access to the period segment, S7, of the LED digit.)

Memory

The connections between the XC95108 CPLD and the 32 KByte static RAM are shown in Figure A.13. Not shown are the connections to ground and 5V on pins 14 and 28 of the RAM package, respectively.

The chip-select is pulled high through a 10KΩ resistor to disable the RAM when the CPLD is being configured. This lets us load values into the RAM and then reconfigure the CPLD without corrupting the RAM data while the CPLD pins are tristated.

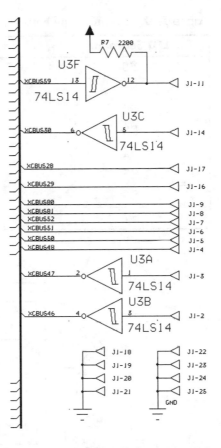

Figure A.11   XS95 Lite Board parallel port interface schematic.

Table A.8   Correspondence between the pins of J1 and the CPLD and the arguments of XSPORT.

J1 Pin	XC95108 Pin	XSPORT Arg.
2	46	B0
3	47	B1
4	48	B2
5	50	B3
6	51	B4
7	52	B5
8	81	B6
9	80	B7

Table A.9 Connections between the CPLD and the LED digit segments.

LED Segment	XC95108 Pin
S0	21
S1	23
S2	19
S3	17
S4	18
S5	14
S6	15
S7	24

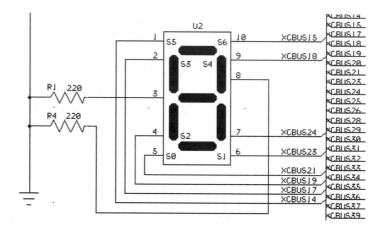

Figure A.12   XS95 Lite Board LED display schematic.

Table A.10 lists the assignment of RAM address, data, and control lines to CPLD pins as viewed by the portion of the XSLOAD program which downloads data to the RAM. Note that the order of the data and address pins in the list differs from the ordering of the pins on the RAM part in the schematic. However, it makes no difference to the actual RAM chip which address or data pin is the least-significant or most-significant as long as every device that accesses the RAM uses the same ordering. The ordering shown in Table A.8 matches the one used in the XS95 Board which was chosen based upon layout considerations for the printed wiring board.

Clock Circuit

The oscillator circuit shown in Figure A.14 generates a 12 MHz clock signal for the XC95108 CPLD. The 12 MHz resonator connected between the input and

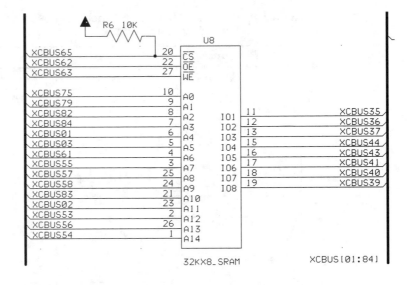

Figure A.13   XS95 Lite Board 32 KByte RAM schematic.

Table A.10   The RAM pin functions assigned to the pins of the XC95108 CPLD.

Pin Function	XC95108 Pin
A0	75
A1	79
A2	82
A3	84
A4	1
A5	3
A6	83
A7	2
A8	58
A9	56

Pin Function	XC95108 Pin
A10	54
A11	55
A12	53
A13	57
A14	61
D0	44
D1	43
D2	41
D3	40
D4	39
D5	37
D6	36
D7	35
/CS	65
/OE	62
/WE	63

output of the first inverter causes it to oscillate. (The two resistors bias the inverter to make sure the oscillation starts up.) The second inverter cleans-up the output of the first inverter and then applies the signal to one of the global clock inputs of the CPLD (pin 9).

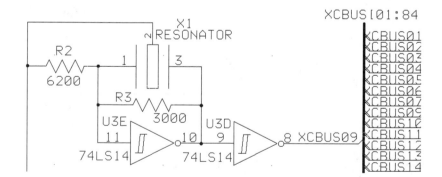

Figure A.14   XS95 Lite Board clock circuit schematic.

Tips for Constructing and Testing the XS40 and XS95 Lite Boards

You may be one of those people who can construct something and have it work the first time, but most of us aren't that lucky. So here are some tips to help you build the Lite boards and test it if it doesn't work:

- It's always difficult to build sturdy connections from the breadboard to the power and parallel port cables. Try soldering 2" wires to the power jack, gluing the jack to the breadboard with epoxy, and then connect the wires to the power buses of the breadboard. For the DB25 connector, solder 25 wires to it and then attach it to the breadboard with a couple of bolts and 1" spacers. Then you can connect the wires where they are needed while the DB25 connector is held firmly by the bolts and spacers.

- Construct the power supply circuitry first and make sure it puts out the correct voltages.

- Place the PLCC adapter on the breadboard next. **Do not put the FPGA or CPLD in it yet!** Connect the VCC and GND pins. Then apply power and make sure the correct voltages appear on each VCC and GND pin. Then turn off the power.

- Build the parallel port interface by placing the 74LS14 chip on the breadboard and connecting the wires from the DB25 connector. **Do not put the FPGA or CPLD in the PLCC adapter yet!** Create a bitstream file using the XILINX Foundation Series tools and then download it using the XSLOAD program. Use a logic probe to verify that the bitstream data and clocks appear on the appropriate pins.

- Place the 7-segment LED on the breadboard and connect it to the PLCC adapter. Then turn on the power and make sure (once again) that the correct voltages appear on each VCC and GND pin of the PLCC adapter. Then turn off the power.

- **Now you can insert the XC95108 CPLD or XC4005XL FPGA in the socket of the PLCC adapter!** Then turn on the power and download a bitstream for a circuit that uses the 7-segment LED (the LEDDCD40 or LEDDCD95 project is a good choice for this). Then use the XSPORT program to test the design and see if it operates correctly.

- If your test isn't successful, check the following items:

  * Check that the 74LS14 and the XC95108 or XC4005XL chips are getting power on the right pins and that all their ground connections are in place.

  * Make sure the 9V DC wall-mounted transformer is plugged in.

  * Verify that the DB25 connector is wired correctly. It wouldn't be the first time a connector was wired backwards.

* Try downloading using different parallel port numbers with the XSLOAD program. If you don't know the parallel port number, try them all: 1, 2, 3, and 4.

* The 74LS14 may be malfunctioning. It's cheap—replace it.

- After proving that you can successfully download to your Lite board, place the clock generation circuitry on the breadboard and verify that it produces a 12 MHz digital signal.

- Finally, place the 32 KByte static RAM on your breadboard and connect it to the PLCC adapter. Don't forget the pull-up resistor on the /CS input of the RAM. Then download the GNOME microcontroller bitstream and a HEX file with a program in it (see Chapter 9). Executing the program will verify that the memory is operating correctly.

Installing and Using the XSLOAD and XSPORT Programs

XESS Corp. provides the following two programs for use with the XS40 and XS95 Boards (and their Lite versions):

XSLOAD.EXE downloads .SVF configuration files for the XC9500 CPLD, .BIT configuration files for the XC4000 FPGA, and .HEX files for the 32 KByte RAM.

XSPORT.EXE forces a byte-value onto the input pins of the XC9500 CPLD or XC4000 FPGA that are connected to the PC parallel port.

Installing the Programs

Installation of these programs is easy:

1. Go to the CD-ROM directory where the tools are stored (D:\XSE).
2. To install the tools for Windows 95, double-click the icon labeled SETUP.EXE. To install the tools for Windows NT, double-click the icon labeled NTSETUP.EXE and then double-click the PORT95NT.EXE icon to install the parallel port driver for NT.

Using XSLOAD to Download Configuration Files

You can download an XC4000-based design into the XS40 Board as follows (assuming the name of the design is CIRCUIT):

```
XSLOAD CIRCUIT.BIT
```

You can download an XC9572 or XC95108-based design into the XS95 Board as follows:

```
XSLOAD CIRCUIT.SVF
```

If you also want to download a HEX file into the RAM of the XS40 or XS95 Board, use one of the following commands (assuming the HEX file is called FILE.HEX):

```
XSLOAD FILE.HEX CIRCUIT.BIT

XSLOAD FILE.HEX CIRCUIT.SVF
```

XSLOAD assumes the XS Board is connected to parallel port #1 of your PC. If you are using another port number, you can specify that like so:

```
XSLOAD -p 2 FILE.HEX CIRCUIT.BIT
```

Using XSPORT for Testing Designs

Assuming your CPLD or FPGA design has its inputs assigned to pins which are connected to the PC parallel port, then you can use XSPORT to force values on the inputs. XSPORT supports up to eight inputs. For example, to force the binary values 1, 0, and 1 onto the three least-significant bits of the parallel port, use the following command:

```
XSPORT 101
```

In the example given above, the upper five bits of the parallel port are forced to logic 0.

Like XSLOAD, XSPORT assumes the XS Board is connected to parallel port #1 of your PC. If you are using another port number, you can specify that like so:

```
XSPORT -p 2 101
```

Appendix B

Using ABEL with Xilinx PLDs

Summary

This application note provides a basic overview of the ABEL language and gives examples showing how to use ABEL to fully utilize the specific features of Xilinx PLDs.

Introduction

ABEL (Advanced Boolean Expression Language), combined with the Xilinx fitter software, provides a complete behavioral development environment for entering, simulating, and implementing designs for Xilinx PLDs. And, because ABEL was developed specifically for programmable logic devices, it provides several important features that support the Xilinx FPGA and CPLD architectures.

ABEL Language Structure

ABEL designs are organized into modules. Each module contains at least one set of declarations, logic descriptions, and an optional set of test vectors. Most designs are completely specified in a single module. However, using the hierarchal feature found in ABEL 6, multi-module designs can also be specified. Figure B.1 shows the distinct sections of code that are necessary to completely specify a design.

The Header Section

The module name is specified by the keyword **Module.** An optional **Title** may also be used after the module name to further describe the design. The double quote or the double slash can also be used to add comments. At the end of each module definition the keyword **End** is used to specify the end of the design.

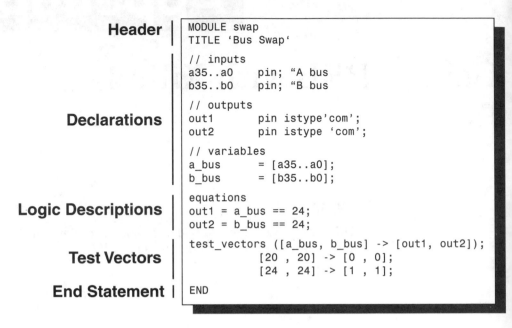

Header

Declarations

Logic Descriptions

Test Vectors

End Statement

```
MODULE swap
TITLE 'Bus Swap'

// inputs
a35..a0     pin;  "A bus
b35..b0     pin;  "B bus

// outputs
out1        pin istype'com';
out2        pin istype 'com';

// variables
a_bus       = [a35..a0];
b_bus       = [b35..b0];

equations
out1 = a_bus == 24;
out2 = b_bus == 24;

test_vectors ([a_bus, b_bus] -> [out1, out2]);
          [20 , 20] -> [0 , 0];
          [24 , 24] -> [1 , 1];

END
```

Figure B.1   ABEL Module Structure

For example:

```
MODULE mydesign

TITLE  'version 1 of mydesign'
"Additionally, comments can be added using "the double quote
// or the double slash

END
```

The Declarations Section

Declarations are used to define constants, signals, and sets, and to pass property statements to the fitter for controlling device specific features which are not directly supported by ABEL. The various elements contained in the Declaration section are:

- Constants—Constants are declared by assigning a value to a constant name.
- Input pins—specified with the **pin** keyword without types.
- Output pins—specified by the **pin** keyword and either registered or combinatorial by using the **istype** keyword.

- Nodes or buried logic—specified by the **node** keyword and can be either registered or combinatorial by using the **istype** keyword.
- Arrays of signals—declared by ending the signal name with a number and using the double period (**..**) syntax.
- Sets—declared to make the ABEL code easier to read and write by replacing long redundant signals with a single reference. In addition, higher level operations performed on sets allow very powerful and complex designs to be specified very quickly.

For example:

```
"Declaring constants
On = 1;
Off = 0;

"Declaring input pins
my_input, my_clk pin;

"Declaring output pins
my_combinatorial_output pin istype 'com';
my_registered_ouput      pin istype 'reg';

"Declaring nodes
my_combinatorial_node    node isypte 'com';
my_registered_node       node istype 'reg';

"Declaring a 5-bit array of nodes
Count4..Count0           node istype 'reg';

"Declaring a set to reference all 5-bits
Counter = [Count4..Count0];

"Property statements
xepld PROPERTY 'fast on';"Sets all outputs          "to fast slew.
```

Note that in the declarations section, the order in which constants and sets are defined is important. If constant X is defined with constant Y, for example, then constant Y must be defined first.

The Logic Description Section

The logic description section specifies the functions of the design. This can be done in three ways: Equations, Truth Tables, and State Diagrams. Equations are useful for designs with regular patterns such as counters or multiplexors. Truth Tables are a good entry method for designs that do not have regular patterns, such as a 7-segment LED decoder. State Diagrams are useful for specifying designs with complex state machines.

Following the Declarations section, the design section is delimited by using the keyword **Equations**, **Truth_Table**, or **State_Diagram.** State_diagrams and truth tables using sequential logic will need an accompanying equations section to define clock signals as well. These keywords must be used when switching between the three design methods.

Equations

Equation design entry primarily consists of assignment statements. These can be combinatorial assignments (**=**), or registered assignments using the delay operator (**:=**). All registered equations using the delay operator (:=) will behave as being implemented as an edge triggered flip-flop. Therefore, they must also have a clock associated with the signal name. This is done by using the **.clk** dot extension.

For example:

```
my_registered_node := my_input;
my_registered_node.clk = my_clk;
```

This is logically equivalent to describing the same logic with detailed dot extensions as follows.

```
my_registered_node.d = my_input;
my_registered_node.clk = my_clk;
```

Using equations to specify a design is similar to programming in other languages, except the context of the equations are evaluated in parallel rather than sequentially. In the following example, the order in which the equations are written is only important in programming languages.

In ABEL, all of the equations are evaluated concurrently, thus, the order in which they are presented is *not* important.

For example, in a normal programming language such as C, the code:

```
x = x + 1;
total = total +x;
```

is not the same as

```
total = total + x;
x = x+1;
```

However, in ABEL

```
x := x+1;
total :=  total + x;
```

is the same as

```
total := total + x;
x := x + 1;
```

In order to process information sequentially in digital logic, registers are used. In the previous example, note that the assignments are made with a ":=" . This means that in order for **x** or **total** to actually change values, a rising edge clock signal must be received by the register. The implementation for this design would look something like the following (line numbers are added to accompany the following explanation):

```
1      MODULE example1
2
3      my_clock        pin;
4      x7..x0          node istype 'reg';
5      total7..total0  pin  istype 'reg';
6
7      x = [x7..x0];
8      total = [total7..total0];
9
10     @carry 4;   "Limit the carry chain to
                   "4-bits
11
12     EQUATIONS   "Signals the beginning of
                   "an equation section
13     [x, total].clk = my_clock;
                   " Set the clock signals for
                   " registers to my_clock
14
15     x := x+1;   "This will implement an
                   "counter counting by 1
16     total := total + x;
                   "Note this is an adder that
                   "uses the @carry directive
                   "to implement the 8 bit
                   "adder with two 4-bit adders
17
18     TEST_VECTORS ([my_clock] -> [x, total])
19
20     [.C.] -> [ 1 , 0];
21     [.C.] -> [ 2 , 1];
```

```
22    [.C.] -> [ 3 ,  3];
23    [.C.] -> [ 4 ,  6];
24    [.C.] -> [ 5 , 10];
25    [.C.] -> [ 6 , 15];
26    [.C.] -> [ 7 , 21];
27
28    END
```

The beginning of this design (as for all ABEL designs) defines the module name. Following the module name are the declarations of the module. On line 3, the clock input is defined. On line 4, the nodes that will contain the information for **x** are declared as an 8-bit register. Line 5 declares the output pins for **total**, an 8-bit register. On line 7 and 8, the 8 bits are renamed to a single variable name, allowing the equations to be written quickly and clearly. Line 10 is an ABEL compiler directive which limits the lookahead carry chain of the adder to 4 bits. On line 12, the **EQUATIONS** keyword signifies the end of the Declarations section and the beginning of the logic descriptions.

Every register must have a clock associated with it and in this design, line 13 specifies that the **x** and **total** registers are clocked by the input **myclock**. The actual logic assignments on lines 15 and 16 determine how the registers are affected when **myclock** goes high. At the rising clock edge, **x** (after clock) will get the value of **x** (before clock) **+ 1**, and **total** (after clock) will get the value of **total** (before clock) **+ x** (before clock). Figure B.2 shows a block diagram of the circuit described by this code.

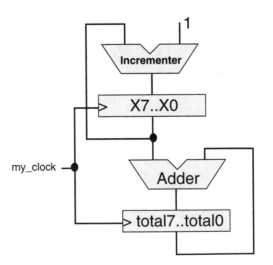

Figure B.2   Block Diagram

Test Vectors

Test vectors are also included at the end of this design. These test vectors show the expected values of the output for each rising clock signal, and are extremely useful for verifying the design. Test vectors are also included in the JEDEC programming file which can be used in the XC9500 family to perform an **INTEST operation** through the **JTAG** port.

Dot Extensions

Dot extensions, illustrated in Figure B.3, give more control over the implementation of the design. Dot extensions such as **.AP** or **.AR** are used to specify asynchronous preset and asynchronous reset for flip-flops. Other common dot extensions are **.OE** used to specify the output enable, and **.PIN** used with bidirectional signals. For a complete set of supported dot extensions, please refer to the documentation for the particular version of ABEL being used.

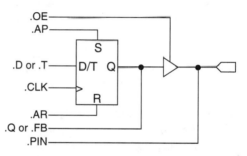

Figure B.3   Directly Supported ABEL Dot Extensions

Logical Operators

Logical operators allow the user to manipulate signals logically and can be applied to both signals and sets. When applied to sets, they are applied bitwise, and the sets must be the same size.

The operators are shown in Table B.1.

Arithmetic Operators

Arithmetic and relational operators can be used on sets to quickly generate adders, counters, and comparators. Large arithmetic functions, such as adders and magnitude comparators, will generate very wide equations to implement the carry look ahead signals. In order to control the width of these lookahead carry equations, the compiler directive, **@carry**, can be used. For example, **@carry 4**, would limit the carry chain to 4-bits. An 8-bit adder therefore, would be implemented as two 4 bit adders. Each adder would perform the carry look-ahead in parallel for its own four bits. However, a carry signal will be generated by the lower four bits that will be cascaded to the higher order adder. Typically, carry chain lengths of 3 or 4 is a good trade-off between speed and density.

Relational Operators

Relational operators are also used for generating the boolean result for conditional equations. These are useful for making the comparisons used in address decoding.   Relational operators are used in conjunction with the conditional equation, **when...then...else** statement. Using conditional equations simplifies the task of building components with control signals.

For example, using operators:

```
"Using the logical OR operator.
my_reg_output := my_input # my_reg_node;
my_reg_output.clk = my_clock;

"Using the arithmetic operator
sum := a + b;
sum.clk = my_clock;

"Using the relational operators with
"conditional equations
 WHEN (a != b) THEN
     c := my_input;
ELSE
     c := my_reg_node;
 c.clk = my_clock;
```

Conditional equations allow the designer to make certain assignments depending upon the result of a relational operator. In the previous example, **c** is assigned **my_input** if **a** did not equal **b**. Otherwise, **c** will be assigned to the value stored in **my_registered_node**. Multiple assignments are made by containing them within the {} symbols.

The **else** statement does not need to be included in every conditional equation. When there are multiple conditions, they can be specified one at a time, however, make sure all possibilities are covered, otherwise unexpected behavior may occur in the design. For example:

```
"Using multiple WHEN without ELSE
"statements, the description of a mux
 "is done by conditionally setting Output
"to Data_A or Data_B. Note the {} are used
```

Using ABEL with Xilinx PLDs   Appendix B

```
"to make multiple assignments for each
"conditional.
Output2.clk = myclock;
WHEN (select == 0) THEN
          {Output = Data_A;
Output2 := ValueA;}
WHEN (select == 1) THEN
          {Output = Data_B;
 Output2 := ValueB;}
WHEN (select == 2) THEN
          {Output = Data_C;
 Output2 := ValueC;}
"Note, select may also equal 3 and not
"defining the behavior for 3 will make
"the condition a don't care.
```

Table B.1 ABEL Operators

	Logical		Arithmetic		Relational
&	AND	-	Twos complement	==	Equal
#	OR	A-B	Subtraction	!=	Not equal
!	NOT	A+B	Addition	<	Less than
$	XOR	<<	Shift left	<=	Less than or equal
		>>	Shift right	>	Greater than
				>=	Greater than or equal

Using Truth Tables

Truth table design entry consists of specifying the input signals and the responding output signals, which can be registered or combinatorial. Truth tables are handy when specifying designs with irregular patterns, such as a 7-segment LED decoder.

A truth table starts with a header specifying all of the input and output signals. A transition specified with the -> syntax is a combinatorial transition, while a :> specifies a registered transition. Registered truth tables must also include a clock equation to specify the clock input to the register. For example:

```
"Syntax for defining a truth table
TRUTH_TABLE ([ my_input ] -> [my_combinatorial output] :>
[my_registered_output])

"A mux described with a truth table
```

```
    EQUATIONS          "Note that we define the clock
                       "with an equation first.
     Output.clk = my_clock;

    "Then, we specify the truth table input
    "and outputs.
    TRUTH_TABLE
    ([select]:> [ Output ])
      [ 0 ]  :> [ Data_A ];
    "If select is 0,then Output gets Data_A
      [ 1 ]  :> [ Data_B ];
    "If select is 1, then Output gets Data_B
```

Entering Test Vectors

Test vectors can optionally be added to perform a functional test of the logic descriptions. For the XC9500 family, they are included in the JEDEC file which can also be used with the JTAG INTEST feature to test the functionality of the physical device in the system.

The **TEST_VECTORS** keyword signifies the end of the previous section and the start of the test vectors that are used to functionally simulate the design. Test vectors are entered in the same format as truth tables, and special signals are defined to help make the process easier. The three most common are **.C.** , **.Z.**, and **.X.**. The **.C.** signal represents a clock that starts from a logical low, goes to a logical high, and returns to a logical low. This is more convenient than entering three test vectors for each clock pulse. The **.Z.** is used to represent signals that are 3-stated, and the **.X.** signal represents a don't care signal. Don't care signals can be used both for inputs and outputs.

Using State Diagrams

State diagrams tend to produce very legible and easy to maintain code. The following steps are used in creating a design using state diagrams.

1. Declare the state bits.

2. Declare a name for the set of state bits.

3. Assign a value for each state.

4. Define the state machine clock signals with equations.

5. Define the state transitions using the name defined in step 2.

Declaring state bits, and defining how each state is represented by those state bits is done by using the sets and constant declarations presented earlier. After defining the state bits and states, an equations section is needed to specify clock signals and other control logic. Using the STATE_DIAGRAM keyword, the designer can now specify assignments and state transitions for each of those states. To specify an unconditional state transition, use the **GOTO** statement. For conditional transitions, use the **IF...THEN... ELSE** statement. Note that this is different from the **WHEN...THEN...ELSE** used in the equations section.

For example:

```
module traf

title 'Traffic Light Controler'

" A controller is needed to control the
" timing of a traffic light.
" The green light should be lit for thirty
" seconds.  Then a yellow light for two
" seconds, and a red light for
" thirty seconds We then repeat the entire
" cycle. This design must run on a clock " which has a period of
1 sec.

"clock in with a period of 1 second.
clk   pin;

"reset to determine initial state.
reset pin;

"Declare some nodes for a counter.
  Count4..Count0  node istype 'reg';
  Counter = [Count4..Count0];

"Step 1.  Declare the state bits.
"These bits are also our outputs
"in this example...
red       pin  istype 'reg';
yellow    pin  istype 'reg';
green     pin  istype 'reg';

"Step 2.  Declare a name for the set of
"state bits.
"Step 3.  Assign a unique value for the
"for each state.
  Light =   [green, yellow, red];
  GO =      [ 1,      0,     0 ];
  CAUTION = [ 0,      1,     0 ];
```

```
    STOP =    [ 0,       0,       1 ];

Equations
"Step 4. Set up the clock and reset lines
"for the state machine.
green.ap      = reset;
red.ar        = reset;
yellow.ar     = reset;
Counter.ar    = reset;
Counter.clk   = clk;
[green, yellow, red].clk = clk;

"Step 5.  Define the state transitions
"using the name defined in step 2.

State_Diagram Light

  state GO:
  IF (Counter < 30) then GO with
     Counter := Counter + 1;
  ELSE goto CAUTION with
     Counter := Counter + 1;

  state CAUTION:
  IF (Counter != 0) then CAUTION with
     Counter := Counter + 1;
  ELSE goto STOP with
     Counter := Counter + 1;

  state STOP:
  IF (Counter < 30) then STOP with
     Counter := Counter + 1;
  ELSE goto GO with
     Counter := 1;

test_vectors
([clk,reset] -> [red, yellow, green])

     [0,1]    ->    [0,0,1];

     [.c., 0]   ->    [0,0,1];
     @repeat 29
{[.c., 0]    ->    [0,0,1];}

  [.c., 0]    ->    [0,1,0];
```

```
[.c., 0]     ->     [0,1,0];

[.c., 0]     ->     [1,0,0];
@repeat 29
{[.c., 0]    ->     [1,0,0];}

[.c., 0]     ->     [0,0,1];
    @repeat 29
{[.c., 0]    ->     [0,0,1];}

[.c., 0]     ->     [0,1,0];
[.c., 0]     ->     [0,1,0];

[.c., 0]     ->     [1,0,0];
    @repeat 29
{[.c., 0]    ->     [1,0,0];}
```

end

Assignments can be made for state diagrams in two ways. If the assignment is made combinatorially in the state, then the outputs will be decoded from the state bits. This will improve the density of the final design, however it will add some delay. As shown in Figure B.4, decoding from the present state adds a level of logic to achieve the desired output. For example:

```
zero_state:
output1 = 0;
GOTO one_state;

one_state:
output1 = 1;
GOTO zero_state;
```

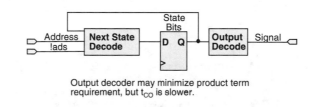

Output decoder may minimize product term
requirement, but t_{CO} is slower.

Figure B.4   Decoding Present State

If the signal is declared as a register type, then while the next state is determined, the output is decoded at the same time, and will transition at the next clock edge. This implementation is shown in Figure B.5. To implement this in ABEL, we use the **with** statement in our state diagram.

For example:

```
zero_state :
GOTO one_state
with output1 := 1;

one_state :
GOTO zero_state
with output1 := 0;
```

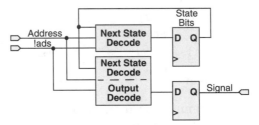

Duplication of the next state decoder may increase product term count, but t_{CO} is faster.

Figure B.5   Decoding Next State

Using Property Statements

Although ABEL was initially developed for PLDs, there are still device-specific features that ABEL does not support directly. Instead, ABEL provides a **property** statement allowing device specific commands to be passed to the fitter software. Property statements must be placed in the declarations section. These property statements allow the user to control the following:

- Slew rates
- Logic optimizations
- Logic placement
- Power settings
- Preload values

fast

The **fast** property controls the output slew rate, and there can only be one **fast** property used in each design. If there are only a few signals that require a fast slew rate, they can be listed individually after the property, and the remaining signals will be slew rate limited. Or, if there are only a few signals that need to be slew rate limited, then those signals can be listed.

```
xepld property 'fast on';
"all pins have fast slew rate

xepld property 'fast on x1 x2';
" only x1 and x2 are fast
  " the remaining pins are slew limited

xepld property 'fast off x1 x2';
"only x1 and x2 are slew limited
"the remaining pins are fast
```

logic_opt

The **logic_opt** property allows the user to control the logic optimization done by the fitter. This should be used on selective nodes, where collapsing those nodes would cause the design to become very large.

```
xepld property 'logic_opt off';
"Preserves all combinatorial nodes

xepld property 'logic_opt off x1';
"preserve x1 and collapse other nodes to fitter limits
```

minimize

The **minimize** property is used to prevent boolean minimization on equations, and is primarily used to prevent removal of redundant product terms in combinatorial logic.

```
xepld property 'minimize off x1 x2';
"keep redundant product terms for x1 & x2
```

partition

The **partition** property is used when specific placement of logic is desired.

```
xepld property 'partition fb1 x1 x2';
"place the functions of x1 and x2 in
"function block 1

xepld property 'partition fb1_2 x1';
"place the function x1 in
"function block 1, macrocell 2
```

pwr

The **pwr** property controls the power settings for individual macrocells.

```
xepld property 'pwr low';
"places all macrocells in low power mode

xepld property 'pwr low x1 x2'
"places x1 and x2 in low power mode
"the remaining in STD power mode

xepld property 'pwr std x1 x2'
"places x1 and x2 in STD power mode
"the remaining in low power mode
```

.prld

The **.prld** property controls the initial state of the registers at power up. Note that because this property is only passed on and used by the fitter, and is not used by ABEL, the preload value will not be reflected in test vectors. The default preload value is 0 for all XC9500 registers. Therefore, only registers that require a value of 1 need to be specified.

For example:

```
xepld property 'equation x1.prld = VCC';
"preload register x1 to a 1
```

Design Examples

The following examples demonstrate some basic, specific design principles.

Bi-Directional pins

This example shows how to implement a bi-directional signal in ABEL. Bi-directional signals are commonly found whenever a bus is being used by several different devices. This usually involves some kind of control signal to allow only one device to drive the bus at a given time. In this example, the input pin **write** is used to control if data is being driven on to the data pins **D7..D0** from an outside source, such as a microprocessor, or if data is being driven from the CPLD to be read from the data pins by an external source

```
module bidi;

"This design will take a value from the
"pins D7..D0 and store it in a Register
```

```
"when the signal, write, is high. When
"write goes low, it will output the
"saved value at pins D7..D0

"inputs
write           pin;
myclock         pin;

"Bi-directional signal also has a register
"associated with it.
D7..D0          pin istype 'reg';

"Define my sets
Data = [D7..D0];

Equations;
Data.oe = !write;      "3-State the data lines
                       "when writing to register

Data.clk = myclock;

WHEN (write==1) THEN Data := Data.pin;
                "When we are writing to the part, read the
                "data pins and save in data register.
ELSE
Data := Data;
                "Else, drive the data pins with the value
                "saved in the register so we can read it "back.
end;
```

Latches

Latches can be implemented in two ways. In the first example, latch_output utilizes the asynchronous set and reset of a flip-flop to implement a latch.

```
module ltest1

input,le        pin;
latch_output    pin istype 'reg';

equations

latch_output.ap = input & le;
latch_output.ar = !input & le;
latch_output.clk = 0; "Clock must be
                          "grounded
```

```
latch_output.d = 0; "D-input must be
                     "grounded

test_vectors    ([input, le] -> [latch_output])
                [0,0] -> [0];
                [1,0] -> [0];
                [0,1] -> [0];
                [1,1] -> [1];
                [1,0] -> [1];
                [0,1] -> [0];

end;
```

Latches can also be implemented combinatorially by using a feedback path and providing a redundant product term to cover glitches. This will require the following code:

```
MODULE comlatch;

le          pin;
input       pin;
latch_out   pin  istype 'com,retain';

// The ABEL compiler will retain redundant
// logic for the latch_out output
// because they have the RETAIN attribute.
// However, the MINIMIZE OFF property
//statement is required to instruct the
// Xilinx fitter to also retain the
// redundant logic.

xepld property 'minimize off latch_out';
" The fitter will retain redundant logic
" for these nodes

EQUATIONS;

    latch_out = input & le
  " latch is transparent high
# latch_out & !le
" latch data on falling edge of le
# latch_out & input;
" Redundant product term

TEST_VECTORS
([ le, input] -> [latch_out]);
```

```
[1 , 0 ]   -> [   0 ];  " transparent
[1 , 1 ]   -> [   1 ];  " transparent
[0 , 1 ]   -> [   1 ];  " latch a 1
[0 , 0 ]   -> [   1 ];  " change input
[1 , 0 ]   -> [   0 ];  " transparent
[0 , 0 ]   -> [   0 ];  " latch a 0
[0 , 1 ]   -> [   0 ];  " change data

END; " All modules must have an END statement
```

Counters

Counters are useful in a variety of applications, such as memory interfaces, generating delay states, or simple state machines. This example shows how to build a loadable up/down counter with a count enable.

```
module counter

"32-bit Up/Down counter with parallel load "and enable

"Outputs
Q31..Q0        pin istype 'reg';

"Inputs
D31..D0        pin;
Load           pin;              " Load Cmd
Count_Enable pin;               " Count Cmd
UpDown         pin;              " Up/Down Cmd
myclk          pin;              " Clock

Counter = [Q31..Q0];
Input   = [D31..D0];

Equations
@carry 4;
Counter.clk = myclk;
WHEN (!Load & Count_Enable & UpDown)
    THEN Counter := Counter + 1
else
WHEN (!Load & Count_Enable & !UpDown)
    THEN Counter := Counter - 1
else
WHEN (Load)
    THEN Counter := Input;
else
```

```
Counter := Counter
end;
```

Multiplexers

Multiplexers can be used to control the data flow of a design. The following
example demonstrates how to implement a 16 bit 4-1 registered multiplexer:

```
module mux2
A15..A0          pin; " Inputs
B15..B0          pin; " Inputs
C15..C0          pin; " Inputs
D15..D0          pin; " Inputs
Q15..Q0          pin istype 'reg'; " Output

Sel1..Sel0       pin;
clk              pin;

Output  = [Q20..Q0];
DataA = [A20..A0];
DataB = [B20..B0];
DataC = [C20..C0];
DataD = [D20..D0];
Select  = [Sel1..Sel0];

Equations

Output.clk = clk;

WHEN Select == 0 THEN Output := DataA
else WHEN Select == 1 THEN Output := DataB
else WHEN Select == 2 THEN Output := DataC
else Output := DataD;

end
```

Conclusion

ABEL allows complex behavioral designs to be easily implemented and simulated. In addition, the special features and capabilities of the device are easily
accessed through ABEL property statements. ABEL is a simple yet powerful
software tool that provides the designer with an efficient language for devel-
oping Xilinx FPGA and CPLD designs.

Index

for XS40 board, 76–82
for XS95 board, 82–84
One-hot encoding, 211, 222
Opcodes, 314
Open button, 38, 43
Operations -> Program menu
item, **JTAG Programmer** window, 56, 83, 92
Options -> Create Netlist menu
item, **Schematic Editor,** 76,
116, 124, 156, 285
Options -> Export Netlist menu
item, **Schematic Editor,** 43, 76,
116, 124, 156, 285
Options -> Integrity Test menu
item, **Schematic Editor,** 43, 76,
116, 124, 285
Options -> Preferences... menu
item, 46, 162
Logic Simulator window, 162
Waveform Viewer window,
44–46
Options button, **Implement
Design** window, 82
Output -> Create SVF File menu
item, **JTAG Programmer** window, 55, 83
Output radio-button, **Design Wizard-Ports** window, 84, 87–88

P

Pad Report, 80–81
PAL structure, 26–28
Parallel port interface:
XS40 Lite Board, 405–7
XS95 Lite Board, 413–14
Parity generator, 73–74
in VHDL, for the XS95 board,
109–13
for the XS40 board, 113–19
for the XS95 board, 103–9
eight-bit even-parity generator, 106–9
four-bit even-parity generator, 106
two-bit even-parity generator, 104–6
partition property, 437
Place&Route Report, 79–80, 94
Place&Route stage, in implementation process when targeting
FPGA, 53

PLA (programmable logic array)
devices, 25–26
Post Layout Timing Report, 81
power supply:
XS40 Lite Board, 403
XS95 Lite Board, 411
Preferences window, **Waveform
Viewer,** 162
-**prid** property, 438
Priority encoders, 72–73
Program counter (PC), 314
Programmable array logic (PAL)
structure, 26–28
Programmable logic:
advantages of, 23–25
architecture, 25–31
design flow, 32–33
design techniques, 23–68
field-Programmable gate array
(FPGA), 28–31
PAL structure, 26–28
PLA devices, 25–26
SPLD (simple Programmable
logic device), 26–28
Trip-Genie project design:
HDL mode, 60–67
schematic mode, 33–60
See also Trip-Genie project
design, HDL mode; Trip-Genie project design, schematic mode
Programmable switch matrices
(PSMs), 28
Programmer's model of memory,
259–60
Project -> Create Macro menu
item, 155
Project Manager window, 34–35,
36–39, 43, 44, 51, 54, 134
Contents tab, 35
Document -> Add menu item,
43, 82, 90, 94, 98, 113, 137,
139, 253, 285
Files tab, 35, 132
Finish button, 85, 88, 96
Flow tab, 35, 39, 50, 61, 67, 78,
86, 113, 137, 247
Implementation -> Options
menu item, 165
Reports tab, 35, 78, 91, 94
**Synthesis -> Force Analysis
of All HDL Source Files,** 65
Synthesis -> New Library
menu item, 132

Synthesis tab, 35
**Tools -> Design Entry ->
HDL Editor,** 36, 61
**Tools -> Design Entry ->
Schematic Editor,** 36, 39
Versions tab, 35
Property statements, 436–38
fast property, 436–37
logicopt property, 437
minimize property, 437
partition property, 437
-**prid** property, 438
pwr property, 438
PSMs (programmable switch matrices), 28
pwr property, 438

R

RAM16x2D symbol, 296
Read mode, memory, 259–60
Register set, DWARF microcomputer, 366, 368–69
Report Browser window, 79, 91
**Asynchronous Delay
Report,** 81
Bitgen Report, 81–82
Map Report, 79
Pad Report, 80–81
Place&Route Report, 79–80,
94
Translation Report, 79
Reports tab, **Project Manager**
window, 35, 78, 91, 94
Return instructions, DWARF microcomputer, 361–62
Run button, 78, 86

S

*.SCH, 43
Schematic Capture window, 115,
161
**Hierarchy -> Create Macro
Symbol from Current
Sheet** menu item, 120
Mode -> Draw Buses menu
item, 121, 124
Mode -> Draw Bus Taps
menu item, 122
Mode -> Symbols menu item,
39, 115, 121, 285